Consuming Cities

This book is about cities as engines of consumption of the world's environ-ment, and the spread of policies to reduce their impact. It looks at these issues by examining the impact of the Rio Declaration and assesses the extent to which it has made a difference.

Consuming Cities examines this impact using three categories of countries for examples: first, four countries from the world's core economies – the USA, Japan, Germany and Britain; second, the experience of the 'giant' states of China and India; third, the contributors consider the case of smaller economies by including two pairs of countries from north and south of the equator – Sweden and Poland, and Australia and Indonesia. Each of these final pairs includes one 'developed' and one 'developing' country.

All the contributors have direct experience of the urban environment and urban policies in the countries about which they write, and offer an author-itative commentary that serves to bring the urban 'consumption' dimension of ecologically sustainable development into sharper focus. It critically eval-uates the success of the Rio Declaration and considers the wider question of global governance for the ecological regulation of cities.

Nicholas Low is Associate Professor in the Faculty of Architecture, Building and Planning, the University of Melbourne, Australia. **Brendan Gleeson** is Senior Research Fellow at the Urban Frontiers Program, University of Western Sydney, Australia. **Ingemar Elander** and **Rolf Lidskog** are Associate Professors in the Centre for Housing and Urban Research, the University of Örebro, Sweden.

Consuming Cities

The Urban Environment in the Global Economy after the Rio Declaration

Edited by Nicholas Low, Brendan Gleeson, Ingemar Elander and Rolf Lidskog

London and New York

First published 2000
by Routledge
11 New Fetter Lane, London EC4P 4EE

Simultaneously published in the USA and Canada
by Routledge
29 West 35th Street, New York, NY 10001

© 2000 Nicholas Low, Brendan Gleeson, Ingemar Elander & Rolf Lidskog

Typeset in Galliard by The Florence Group, Stoodleigh, Devon
Printed and bound in Great Britain by TJ International Ltd, Padstow, Cornwall

British Library Cataloguing in Publication Data
A catalogue record for this book is available from the British Library.

Library of Congress Cataloging in Publication Data
Consuming cities: the urban environment in the global
economy after the Rio Declaration/edited by Nicholas Low . . . [et al.]
    p. cm.
    1. Urban ecology 2. Sustainable development. I. Low, Nicholas.
    HT241.C66 1999
    304.2′09173′2–dc21   99–18549
                            CIP

ISBN 0–415–18768–0 (hbk)
ISBN 0–415–18769–9 (pbk)

Contents

Figures

Tables

Contributors

Professor Andrew Blowers, Faculty of Social Sciences, The Open University, UK.

Mr Peter Christoff, Department of Geography and Environmental Studies, The University of Melbourne, Australia.

Dr Ingemar Elander, Centre for Housing and Urban Research, The University of Örebro, Sweden.

Dr Brendan Gleeson, Urban Frontiers Program, University of Western Sydney, Australia

Mr Martin Gürtler, Internship: Wuppertal Institute for Climate, Environment and Energy, Wuppertal, Germany.

Professor Toshio Hase, Tokyo International University, Japan.

Dr Lea Jellinek, Faculty of Architecture, Building and Planning, The University of Melbourne, Australia.

Professor Arvind Krishan, Centre for Advanced Studies in Architecture, School of Planning and Architecture, Delhi, India 110 002.

Professor Robert Lake, Center for Urban Policy Research, Rutgers University, New Brunswick, USA.

Dr Rolf Lidskog, Centre for Housing and Urban Research, the University of Örebro, Sweden

Associate Professor Nicholas Low, Faculty of Architecture, Building and Planning, The University of Melbourne, Australia.

Professor Timothy Luke, Department of Political Science, Virginia Polytechnic Institute and State University, USA.

Professor Asesh Kumar Maitra, Centre for Advanced Studies in Architecture, School of Planning and Architecture, Delhi, India 110 002.

Professor Tadeusz Markowski, Department of City and Regional Management, Faculty of Business Administration, the University of Łódź, Poland.

Dr Helena Rouba, Department of City and Regional Management, Faculty of Business Administration, the University of Łódź, Poland.

Mr Joachim Spangenberg, Program Director 'Sustainable Societies', Wuppertal Institute for Climate, Environment and Energy, Wuppertal, Germany.

Professor Fukashi Utsunomiya, Department of Political Science and Economics, Tokai University, Japan.

Ms Anke Valentin, Junior Research Fellow, Wuppertal Institute for Climate, Environment and Energy, Wuppertal, Germany.

Dr Mark Wang, Department of Geography and Environmental Studies, The University of Melbourne, Australia.

Dr Yongyuan Yin, Senior Associate, Environmental Adaptation Research Group, Sustainable Development Research Institute, University of British Columbia, Vancouver, Canada.

Dr Stephen Young, Department of Government, The University of Manchester, UK.

Preface

This book is about cities as engines of consumption of the world's environment, and the spread of policies to reduce their impact. In the first part of the book we set the scene by discussing the question of cities as environmental systems for consumption. We consider the global setting of the environmental crisis, and the possibility of moving beyond current transnational institutions and environmental regimes (within the United Nations framework) for resolving questions of justice and the environment. The second part, which is the core of the book, is a collection of separately authored national studies under the rubric of ecological sustainability and environmental justice which seek to assess the impact of the Rio Declaration and Agenda 21 on the environmental regulation of cities in different nations. The national studies enable us to draw some tentative general conclusions in the final chapter about 'consuming cities', their governance, and the discourse of ecologically sustainable development.

Until recently there was relatively little reference in environmental discourse to the role of cities in the future of the global environment. The Rio conference made an important contribution to changing that lack of focus. Nature was no longer viewed either as a resource supporting the world's population or as an idealised 'other' – other, that is, than urban. The urban and natural environments were rather seen as an indivisible matrix in which human and natural processes interact. Increasingly the world order governed by the competitive market pits cities and their governments against one another. In this world of competing cities the place of the nation-state is equivocal, yet the state remains the primary source of authoritative regulation.

Following the Rio conference it has become even clearer that the environmental regulation of cities is crucial to the future of the planetary ecosystem. The Toronto-based International Council for Local Environmental Initiatives is promoting adoption of local Agenda 21s by city governments. However, the momentum generated by the United Nations Conference on Environment and Development (UNCED) is flagging, the successful negotiation of an effective regime to phase out production of ozone depleting gases has not been repeated in other critical areas such as biodiversity and climate change.

The aims of the book are as follows:

1   to bring the urban 'consumption' dimension of ecologically sustainable development into sharper focus;
2   to evaluate critically the success of the Rio Declaration and Agenda 21 as a means of bringing about proper regulation (in the urban domain) of the global environment for a sustainable society;
3   to consider the wider question of global governance for the ecological regulation of cities, and what further development of global institutions might be needed.

The national coverage is necessarily selective. The key assumption governing the selection of national studies is that the relationship between a national state and the global economy is an influential factor (though by no means the only or even necessarily the most significant factor) in the nation's response to the Rio Declaration. Underlying this assumption are two questions: to what extent is urban environmental regulation today a matter for intervention at global level? And what form might the institutions developed for this purpose take?

Following the above assumption, four countries were chosen from the world's core economies: the USA, Japan, Germany and Britain. The experience of giant states – in both population and economic potential – is then considered: China and India. Whatever happens in these countries is of pivotal importance for the urban success of global initiatives like the Rio convention. The case of the smaller countries is included by considering two pairs of countries from the North and the South: Sweden and Poland, Australia and Indonesia. Each pair includes one 'developed' and one 'developing' country. In all cases the authors have direct experience of the urban environment and urban policies in the countries concerned.

Large areas of the world have of course been left out. Limited space in the book meant that only a small sample of countries could be covered. For instance we have not been able to cover the regions of Africa, the Commonwealth of Independent States (Russia and associated countries), or South America. Each of these regions has a different experience from our small sample of countries, and each region's experience is different from the others. Experiences of different countries within regions also varies: we are not suggesting here that Sweden's experience is the same as all Scandinavian countries, nor that Germany and Britain are similar to France and Italy, nor that Indonesia's experience is common to all South East Asian nations. Commentaries on such vast and complex nations as China and India must be regarded as partial surveys. Another limitation of the book is the different perspectives which the authors bring to bear on the questions posed. Though they are all urbanists, we cannot say that the authors of national studies view the world through the lens of a common political perspective or academic discipline. But this variation may also be seen as representing one aspect of reality – in which perspectives differ. Commonality of purpose was to some degree assisted by a face-to-face workshop in 1997. Despite these limitations, however, we think our main intention has been achieved: namely, to bring the question of consuming cities into focus and to test some assumptions against the experience of distinguished observers of urban policy around the world.

The book is truly a co-operative international effort, not only among the editors but with all the contributors. The idea for the book was first mooted during a visit by Nick Low in 1996 to the Centre for Housing and Urban Research at the University of Örebro. It was followed up in meetings between Nick and Brendan Gleeson at the Urban Research Program of the Australian National University (Canberra) and further developed in the following year when most of the authors came together at a workshop of the 1997 conference on Environmental Justice at the University of Melbourne (http://www. arbld.unimelb.edu.au/envjust/). Thanks are due both to the Centre for Housing and Urban Research and the Faculty of Architecture, Building and Planning of the University of Melbourne for hosting these occasions. The British Council through its New Images programme generously supported the visit to Melbourne by Professor Andrew Blowers. Preparation of Chapter 4 on the USA would not have been possible without the research assistance provided by Anne Leavitt-Gruberger and Liesje DiDonato. Helpful comments on earlier versions of this chapter were provided by participants in the Melbourne workshop, the Urban Affairs Colloquium at the University of Delaware, and the CUPR faculty seminar at Rutgers University. The authors of Chapter 6 on Germany and the editors wish to acknowledge the help of Professor Klaus Kunzmann of the Faculty of Spatial Planning of the University of Dortmund for his earlier draft, and especially for his comments on the Ministry of Regional Planning in the sustainability process and the results of the United Nations Habitat II conference. The editors also wish to acknowledge the participation of Dr Renat Perelet of the Institute for Systems Analysis, Russian Academy of Sciences (Moscow), for his participation in the Melbourne workshop and to thank him for his paper on Russia which failed to be turned into a chapter only because of technical problems of communication.

Most especially the editors would like to thank the authors of the national studies for their work and for their patience in the editorial process.

The editors
November 1998

1 Cities as consumers of the world's environment

Brendan Gleeson and Nicholas Low

Since most of us spend our lives in cities and consume goods imported from all over the world, we tend to experience nature merely as a collection of commodities or a place for recreation, rather than the very source of our lives and well-being.

(Wackernagel and Rees, 1996: 7)

Introduction

In the past, cities have borne witness to humanity's struggle for ascendancy over nature. The city has been a haven where humanity could escape the caprice of the natural world, a vantage point from which nature could be safely consumed, experienced and enjoyed. But escape from nature assumed an infinite cornucopia. In the world which is fast taking shape cities as engines of consumption are turbocharged – world consumption expenditures rose six times between 1950 and 1998 (UNDP, 1998a). Yet at the same time it is increasingly understood that 'nature' is finite, the cornucopia may empty, and the social health of the city depends upon its integration within natural ecological systems (Breheny, 1990: 9.4 cited in Haughton and Hunter, 1994: 10; Urry, 1995; Wackernagel and Rees, 1996; Bell, 1997).

This growing appreciation of the interrelationship between cities and the 'natural' world is encouraging a convergence of ecology with city planning. This convergence has brought with it new perspectives. The idea of social justice grew up with the industrial city and the close juxtaposition of great wealth and great poverty. The city became the site of constant struggles between conflicting groups and classes. Of course that struggle was always in large part over the environment and its qualities. But the environment was 'urban', localised and subdivided, and the struggles were contained within city administrations and nation-states. Today's environmental struggles sited in the city cannot be contained in this way. Not only has a globalised corporate sector pitted cities (as well as nation-states) against one another in a nexus of market relations, but the entire global system of corporate relations, of which city administrations are increasingly part, is responsible for the health or sickness of the whole planetary environment. Just as urbanisation brought into focus the question of social justice at the national level, so today's global processes of development bring into focus that of environmental justice at global level (see Low and Gleeson, 1998).

As argued by Habermas (1995) and Dryzek (1997), discourses co-ordinate social action:

> A discourse is a shared set of assumptions and capabilities embedded in language that enables its adherents to assemble bits of sensory information that come their way into coherent wholes. So any discourse involves a shared set of basic, often unspoken, understandings.
>
> (Dryzek, 1999: 268)

Discourses shape the development of institutional rules. But discourses are also embedded in other relationships of power, notably economic power. There are, of course, many different discourses at work in the political world today, and the particular mixture of discourses and its relationship with economic and institutional power structures amounts to a model of development.

In the first section of this chapter we interrogate the discourse of urban ecological sustainability. We outline the scope of the discourse and its principles for protection of the global environment. We then consider the model of development which is dominant today. We argue that, while sustainable development is a contested concept and problematic in implementation, the discourse has the potential to guide development in a benign direction. However, sustainable development today appears to be playing only a supporting and apologetic role in a model of development which is increasingly dominated by a discourse of market co-ordinated, competitive growth. We raise the question on which this collection of essays is intended to throw some light: *how far has the discourse of urban sustainability, promoted at the Rio 'Earth Summit' of 1992 and at subsequent international fora, penetrated the world's national urban policy systems?*

In the second section we show why there is reason to question whether urban sustainability has made much headway. The evidence suggests that the world of cities is not moving towards ecological sustainability. Quite the contrary. Cities throughout the world appear to be faced with a forced choice between economic prosperity and environmental responsibility. We question the spatially fetishised concept of the 'individual' city handed down from antiquity and now part of a discourse of 'competing cities' and 'city-marketing'. We argue that cities must be viewed in their global economic context. Such a view demands that the problematic of cities be considered anew, and inevitably within a framework of global governance.

Sustainable development and the changing model of growth

The Fordist prospectus of endless economic growth which accompanied the post-war 'long boom', was challenged in the 1970s by ecologists who pointed to the Earth's limited capacity to service economic development (Hardin, 1968; Ehrlich, 1971; *The Ecologist*, 1972; Meadows *et al.*, 1974). The oil shocks of the 1970s reinforced this challenge by demonstrating in practice how dependent whole economic systems were upon a single natural resource. Dependence coupled with limits added up to a serious threat to prosperity.

The continuing agenda of the United Nations since the Second World War was the extension of the current development model to the so-called 'Third World'. So any abatement of the growth model, just as it was beginning to spread outward, would prevent the many from joining the prosperity of the few developed core nations. As is well known, the General Assembly of the United Nations in 1983 established an independent commission chaired by Gro Harlem Brundtland, Prime Minister of Norway, to find ways in which economic development and growth could be reconciled with conservation of the Earth's environment and ecological systems. The full text of the oft quoted definition of sustainable development reveals the UN agenda:

> Humanity has the ability to make development sustainable – to ensure that it meets the needs of the present without compromising the ability of future generations to meet their own needs. The concept of sustainable development does imply limits – not absolute limits but limitations imposed by the present state of technology and social organization on environmental resources and by the ability of the biosphere to absorb the effects of human activities. But technology and social organization can be both managed and improved to make way for a new era of economic growth. The Commission believes that widespread poverty is no longer inevitable. Poverty is not only an evil in itself, but sustainable development requires meeting the basic needs of all and extending to all the opportunity to fulfill their aspirations for a better life.
> (United Nations World Commission on Environment and Development, 1987: 8)

The idea of sustainable development was from the start embedded in a particular model of growth with specific ethical content. That model has now changed. The 1970s model is based on the assumption of a virtuous form of growth in which the masses come to share in a general prosperity in which everyone's needs are met. Sustainable development originally expressed the concern to continue with this model while acknowledging limits to natural environmental resources. The model was carried forward in a number of UN fora and institutions (Habitat, UNEP, UNDP, the Rio Earth Summit and Rio Plus Five).

The 1970s model portrayed a socially sustainable capitalism. Where markets and enterprises did not meet human need, then governments would step in. This is the model of the welfare state explicated in the precepts of Keynes (1931, 1936), Myrdal (1956), Rawls (1971) and Shue (1980). A model which is both socially and ecologically sustainable has not yet emerged. Meanwhile the prospect of social sustainability has radically changed. Globalisation of the economy imposes a discipline which is patently no longer consistent with the spread of equality and the meeting of need. Since the 1970s the world has moved significantly backwards to an earlier model of capitalism, and this move has made 'ecologically and socially sustainable development' a much more distant prospect.

The discourse of urban sustainable development

What then is the scope of the discourse of urban sustainable development? The term 'sustainable development' can be understood in different ways. Thus, for example, Munasinghe (1993: 3) finds three approaches to sustainable development: the *economic approach* based on the idea of maximising income while sustaining 'natural capital', the *ecological approach* focusing on stabilisation of biological and physical systems and maintaining biodiversity, and the *socio-cultural approach* which stresses equity within and between generations. In this latter approach, 'preservation of cultural diversity across the globe, and the better use of knowledge concerning sustainable practices embedded in less dominant cultures, should be pursued' (ibid.). Mitlin and Satterthwaite (1996: 24) observe that, 'for many people writing on sustainable development, it is different aspects of development or of human activities that have to be sustained – for instance sustaining economic growth or "human" development or achieving social or political sustainability'.

In these authors' view, 'meeting the needs of the present' includes *economic need*: access to an adequate livelihood, economic security, social, cultural and health needs, *environmental need*: healthy, safe and affordable shelter, homes, workplaces and living environments free from environmental hazards, and the need for choice and control of homes and neighbourhoods; and finally *political needs*: participation in political decision-making within a broader framework guaranteeing civil and political rights 'and the implementation of environment legislation'. '*Without compromising the ability of future generations to meet their own needs*' means minimising the use or waste of non-renewable resources, the sustainable use of renewable resources, and ensuring that wastes from cities are kept within the absorptive capacity of local and global sinks (Mittlin and Satterthwaite, 1996: 31–2). In a normative analysis, Haughton and Hunter (1994: 17) identify three principles for sustainable development: intergenerational equity, social justice and transfrontier responsibility. Essentially these are principles of justice – between generations, between social classes and strata, and between places: 'where feasible, the impacts of human activity should not involve an uncompensated geographical distribution of environmental problems' (ibid.).

There is a difference between anthropocentric positions emphasising the meeting of human needs both now and in the future, and the more ecocentric positions of those who believe that the earth and its ecosystems have intrinsic value and should be allowed to develop in their own manner without separate reference to the needs of humans. To capture this dimension of sustainable development Baker *et al.* (1997: 8–18) erect a ladder of sustainable development. The bottom rung they term the 'treadmill' of development (Simon and Kahn, 1984). In this approach there is faith that technological development will solve all future environmental problems, the natural environment is viewed exclusively as the resource base for development. The further spread of international capitalism is given an unqualified welcome. On the next rung, 'weak sustainable development', we typically find the work of Pearce *et al.* (1989) who argue that environmental problems emerging from capitalist markets can be solved through the appropriate application of the principles of neoclassical economics. Both these rungs we have elsewhere termed 'market environmentalism' (Low and Gleeson, 1998: 160). The next rung is 'strong sustainable development'. Here the model of development must be politically adapted to ensure that production is geared towards environmental protection which is viewed as a precondition of development. This is akin to Blowers's (1996) interpretation of 'ecological modernisation' which demands the reduction of waste and pollution through technological improvements, the refinement of markets and regulatory frameworks to better reflect ecological priorities, and the 'greening' of social and corporate values and practices. In similar vein Weale (1992) envisages a highly interventionist state and commends, for example, the Swedish and Netherlands practices. Christoff (1996) divides ecological modernisation itself into 'weak' and 'strong' variants placing at the 'strong' end the work of Beck (1995) which demands a 'reflexive' modernity in which governments learn from experience (see also Dryzek, forthcoming).

At the top of the ladder Baker *et al.* (1997) locate the 'ideal model' of sustainable development. We think 'radical' is a more accurate term than 'ideal' since what is demanded is deeply rooted change in human values and institutions. This is the 'deep ecology' approach advocated by those such as Naess (1989), Ekins (1992), Goldsmith (1992) and Echlin (1996). In this ecocentric approach both humans and non-humans have intrinsic value and the latter cannot be reduced to mere instruments for the use of the former. What is to be sustained is not the spread of 'development' as in the Brundtland model or the global capitalism of market environmentalism, but the integrity of the Earth's ecosystems. There is, however, more than one model at the top of the ladder. Ecosocialism (Pepper, 1993; O'Connor, 1994; Schwartzman, 1996) and ecoanarchism (Bookchin, 1990, 1995a, 1995b) are two radical models with an anthropocentric stance. Ecofeminist models (Mathews, 1991; Merchant, 1992, 1996; Plumwood, 1993, 1997) also have a strong claim to a place on the radical top rung and, though generally ecocentric, must be distinguished from the models of the deep ecologists.

Sustainable development, then, is about the achievement on a global scale of three principles: economic development, social justice and ecological responsibility. These principles exhibit a dialectical tension. Sustainable development is in practice always likely to be a shifting compromise among them. The weight given to each of these principles in different philosophical approaches varies greatly and it may be argued that in some variants only two are present: for instance economic development and ecological responsibility in market environmentalism, and ecological responsibility and social justice in the ecocentric model. The common element is 'ecological responsibility' without which the discourse of sustainability cannot be distinguished from its antecedents.

Much of the literature explores the nature of the goals to be pursued, the policies and programmes (economic and institutional) to be formulated, the problems of implementation of sustainable development and difficulties with the interpretation of knowledge in a highly politicised discourse (see for example Blowers, 1993; Lidskog, 1996). The interpretive and implementational difficulties are considerable but they do not seem insuperable (any more than other contested political principles such as participation, justice or democracy). There seems little doubt that the three principles *can* be reconciled. The question, though, is *will* they? More precisely, will the current model of economic development associated with globalisation deliver all three?

The model of economic development

At first the trajectory of world capitalist expansion seemed destined to fulfil the agenda of development. Definite improvements in Third World living standards were noted (World Bank, 1991, 1992). The collapse of communism and the revelation of immense environmental problems in the former command economies made the capitalist growth model seem comparatively attractive (Carter and Turnock, 1993; Kapuscinski, 1994). In recent writing on sustainable development, however, there is growing concern that the model of global development which has sprung from the demise of Fordism is restricting the spread of both social justice and ecological responsibility (Lipietz, 1992). Rees (1992), argues that the economic logic of globalisation of production and the expansion of trade means that cities no longer feel the effects in any direct sense of their impact on the earth's environment. The governments of highly developed cities can import all the necessities – and luxuries – of life from far afield and export most of their wastes. Stephen Lewis, former Canadian ambassador to the United Nations, noted that the message of the Brundtland Report, which spoke to issues of the poor and homeless, the vulnerable and disadvantaged, 'has largely been disregarded by much of Western society, interested as it is in the balance sheets and in whether a given project results in more or less of the "permissible" levels of environmental degradation' (cited in Stren *et al.*, 1992: 5–6).

The 1996 report of the UN Centre for Human Settlements (Habitat, 1996) expresses an increased sense of urgency and even impatience: 'Already more than 600 million people in cities and towns throughout the world are homeless or live in life- and health-threatening situations. Unless a revolution in urban problem solving takes place, this numbing statistic will triple by the time the next century passes its first quarter' (ibid: xxi). The report argues that the problems of cities arise from inadequate planning: 'The need for planning becomes ever more necessary in the light of the increased social, economic, and environmental impacts of urbanization, growing consumption levels and renewed concerns for sustainable development since the adoption of Agenda 21' (ibid: xxxi). But where, in the current model of economic development, do we find demands for more *planning*, which, of necessity, entails more *intervention* by governments?

The Habitat report draws attention to a number of global changes which make adequate planning highly problematic. Debt burdens have meant that more and more countries have had to curtail expenditure on essential human services and physical infrastructure (Habitat, 1996: 8). Structural adjustment programmes (often enforced by the IMF) have made it increasingly difficult for national governments to find the funds to maintain, let alone improve, the quality of life of city populations (ibid: 162). The control of world trade has become much more centralised and 'national and city governments have increasingly sought to ensure their countries or cities remain competitive' (ibid: 9). While there have been substantial gains world-wide over the last 30 to 40 years, the report remarks, 'There is a growing body of data showing a slowing in social progress or even a halt or decline in some countries during the 1980s' (ibid: 99). The world's wealthiest cities 'have, to a considerable degree, transferred the environmental costs that their concentration of production and consumption represents from their region to other regions and global systems' (ibid: 155).

Discussions of the normative meaning of sustainable development, of programmes and policies addressing both markets and government institutions, of implementation issues and empirical evidence of failure to meet underlying principles remain today important, but their focus has been directed more towards the efforts of individual cities, nations and localities and at the positive requirements of sustainability than to the underlying global conditions in which these activities take place. If the basic model of global economic development being pursued by the powerful governors of the world economy is blocking sustainable development, then it is this model which must now become the focus of attention.

The period in which the discourse of sustainable development flourished and matured was also a period in which the model of global development changed very rapidly. Among the first to conceptualise this change were sociologists, human geographers (Harvey, 1987, 1989b; Lash and Urry, 1987) and neo-Marxist regulation theorists (Aglietta, 1974; Lipietz, 1985; Leborgne and Lipietz, 1988; Hirsch and Roth, 1987; Esser and Hirsch, 1989). In reality the so-called Fordist model was neither a global model – that is, a model for the whole matrix of the global economy – nor a universally adopted national model. There were many 'Fordisms'. Some contained a high degree of egalitarian democracy (Sweden, Germany), others were in varying degrees authoritarian (Singapore) or libertarian (USA, UK). Yet they all shared a concern for social security as a condition of economic development. These models are now being transformed by a wilder capitalism which erodes social security.

The shape of the emerging model has, since the early 1990s, become clearer. Lipietz as early as 1989 characterised the new 'societal paradigm' as 'liberal-productivism'. Deregulation, free trade, technological change, each justifying the other two 'like three mice chasing their tails' (Lipietz, 1992: 31). The consequences of this paradigm are intensified social polarisation, economic instability and ecological destruction. Certainly Lipietz's characterisation has a particular French flavour to it. But there are definite commonalities in observations from Britain (Hutton, 1995), the USA (Harrison and Bluestone, 1990; Athanasiou, 1998), Ireland (Douthwaite, 1992) and Germany (Martin and Schumann, 1997). The United Nations has added its own analysis to the literature of social and economic polarisation in the global economy (United Nations Development Programme, 1998a, 1998b).

Though these critiques vary in analytical focus and normative stance, they tell basically the same story. Since the early 1980s the gap between rich and poor has expanded in the developed world. Average living standards have declined, especially real wage levels (Douthwaite, 1992). Social polarisation is growing fast. The earlier analyses speak of an 'hour glass society' (Lipietz, 1992: 35) or a 'two thirds/one third society' (Hutton, 1995). Later work speaks of a 'twenty/eighty society' (Martin and Schumann, 1997). In this social prospectus 20 percent of the society has at least a comfortable, and at most a highly luxurious lifestyle derived in various ways from the profits of the corporate sector and financial institutions, 80 percent of the society either scrapes a living with temporary and casual work or, at worst, falls into periods of unemployment poorly compensated by the state, and health-threatening poverty. Twenty percent of the population consume 80 percent of the available environmental resources. 'Inequalities in consumption are stark. Globally, the 20% of the world's people in the highest-income countries account for 86% of total private consumption expenditures – the poorest 20%, a minuscule 1.3%' (United Nations Development Programme, 1998a, Summary, p. 2). Wackernagel and Rees (1996: 13–14) observe that the share of consumption of the earth's resources by the rich nations has been steadily increasing; North Americans now consume three times the global average: 'If everyone on Earth lived like the average Canadian or American, we would need at least three such planets to live sustainably' (ibid.: 13, and see Habitat, 1996: 155).

Signs that a country is doing less than it might to serve the short term profit needs of the major global financial institutions now bring instant retaliation from the markets in the form of speculation against the currency. What Martin and Schumann (1997: 41–6) describe with a sense of horror, Mexico 1995 – devaluation followed by the biggest financial rescue package in history – was repeated in double measure less than three years later for the mighty economic 'tigers' of East Asia. The impact of this financial meltdown on East Asian societies is incalculable. The new middle class has been badly hurt (see Chapter 13, this volume). During the crisis the world saw pictures of Koreans giving away their life savings in gold to the Korean government to pay back the IMF loans which, as the Harvard economist Jeffrey Sachs has acknowledged, are designed to help a few dozen international banks escape losses on risky loans (Hewett, 1998). Just as social strife increased in Mexico following the 'rescue', so it has followed in Indonesia, which was already moving into a period of political instability with the ageing of the Suharto regime, and Malaysia. The Asian crisis has now left the US dollar and European currencies bloated against Asian currencies. The crisis has spread to Russia and is propagating into Latin America, Canada and Australia. As we write, the US banking system is beginning to feel the effects. China cannot be immune.

This is a very different picture of development from the optimistic one promulgated in the Brundtland Report. We have seen the diminishing size of the middle band in society. Are we also now beginning to see the diminishing size of the top? Will the prospectus stop at 20:80 or move in ten years to 10:90 or yet further? Such a scenario bespeaks world deflation and depression. The question of the environment, both 'green' and 'brown', has now been problematised in global terms (Douthwaite, 1992; Lipietz, 1992, 1995, 1996; Martin and Schumann, 1997; Athanasiou, 1998). Let us now turn to the environmental question which is central to this book, the environment of cities, in this apparently new (since the late 1970s) but perhaps merely geographically restructured global system. In what follows we postulate a nexus of contextual issues which will later be reviewed in the light of commentaries on national urban policy systems (Chapters 3 to 13).

The urban environment in the global economy

Certain urban sustainability discourses have tended to assume that the city is an autonomous entity whose ecological health can be ensured through the regulation of endogenous social processes and the improvement of physical infrastructure. We argue against any such attempt to divorce the city from the broader contexts that sustain it. Indeed, as Harding and Le Galès (1997) have argued, countering the new urban problems and polarising tendencies thrown up by competitive entrepreneurialism is increasingly understood by urban governments in Europe to require national (and supra-national) interventions. In the contemporary world, the processes of economic and political globalisation have extended the horizons of the social and environmental systems that support individual cities. Indeed, globalisation has enhanced the integration of cities and their supporting systems within new regional, national and supra-national networks and alliances.

Increasingly, social scientists speak of urbanisation as a global phenomenon, which is to say a process that must be understood by reference to its wider systemic causes. Yet today's dominant discourse of 'planning' for cities is marked by two contradictory propositions. On the one hand we are invited to believe in the autarchy of cities as competing corporations – a concept which perhaps parallels that of the 'sovereignty' of consumers – and on the other hand we are told that global market relations are the principal determinants of city fortunes and that city-corporations must now conform or perish.

From the perspective of sustainability or 'ecological rationality' (Dryzek, 1987), it is of critical importance to understand how political and economic globalisation have changed the systemic context for urbanisation. Such an understanding will help to clarify both the ecological significance of cities and the appropriateness of the many strategies and theories that have been forwarded under the rubric of 'urban–sustainability'. In what follows, we shall examine this broader context of urbanisation from five systemic perspectives: demographic, economic, political-cultural, environmental and ecological. We emphasise the dynamic instability of this set of contexts and conceive cities as nodes with both centrifugal and centripetal functions – within these overlapping, interdependent, and sometimes contradictory, systems. Our key interest here is in the ecological consequences of the contemporary global context of urbanisation.

The demographic context: destabilised population growth

Cities are demographic phenomena: concentrations of population marked by various age, sex and health characteristics. Although cities are also concentrations of animal life, we are here concerned primarily with people. Although the rate of world population growth is diminishing, the number of humans continues to rise (from about 4.2 billion to about 5.7 billion over the last two decades). By one recent estimate, the world population will reach 7.2 billion by 2010, though annual growth rates will differ greatly among regions, ranging from 2.5 percent for Africa to –0.1 percent for Europe (UNDPCSD, 1997a). Moreover, urbanisation is continuing rapidly, meaning that this expanding population will increasingly reside in cities. In 1950 just 29 percent of the world's population lived in urban areas; by 1994, the proportion had risen to 45 percent and, by 2005, the United Nations expects that the majority of the world's population will live in urban areas, and approximately 40 percent of this urban population will be children (UNCHS, 1996: 20).

Not surprisingly, the United Nations sees human settlements in particular cities as key influences on humanity's environmental and social well-being in the coming century. The contemporary human settlement pattern is marked by rapid urbanisation, an increasing land area taken up by cities, and the growth of mega-cities, especially in the developing world. The term 'wild urbanisation' has been coined to describe the explosive growth of developing cities (Altvater, 1997). Of the fifteen largest urban agglomerations in 1950, four were in developing countries. In 1997 eleven out of fifteen are located in developing countries. These cities are already facing severe problems of urban degradation, industrial pollution, waste generation and general congestion (UNDPCSD, 1997a: 4). The United Nations believes that in 2015, thirteen of the fifteen largest mega-cities will be in developing countries: ten in Asia, two in Latin America and one in Africa (UNDPCSD, 1997a: 4). Just two, Tokyo and New York, will be located in the developed world.

City population growth is concentrated in the developing world, where it averages 5 percent per annum, compared to about 0.6 percent in industrialised countries (UNCHS, 1997a). Approximately 55 million people are currently added to the population of developing cities each year. On the face of it, this modernisation of developing countries is repeating the pattern of rural–urban population shifts which characterised historical urbanisation in the West. However, the rural–urban transition in developing countries has departed from the Western historical experience by occurring both at a greatly increased rate and often in the absence of broad-based economic growth (UNDPCSD, 1997b). In the larger cities of the developing world: São Paulo, Mexico City, Cairo, Lagos, Bombay, Shanghai and Beijing, dizzying rates of urbanisation mean that these urban areas are experiencing in just one generation what London went through in ten or Chicago in three (Harvey, 1996: 43). Urban growth is 'destabilised' by the 'combination of poverty, rapid population growth and environmental damage' (UNDPCSD, 1997a: 4).

In addition to economic imperatives, a succession of local disturbances, civil wars, regional conflicts and rural ecological degradation have combined to encourage migration to cities, both in the developing world and in former Soviet bloc states. Indeed, ecological problems are emerging as a major source of forced migration and urbanisation. In 1996, the International Organisation for Migration (IOM) estimated that 25 million persons are environmentally displaced world-wide (UNDPCSD, 1997a: 9). Slums and squatter settlements are now home to an estimated 25–30 percent of the urban population in the developing world (UNDPCSD, 1997b). At the household level, there are also major gulfs between the situations in rich and poor cities: the average household in developing cities contains 6.5 persons, whilst, in developed cities, the figure drops to an average of 2.5 persons (UNCHS, 1997a). The United Nations believes that the situation is deteriorating in the developing world, especially in African cities where household sizes are rising as real incomes fall.

This is not the place for a full analysis of global migration patterns and their connection to urbanisation processes. What we wish to emphasise here is the increasing movement of humanity towards city living, and the extra-ordinary diversity of demographic contexts in which this movement is realised.

The economic context: marketisation, poverty and spatial polarisation

Since the Second World War, the increasing globalisation of the market economy has furthered the integration of cities within flows and circuits of trade and investment. This tendency to integration has been riven by increasing contradictions at the urban scale, where circuits traced by various fractions of capital: financial, merchant, industrial-manufacturing, property, statist and agro-business, intersect in competition for shares of local economies. Within the global urban system, an upper echelon of world cities (typically, London, Tokyo and New York) has emerged which is more attuned to the rhythms of the global economy than to those of the nation-state in which it is located (Sassen, 1991; Warf, 1996: 40). These urban centres have played a disproportionate role in the production and transformation of international economic relations in the late twentieth century. However, no major city remains completely outside the expanding sphere of market relations, and all are drawn increasingly within the processes of capitalist urbanisation (Harvey, 1996: 46).

Cities may be now linked, as never before, in a global economic system, but this structural unity coexists with, and indeed fosters, a widening gulf between wealth and social conditions in different urban areas. Some sections of society in developing nations became enormously enriched, especially in East Asia (see Chapter 13). Other regions, such as sub-Saharan Africa, suffered terribly in recent decades as national debt ballooned and real incomes fell, catalysing political instability and war. In the former Soviet bloc countries, the introduction of market relations has brought impoverishment and insecurity for many, creating new elites and power alliances, and fostering the growth of informal economic sectors. Globalisation, then, has been a deeply duplicitous process causing both heightened integration of national economies and international policy convergence on the ideal of free trade, whilst at the same time encouraging new forms of political instability, social fragmentation and environmental stress – which we discuss below (see also UNCHS, 1996: 2).

Within the Western world, social commentary points to the 'postmodernisation' of cities, meaning *inter alia* increases in both cultural pluralisation and social polarisation, and the rise of new, flexible patterns of consumption and production that have served in recent decades to renew the processes of capitalist accumulation (Harvey, 1989a). As the key intersection points for ever expanding global flows of traded goods, money and services, Western cities can now offer their citizens (at least those who can pay) new consumption patterns drawn from nearly every imaginable geographical and cultural context. The fabulous expansion in the range and origins of traded goods results both from the proliferation of tastes, associated with complex changes to culture, mobility and production, and the increasing availability of cheap commodities produced in developing countries. But, as Harvey puts it: 'urbanization in the advanced capitalist countries . . . has not in recent history been about sustaining bioregions, ecocomplexes, or anything other than sustaining the accumulation of capital' (1996: 46).

This contemporary pattern of renewing accumulation through rapid urbanisation increases the entropic, and fundamentally anti-ecological, tendencies of the market: 'beyond a certain point, the continuous growth of the economy . . . can be purchased only at the expense of increasing disorder (entropy) in the ecosphere' (Wackernagel and Rees, 1996: 43, and see Altvater, 1993, 1997). It is often claimed that new communication technologies slow the rate of entropy by reducing the need for travel and the transport of goods. However, this claim is doubtful given the demonstrated power of these new technologies to reshape cultural horizons and thus encourage greater mobility by the many who can now afford to travel by air and motor transport (see UNCHS, 1997b: 13).

As the United Nations Development Programme (UNDP) put it recently, 'Today's consumption is undermining the environmental resource base. It is exacerbating inequalities. And the dynamics of the consumption–poverty–inequality–environment nexus are accelerating' (UNDP, 1998a: 1). The gulf between the developed and developing world is widening. Whilst consumption per capita of industralised countries has been steadily rising over the last quarter of a century, at about 2.3 percent annually, growth has been slow or stagnant in many parts of the developing world. According to the UNDP, 'The average African household today consumes 20% less than it did 25 years ago' (1998a: 2). Overall, the amount of people in the world living in poverty is increasing steadily (UNDPCSD, 1997c: 8). The absolute poor now number about 1.3 billion people. Noting World Bank data on poverty and living conditions for the period 1987–93 (Table 1.1), there appears to have been some improvement in one important indicator of well-being, infant mortality, in even the poorest regions, such as sub-Saharan Africa. However, this welcome finding must be measured against the other data which show a significant rise in the absolute numbers of poor globally, concentrated in certain regions (Latin America/Caribbean, sub-Saharan Africa and South Asia) (UNDPCSD, 1997c: 8). Moreover, mortality rates remain catastrophically high in much of sub-Saharan Africa. In Sierra Leone, for example, the UNDP estimated in 1995 that 50 percent of the national population would die before reaching the age of 40 – Uganda, Malawi, Zambia, Guinea Bissau and Chad all had similarly high mortality rates (UNDP, 1998b).

In developed countries there has also been an appreciable increase in polarisation as a result of cuts to social protection systems developed in the aftermath of the Second World War (UNDPCSD, 1997c: 9). One major study of international income inequality trends in developed countries over the last two decades found that increases in polarisation had been most

Table 1.1   Trends in world poverty

	Incidence of poverty %			Number of poor (millions)			Infant mortality (per 1000)	
	1987	*1990*	*1993*	*1987*	*1990*	*1993*	*1987*	*1993*
East Asia and Pacific	28.2	28.5	26.0	464	468	446	44	35
Latin America and Caribbean	22.0	23.0	23.5	91	101	110	49	43
Middle East and North Africa	4.7	4.3	4.1	10.3	10.4	10.7	67	53
South Asia	45.4	43.0	43.1	480	480	515	97	84
Sub-Saharan Africa	38.5	39.3	39.1	180	201	219	103	93
Total	33.3	32.9	29.4	1,225	1,260	1,301	63	54

Source: United Nations Department for Policy Co-ordination and Sustainable Development (1997c)

pronounced in countries that have implemented a neo-liberal political–economic agenda – notably New Zealand and the United Kingdom (Rowntree Foundation, 1995). The UNDP's new human poverty index indicates that contemporary poverty rates range from 7 to 17 percent in developed nations (UNDP, 1998b: 2).

The increasing disparity in the share of wealth between and within countries partly explains the explosive rates of urbanisation seen in African, South Asian and South American countries (UNDPCSD, 1997a: 4). A major source of this growing disparity is the indebtedness of poor countries to wealthier nations. The present debt crisis of the developing world, which has lasted for about ten years, has both eroded living standards and caused a prodigious overexploitation of natural resources (Altvater, 1993: 125). Between 1982 and 1989, the poorer countries transferred a net total of US$236.2 billion to rich nations. By contrast, between 1992 and 1996 total development aid provided by countries of the OECD Development Assistance Committee declined in real terms from US$62 billion to US$50 billion. Moreover, the share of this aid provided to the least developed countries also declined during the same period (UNDPCSD, 1997c).

In *developing* cities, social asymmetry is reflected in rigid, and frequently policed, separations of wealthy and poor residential zones, though extremes often exist in close proximity. However, Western perspectives of the new polarisation (e.g. Sudjic, 1996: 36; Atkinson, 1996: 6) may conceal emergent indigenous social and environmental models of development as Jellinek (Chapter 13) reminds us. Within *developed* cities: certain life-threatening forms of poverty, notably homelessness and sweated labour, have increased in recent decades and diseases of poverty such as tuberculosis have reappeared (Badcock, 1995; Wolch and Dear, 1993; Sassen, 1991). These trends embody many socio-structural, regional and environmental changes too complex to detail here. Suffice it to say that the general worsening of income polarisation in many developed countries in recent decades has been paralleled by increasing residential differentiation in most major cities (O'Loughlin and Friedrichs, 1996).

The political-cultural context: neo-liberal ascendancy and 'competitive cities'

As Sjölander tells us, globalisation brings with it the internationalisation of the state and the globalised diffusion of neo-liberalism (cited in Altvater, 1997: 1). Neo-liberalism embodies many nostrums of neo-classical economics (e.g. the unquestioned benefits of free trade and the verities of competitive advantage) and combines them with more recent strands of liberal thought, notably Hayek's hostility to social justice and Nozick's erstwhile enthusiasm for the minimalist state and an etiolated public sphere. Coupled with these simplistic nostrums goes rhetoric about the generative and supportive effect of the market for democracy, and the leading effect of the market in the eradication of poverty. These lies and half-truths are thrust unceasingly before governments by a world-wide network of 'independent' foundations funded by the transnational corporations (see Self, 1993; Cockett, 1996).

The collapse of the Long Boom in the mid-1970s, and the resulting fiscal crises of many developed nation-states, ensured a receptive ideological climate for neo-liberalism. Across the Western world, trade problems mounted, unemployment and inflation soared, and pressure increased dramatically on state finances. In a climate of stagflationary recession neo-liberal politics gained ascendancy in the English-speaking world, offering bold solutions – monetarism, deregulation and privatisation – to the twin crises that had beset the state and the economy. The general effect of these policy shifts was a progressive withdrawal of supportive intervention on the part of the state to compensate for 'market failure' and increased intervention to provide the 'firm government' needed to ensure increasing reliance upon the market. At the same time, the increasing (though often rhetorically exaggerated) importance of globalisation added weight to the central argument of neo-liberals that states must surrender their attachment to national economic planning. However, unemployment levels continue to be high, the quality of working life has deteriorated, income polarisation has grown, and access to valued consumption goods and services, including decent housing, has declined for many. In developed cities, these economic and social changes have been revealed in new geographic patterns of employment and industry, and, as explained earlier, rising levels of residential segregation.

Whilst long seen as the English-speaking disease (Bennetts, 1997), neo-liberal ideology and policies have in the 1990s gained new purchase in Continental Europe, former Soviet bloc states and the developing world. Although often encountering fierce political resistance in many of these regional contexts, sponsorship by powerful international bodies, such as the OECD and the World Bank, has seen neo-liberalism make steady advances in most national policy frameworks. In some regions, notably the EU, neo-liberalism has percolated deeply through layers of national and regional governance, but has encountered increasing popular antipathy, and in some cases, dramatic and widespread resistance (e.g. trade union led protests in France and Germany during 1996). The ideology's power derives partly from its apparent common-sense affirmation of the enormous concrete changes that globalisation has brought to most nations, including exposure to new technologies that speed the process of communication and trade and inflate corporate managerial structures whilst reducing the effectiveness of national regulatory systems. Neo-liberalism appears as the ideological corollary of globalisation by insisting that market relations are spaceless as well as timeless, and can neither be contained within national policy frameworks nor avoided.

Like most other domains of public policy, city planning and urban servicing have increasingly been made servants of neo-liberal political economy in many Western countries (see for example Forster, 1995: 70). The processes of economic globalisation, technological change, state fiscal crisis and neo-liberal politics have combined to encourage a spiralling place competition between cities eager to attract investment capital. Harvey (1989b) tells us that in Western (particularly English-speaking) cities, the character of urban governance has shifted from managerialism to entrepreneurialism, though arguably managerialism has merely changed to reflect the corporate form. Cities have been recast as players in a rough and tumble pursuit of highly mobile capital, a game played both in national and international leagues (Warf, 1996).

Allied with this new urban entrepreneurialism has been an increasing emphasis on deregulation, the removal of land-use and other spatial controls which supposedly cause frictions for development (e.g. see Fainstein, 1991; also Gaffikin and Warf, 1993, on the United States and Britain, and Badcock, 1995, and Stilwell, 1993, on Australia). Throughout the West, by the 1990s, orthodox welfarist prescriptions to combat urban decay had been greatly discredited, including even the bastions of state welfarism in Scandinavia (see Chapter 10). The new urban agenda heralded the saving graces of an 'enterprise' culture which alone would liberate marginalised populations from their demoralising dependencies (Warf, 1996: 41).

In summary, it can be said that urban policy has failed to prevent, and perhaps even abetted, the worsening spatial polarisation evident in Western cities. Planning in many contexts has been given over to trickle down economics that have clearly failed to enrich any but the wealthy, the cultural elites, and the locationally advantaged. One notable departure from this depressing assessment is the recent urban and environmental policy thrust of the EU, where there has been a noticeable shift towards new and strengthened forms of spatial regulation (Newman and Thornley, 1996).

What are the possible consequences of the shift to urban entrepreneurialism? In particular, what prospects does the increasingly obligatory game of place competition hold for the populations of developing cities? A glimpse can be caught in United Nations data that compare public expenditure on basic settlement infrastructure across a range of global regions defined both by wealth and geography (Table 1.2). The data reveal shocking disparities in urban investment between rich and poor countries. Spending on infrastructure by High Income Nations, for example, is more than fifty times that of the poorest group of countries. As the United Nations notes, 'All indications are that countries that do not have sufficient levels of urban infrastructure and services, as well as good urban management, are being sidelined by the economic changes and globalization processes under way' (UNCHS, 1997b: 13).

It seems that in the West, spending on urban infrastructure remains comparatively high, though increasingly maldistributed and geared to provision for elites and foreign capital. In the Asia-Pacific region some cities, Jakarta (up to October 1997) for example, sought to copy the model of Singapore by drafting international capital into the creation of vast physical development schemes. The Singapore model combines a narrow bureaucratic focus on engineering works (notably high-rise housing) with authoritarian governance, social rationing according to wealth, and the legitimacy stemming from genuine improvements in provision for basic needs (sanitation and water supply). In the developing world, however, overall investment remains low, disastrously so in many countries, and here the adoption of neo-liberal growth machine politics can only worsen access to services for the poor.

This pattern of inequality of urban development between rich and poor nations is confirmed by another United Nations study which calculated a City Development Index (CDI) for 236 cities, grouped by region (UNCHS, 1997a; see Table 1.3). The CDI is an aggregate indicator of urban social development containing variables that measure child mortality, city product and investment in facilities. From these data, the United Nations draws the obvious conclusion that, 'policy is important. If countries have invested their income in physical and social infrastructure, then dividends will be received in social welfare' (UNCHS, 1997a: 9).

Table 1.2 Government expenditures per person on water supply, sanitation, drainage, garbage collection, roads and electricity

Income grouping (cities in)	*$US per person*	*Regional grouping (cities in)*	*$US per person*
Low-income countries	15.0	Sub-Saharan Africa	16.6
Low-mid-income countries	31.4	South Asia	15.0
Middle-income countries	40.1	East Asia	72.5
Mid-high-income countries	304.6	Latin America and the Caribbean	48.4
High-income countries	813.5	E. Europe, Greece, Nth Africa and Mid East	86.2
		W. Europe, Nth America, Australia	656.0

Source: UNDPCSD (1997c: 12).

Table 1.3 City Development Index (CDI) ranking for 236 cities, totals by region

	Africa	Arab	Asia/Pacific	HIC[a]	LAC[b]	Trans[c]	Total
CDI							
High	1	1	0	34	2	19	57
Med.	5	6	16	0	27	12	66
Low	81	3	26	0	3	0	113
Total	87	10	42	34	32	31	236

Source: UNCHS (1997a: 9).
Notes: [a]Highly Industrialised Countries.
         [b]Latin America and Caribbean.
         [c]Transitional (former socialist states).

The tendency to greater inequality of access to basic urban services, both between rich and poor countries and within all cities, entrenches locational disadvantage and social exclusion. Observing this, the United Nations recently called for policies which increase such inequality to be replaced by the promotion of social cohesion and human solidarity (UNCHS, 1997b: 13). However worthy these sentiments, the United Nations, like other mainstream global institutions, seems unwilling and/or unable to challenge the process of globalisation as it is presently constituted.

The tendency of globalisation, in conjunction with neo-liberal policies, to disenfranchise whole social strata economically and politically is hard to deny. Left to auction themselves to the highest corporate bidder, localities find themselves in a race to the bottom in which entrepreneurial states promote growth without regulating its aftermath (Warf, 1996: 43). The marketisation of all social relations depoliticises democracy by constraining the political–economic, and therefore social, choices open to polities at any level (Altvater, 1997: 5; Lindblom, 1982; Saul, 1997: 115). For Safier (1996), globalisation and the increasing marginalisation of the poor and certain cultural groups (notably immigrants) from the mainstreams of political–economic power are raising the prospects of a dramatic rise in urban social conflict throughout the world.

Safier, like a number of observers (e.g. Sandercock, 1998), welcomes the tendency to cultural pluralisation that has in part been encouraged by globalisation and attendant social changes (notably, cultural postmodernisation) and population shifts (such as, migration and rising mobility). Amidst the contemporary shifts in culture and demography he reads the potential for a global cosmopolitan democracy which could transcend the antagonisms of the past that have been based on notions of cultural purity and superiority (see also Rosenau, 1998). However, at the same time, neo-liberalism and marketisation have exacerbated socio-cultural divisions and eroded the capacity of states to promote the cohesion and social solidarity that the United Nations so desires. The abandonment of many social strata is reflected in new racial, religious, ethnic and economic tensions, all of which are contributing to an increasing disorder across the urban world. New political–economic and cultural relations are needed to counteract the destructive tendency to social atomisation and to avoid the Hobbesian scenario of urban disorder that Safier fears is nearly upon us.

Environmental distribution

In the past, Western political frameworks have valued the environment in instrumental terms, as a resource to be exploited for the production of use values which can then be distributed amongst communities and within humanity in general. But we are well aware now of the inadequacy of this ethical viewpoint and the disastrous environmental consequences of the industrial transformation of nature over the past two centuries. Local community conflict has emerged over the sharing out of environmental benefits and burdens. An 'environmental justice' movement has grown up in the USA around the tendency of high risk industrial processes to be located in places occupied by the poor and the coloured (African American, Latino). But there is nothing peculiar to the USA, or even to the developed world, about such events. Nor are they new. In the past governments have stepped in to resolve such conflicts – through health, sanitation, building and planning regulations (Harvey, 1996) Now these historical gains are threatened in developed cities and, more worryingly, there is a world-wide dimension to the problem. As Ulrich Beck (1992, 1995) has explained, capitalist modernity has produced potent industrial residuals which threaten human and non-human life at every geographic scale.

Pulido (1994) has observed that the political successes of the environmental movement in developed countries may actually accelerate the relocation of hazardous industries to developing nations. This traffic in risk may offer poorer nations the opportunity for modernisation and economic development in the short run, but at the cost of entrenching the injustice of global uneven development. Environmental regulations are increasingly cited by US firms as a reason for their flight to more 'business friendly' countries, such as Mexico. The profits from industrial plants, as well as their products, are largely exported to the country of the operating firm. The host nation incurs the risks which attach to the hazardous industry. This system permits developed countries to externalise industrial risks by moving hazardous forms of production beyond their borders. Firms enhance their profits through the imposition of 'cross-border externalities', given that the nations which host hazardous production may never be fully compensated for the spillover effects of these activities (i.e. environmental degradation, social dislocation).

Another dimension of the traffic in risk is the toxic waste trade. Indeed, Beck (1995: 134) believes that the world-wide traffic in toxic and harmful substances is a defining characteristic of the present age: the 'risk society'. As he puts it so evocatively, 'Supranational groups of regions and countries swallow poisons and waste on others' behalf' (Beck, 1995: 154). In 1990 the United Nations estimated that the world was producing between 300 and 400 million tonnes of hazardous wastes annually, about 98 percent of which was generated by OECD countries, much of it sourced in urban areas (Greenpeace International, 1994). Some of the trade occurs within the developed world. In the late 1980s, for example, it was estimated that 100,000 waste transfers occurred annually within Europe (Smith and Blowers, 1992). However, a significant amount of the commerce in wastes involves transfers of domestic and industrial refuse (both toxic and non-toxic) from developed nations to poorer countries (ibid.: 212). According to Greenpeace, Germany is the largest waste exporter in the world, and in 1993 shipped over 600,000 tonnes of hazardous wastes to ten different countries in Europe (including former Soviet bloc nations) and to the developing world (Edwards, 1995a). The United States in 1992 exported over 145,000 tonnes of toxic wastes abroad, with large amounts being shipped to Canada and Mexico (Edwards, 1995b). Smith and Blowers (1992) detail the export of wastes, some of which included radioactive materials, by both the United States and European countries to Africa during the late 1980s with Guinea-Bissau as a major destination, that country being offered the equivalent of its then existing GNP (some $US120 million) to dispose of European hazardous waste in landfills (ibid.: 212).

The traffic in waste and other environmentally injurious development is worsening international inequity and helping to sustain risky industry throughout the globe. The absence of a supervising state (and the United Nations in its present form cannot yet perform this role) means that a distributional framework cannot be readily applied to the international traffic in risk. Those international agreements which have sought to control aspects of the traffic in risk, such as the Basel Convention, have been shown both to be vulnerable to political attacks by recalcitrant states and difficult to enforce. Even so, notwithstanding the limitations of the Basel Convention, this attempt at an inter-regional regulation of the waste trade foreshadows the potential for more effective controls within the context of global-institutional regulation. Epochal structural changes, including economic globalisation, the mobility of capital (and risk), and the collapse of Cold War antagonisms, have created a new geopolitical context for ecological politics:

> We are on the threshold of a new phase of risk-society politics; in the context of disarmament and the relaxation of the East–West tension, the apprehension and practice of politics can no longer be national but must be international, because the social mechanism of hazard situations flouts the nation-state and its systems of alliance.
>
> (Beck, 1995: 162)

Indeed, we argue that this new international political practice, of which Beck speaks, must seek to eliminate the flourishing traffic in risk which is already worsening the legacy of global uneven development bequeathed by centuries of colonialism and capitalism. This new ecological politics requires a new global institutional context which can both problematise the production of risk and regulate the distribution of hazards between states.

Ecological distribution

In the present context of globalisation, cities are 'growth machines' geared to an ever-expanding consumption of nature (this apt term stems from Molotch, 1976). As we have argued earlier, the urban sustainability discourse is limited by its failure to confront the deeper causes of unsustainability. Most important among these causes are the current growth model of global capitalist economy and its entropic logic of accumulation. Linked to this is the failure of modernisation to accord moral significance, and therefore rights, to non-human nature.

Major global institutions promote growth, indeed accumulation, as the necessary precondition for correcting all unsustainable development patterns (UNDPCSD, 1997b: 31). But capitalist growth, as Altvater (1993, 1997) has shown, is fundamentally entropic, and therefore anti-ecological, because it is geared to the endless expansion of value, a process predicated on the increasing consumption of nature. Productivity increase is one of the main features of industrial capitalism in general and of the Fordist system in particular (see Lipietz, 1992, 1996). Increase of productivity only becomes possible by using more fixed capital – and consuming growing quantities of matter and energy (Altvater, 1997: 14). Conventional urban sustainability analyses, trapped in the entropic logic of capitalist accumulation, can at best formulate schemes whereby the exhaustion of energy and corruption of nature is slowed. This merely delays the moment when globalisation finally overreaches the ecological frontiers laid down by the finite quantity of materials and energy.

Altvater does not argue for a simplistic 'limits to growth' thesis, for not all growth is consumptive (see Jacobs, 1991, 1995). Rather he posits that we must cease to regard the market as the 'natural' mechanism for defining and fulfilling humanity's environmentally patterned needs. If the process of defining and satisfying human values were changed fundamentally through its democratisation there need be no reason to prevent growth remaining as a key social aspiration, albeit in a form very different to the present logic of accumulation with consumptive growth crudely measured by business activity–GDP.

Looking more closely at patterns of growth under the present development model, it is of course clear that not all cities consume nature in similar ways or at the same levels. Defining the 'ecological footprint' of a city or country is a way of accounting for the flows of energy and matter to and from its economy and 'converting these into the corresponding land and water area required from nature to support these flows' (see Wackernagel and Rees, 1996: 3). Table 1.4 shows the footprints of different countries according to the calculations of Wackernagel and Rees. However, the dissimilarity of consumption patterns is reducing as globalisation encourages the

Table 1.4 Ecological footprint analysis of selected countries following Wackernagel and Rees (1996: 97–9)

	Ecological footprint[a] (ha. per capita)	Ecologically productive land within the national territory[b] (million ha.)	Land surface consumed[c] (million ha.)
USA	5.1	725.6	1,315.8
Britain	3.0	20.3	174.0
Germany	3.0	27.7	243.9
Japan	2.0	30.4	250
India	0.38	250.0	345.8
Australia	3.74	575.9	66.9

Notes:
a Calculated by assessing the amount of land surface area of the earth for all uses (including, for example, food supplies, energy production, housing, transport, carbon sinks) per capita.
b Land which can be used for the 'ecological services' required by people (cropland, permanent pastures, forests); does not include built-up areas, roads, deserts, wilderness.
c Ecological services consumed (ecological footprint × population) expressed in terms of land surface area.

growth of neo-Fordist consumption, and neo-Taylorism in the developing world. For the Fordist growth model 'must of necessity emancipate itself from human limitations and initiate and maintain a . . . spiral of accumulation and spatio-temporal expansion' (Altvater, 1993: 51). An increasing proportion of our expanding humanity want what is so immeasurably praised: the material gratifications promised by the Fordist production–consumption model (Altvater, 1993: 52). Here, Western ideals of sustainability may well melt quickly in the heat of rising expectations and desires.

None the less, whilst growth patterns in developed and developing countries are converging, at least materially, there remain substantial differences in the ecological demands made by cities in both regions. Developed cities are still, by far, the predominant consumers of nature (see Table 1.4). In the three decades between 1973 and 1993, world energy consumption increased by 50 percent. However, in 1993 industrialised countries still accounted for over 60 percent of total energy consumption, though their share continues to decline as the Fordist system extends its spatial reach into the developing world (UNDPCSD, 1997b). The costs of this system – ecological, social and financial – must in time be paid in full.

The progress of capitalist expansion, it seems, can be read directly in the rising concentrations of greenhouse gases in the earth's atmosphere (see Houghton *et al.* 1996). Consequent global warming threatens major disruptions – indeed catastrophes – within human and non-human systems. The twin dynamics of ecological exhaustion and social deprivation (relative and absolute) can be expected to catalyse yet further migrations to the wealthy heartlands, and yet more efforts in those privileged nations to keep the mass of deprived and threatened humanity at bay. Altvater speaks of an impending crisis of civilisation as eco-refugees join industrial (economic) refugees in streaming from areas of ruin and risk to nations which have established sustainable alternatives to Fordism's massive discharge of useless and toxic wastes. Indeed, 'humanity . . . is becoming an antiquated being precisely in a self-created environment, whose threat to the foundations of life is also destroying the natural . . . conditions underpinning civilization' (Altvater 1993: 51).

The main institutional ideologies within Western environmentalism, notably free-market environmentalism and ecological modernisation, continue to deny the real systemic origins of the crisis facing the globe. Moreover, these discourses, and the regional and global policy agendas they inform, fail to apprehend the singular ecological issues facing the impoverished masses of developing cities – including, of course, those of the so-called transitional former Eastern bloc (Carter and Turnock, 1993; Harvey, 1996). Here life, not merely amenity, literally hangs in the balance, suspended within critical household-level problems, such as indoor air quality and sanitation. The most immediate threat to health and quality of life comes from forms of household pollution – bad air and water – that receive little attention from Western environmentalists (see Chapters 8, 9 and 13).

> We must recognize that the distinction between *environment* as commonly understood and the *built environment* is artificial and that the urban . . . is as much part of the solution as it is a contributing factor to ecological difficulties. The tangible recognition that the mass of humanity will be located in living environments designated as urban says that environmental politics must pay as much if not more attention to the qualities of those built and social environments as it now typically does to a fictitiously separated and imagined 'natural' environment.
>
> (Harvey,1996: 60, original emphasis)

We share this vision for an ethically honed and critically aware urban environmentalism. An environmentalism which joins the 'natural' with the humanly constructed (both materially and ideally) and conceptualises their relationship must now, we think, inform ecological ethics and politics.

Conclusion

The aim of ecological sustainability for the city is wholly laudable but, as the global norm it must become, it seems to be a distant prospect. The global picture is far from encouraging. Economic forces at present beyond the control of any conscious human processes are working precisely in the opposite direction. A world of individual cities, tightly constrained by their geographic boundaries, competing for 'growth' (defined purely as business activity) within a global market framework is not one which, in the long term, appears capable of ecological sustainability.

The hope of a world economy formed around the implementation of 'clean' production is vain if there is no force to ensure that economic incentives in favour of 'dirty' production are reversed. Incentives need to be applied not only to industry but to governments to implement clean production legislation, policies and programmes. At present the incentives point in the other direction. The 'consuming city' of the developed world is merely the counterpart of the new 'producing cities' of those regions of the global economy to which manufacturing capital has moved. Such cities are attractive to investment precisely because they offer the least regulatory friction to dirty production. Many of the new producing cities suffer from environmental problems as severe as those of the old producing cities of the late nineteenth century. The consuming city on the other hand also creates enormous conflicts over resources and environments, and the UN has admitted that the developed world, under the present global regime, is settling into a condition of social polarisation with a permanently impoverished underclass.

Creating the conditions which will reverse the economic incentives acting on governments is a task which will require re-regulation of the global economy. We doubt if this can be done with the present system of international governance. There are too many loopholes and not enough locations for real political pressure to be applied. Martin and Schumann (1997) argue for democratisation of European government. We have elsewhere argued for democratisation of global governance (Low and Gleeson, 1998). What is really needed is a democratisation at all scales, 'cosmopolitan democracy', in which global democratic institutions function to create or restore real democracy at regional, national and local levels, the restoration of true citizenship at every level (Held, 1995).

But the dismal hypothesis that the urban world as currently governed cannot move towards ecological sustainability needs to be more carefully examined. Despite the thrust of much rhetoric about the independence of city governments, nations and their states remain the locus of political regulatory power in the world. We need to discover if nations have taken up the challenge laid down at the Rio Summit for an 'Agenda 21', an agenda for ecological sustainability to which the cities of the nation will contribute in a systematic way. We need to know if there is indeed a prospect of the regulatory power of the nation-state being brought to bear to create a national framework within which cities can formulate policies for ecological sustainability. If the dismal hypothesis proves correct then the world will have to look beyond the nation-state to make changes to the governance of the global context within which nation-states formulate policy.

References

Aglietta, M. (1974) *Accumulation et régulation du capitalisme en longue periode. Example des Etats Unis (1870–1970)*, Paris: INSEE.
Altvater, E. (1993) *The Future of the Market*, London: Verso.

—— (1997) 'Restructuring the Space of Democracy: the Effects of Capitalist Globalization and of the Ecological Crisis on Form and Substance of Democracy', Paper presented to the Environmental Justice, Global Ethics for the 21st Century Conference, University of Melbourne, Melbourne, October 1–3.

Athanasiou, T. (1998) *Slow Reckoning, The Ecology of a Divided Planet*, London and New York: Verso.

Atkinson, A. (1996) 'Sustainable Cities: Dilemmas and Options', *City*, 3/4: 5–11.

Badcock, B. (1995) 'Towards More Equitable Cities: a Receding Prospect?' in Troy, P. (ed.), *Australian Cities: Issues, Strategies and Policies for Urban Australia in the 1990s*, Melbourne: Cambridge University Press, 196–219.

Baker, S., Kousis, M., Richardson, D. and Young, S. (1997) 'The Theory and Practice of Sustainable Development in EU Perspective', in Baker, S., Kousis, M., Richardson, D., and Young, S., *The Politics of Sustainable Development: Theory, Policy and Practice within the European Union*, London: Routledge.

Beck, U. (1992) *Risk Society, Towards a New Modernity* (tr. Mark Ritter), London: Sage.

—— (1995) *Ecological Politics in an Age of Risk*, Cambridge: Polity Press.

Bell, D. (1997) *Consuming Geographies: We are where we eat*, London: Routledge.

Bennetts, S. (1997) 'New Project to Spark Debate on English-Speaking.

Disease', *Uniken* (University of New South Wales), 426: 1, 4.

Blowers, A. (1993) 'Environmental Policy: the Quest for Sustainable Development', *Urban Studies*, 30 (4/5): 775–96.

—— (1996) 'Environmental Policy – Ecological Modernisation or the Risk Society?', unpublished paper, copy obtained from author, The Open University, Walton Hall, Milton Keynes, United Kingdom.

Bookchin, M. (1990) *Remaking Society, Pathways to a Green Future*, Boston: South End Press.

—— (1995a) *From Urbanization to Cities, Towards a New Politics of Citizenship*, London: Cassell.

—— (1995b) *Re-enchanting Humanity, A Defense of the Human Spirit against Anti-humanism, Misanthropy, Mysticism and Primitivism*, London: Cassell.

Breheny, M. (1990) 'Strategic Planning and Urban Sustainability', *Proceedings of the Town and Country Planning Association Annual Conference*, London: TCPA.

Carter, F.W. and Turnock, D. (1993) *Environmental Problems in Eastern Europe*, London: Routledge.

Christoff, P. (1996) 'Ecological Modernisation, Ecological Modernities', *Environmental Politics*, 5(3): 476–500.

Cockett, R. (1996) *Thinking the Unthinkable: Think Tanks and the Economic Counter-revolution, 1931–1983*, London: HarperCollins.

Dryzek, J. (1987) *Rational Ecology, Environment and Political Economy*, Oxford: Blackwell.

—— (1997) *The Politics of the Earth: Environmental Discourses*, Oxford: Oxford University Press.

—— (1999) 'Global Ecological Democracy', in Low, N.P. (ed.) *Global Ethics and Environment*, London: Routledge.

Douthwaite, R. (1992) *The Growth Illusion, How Economic Growth has Enriched the Few, Impoverished the Many, and Endangered the Planet*, Bideford, UK, Green Books.

Echlin, E.P. (1996) 'From Development to Sufficiency', *The Aisling*, 18: 32–4.

The Ecologist (1972) *A Blueprint for Survival*, Harmondsworth, London: Penguin Books.

Edwards, R. (1995a) 'Dirty Tricks in a Dirty Business', *New Scientist*, 18 February: 12–13.

—— (1995b) 'Leaks Expose Plan to Sabotage Waste Treaty, *New Scientist*, 18 February: 4.

Ehrlich, P. (1971) *The Population Bomb*, London: Ballantine.

Ekins, P. (1992) *Wealth Beyond Measure*, London: Gaia Books.

Esser, J. and Hirsch, J. (1989) 'The Crisis of Fordism and the Dimensions of a "post-fordist" Regional and Urban Structure', *International Journal of Urban and Regional Research*, 13: 417–37.

Fainstein, S. (1991) 'Promoting Economic Development: Urban Planning in the United States and Great Britain', *Journal of the American Planning Association*, 57: 22–33.

Forster, C. (1995) *Australian Cities: Continuity and Change*, Melbourne: Oxford University Press.

Gaffikin, F. and Warf, B. (1993) 'Urban Policy and the Post-Keynesian State in the United Kingdom and the United States', *International Journal of Urban and Regional Research*, 17: 67–84.

Goldsmith, E. (1992) *The Way*, London: Rider.

Greenpeace International (1994) 'Asia Toxic Trade Patrol', Unpublished bulletin, Amsterdam: Greenpeace International.

Habermas, J. (1995) *Between Facts and Norms: Contributions to a Discourse Theory of Law and Democracy*, Cambridge: Polity.

Habitat (1996) *An Urbanizing World: Global Report on Human Settlements, 1996*, Oxford: Oxford University Press for the United Nations Centre for Human Settlements.

Hardin, G. (1968) 'The Tragedy of the Commons', *Science*, 162: 1243–8.

Harding, A. and Le Galès, P. (1997) 'Globalization, Urban Change and Urban Policies in Britain and France', in Scott, A., *The Limits of Globalization, Cases and Arguments*', London: Routledge.

Harrison, B. and Bluestone, B. (1990) *The Great U Turn, Corporate Restructuring and the Polarizing of America*, New York: Basic Books.

Haughton, G. and Hunter, C. (1994) *Sustainable Cities*, London and Bristol: Jessica Kingsley and Regional Studies Association.

Harvey, D. (1987) 'Flexible Accumulation through Urbanisation: Reflections on "post-modernism" in the American City' *Antipode*, 19(3): 260–86.

—— (1989a) *The Condition of Postmodernity*, Oxford: Blackwell.

—— (1989b) 'From Managerialism to Entrepreneurialism: Transformation in Urban Governance in Late Capitalism', *Geografiska Annaler*, 71: 3–17.

—— (1996) 'Cities or Urbanization?', *City*, 1/2: 38–62.

Held, D. (1995) *Democracy and the Global Order*, Stanford, Calif.: Stanford University Press.

Hewett, J. (1998) 'The IMF Pill May Be More Curse than Cure', *The Age*, Melbourne, 14 January, p. A11.

Hirsch, J. and Roth, R. (1987) *Das neue Gesicht des Kapitalismus* (The New Face of Capitalism), Hamburg: VSA Verlag.

Houghton, J.T., Filho, L.G.M., Callander, B.A., Harris, N., Kattenberg, A. and Maskell, K. (eds) (1996) *Climate Change 1995, The Science of Climate Change, for the Intergovernmental Panel on Climate Change*, Cambridge: Cambridge University Press.

Hutton, W. (1995) *The State We're In*, London: Jonathan Cape.

Jacobs, M. (1991) *The Green Economy: Environment, Sustainable Development and the Politics of the Future*, Boulder Col., and London: Pluto Press.

—— (1995) 'Sustainability and "the Market": a Typology of Environmental Economics', in Eckersley, R. (ed.) *Markets, the State and the Environment*, Melbourne: Macmillan,46–72.

Kapuscinski, R. (1994) *Imperium*, London: Granta Books.

Keynes, J.M. (1931) *Essays in Persuasion*, London: Macmillan.

—— (1936) *The General Theory of Employment, Interest and Money*, London: Macmillan.

Lash, S. and Urry, J. (1987) *The End of Organized Capitalism*, Cambridge: Polity Press.

Leborgne, D. and Lipietz, A. (1988) 'New Technologies, New Modes of Regulation: Some Spatial Implications', *Environment and Planning D: Society and Space*, 6: 263–80.

Lidskog, R. (1996) 'In Science we Trust? On the Relation Between Scientific Knowledge, Risk Consciousness and Public Trust', *Acta Sociologica*, 39(1): 31–56.

Lindblom, C.E. (1982) 'The Market as Prison', *Journal of Politics*, 44(1–2): 324–36.

Lipietz, A. (1985) *The Enchanted World: Inflation, Credit and the World Crisis* (tr. I Patterson), London: New Left Books.

—— ([1989] 1992) *Towards a New Economic Order* (tr. M. Slater), Cambridge: Polity Press.

—— (1995) *Green Hopes: The Future of Political Ecology*, tr. M. Slater Cambridge: Polity Press.

—— (1996) 'Geography, Ecology, Democracy', *Antipode*, 28(3): 219–28.

Low, N.P. and Gleeson, B.J. (1998) *Justice, Society and Nature: an Exploration of Political Ecology*, London: Routledge.

Martin, H.-P. and Schumann, H. (1997) *The Global Trap, Globalization and the Assault on Democracy and Prosperity* (tr. P. Camiller), London: Zed Books, and Leichardt, Sydney: Pluto Press Australia.

Mathews, F. (1991) *The Ecological Self*, London: Routledge.

Meadows, D.H., Meadows, D.L., Randers, J. and Behrens, W.W. (1974) *The Limits to Growth*, London: Pan Books.

Merchant, C. (1992) *Radical Ecology: the Search for a Livable World*, New York: Routledge.

—— (1996) *Earthcare: Women and the Environment*, London: Routledge.

Mitlin, D. and Satterthwaite, D. (1996) 'Sustainable Development and Cities' in Pugh, C., *Sustainability, the Environment and Urbanization*, London: Earthscan Publications, 23–61.

Molotch, H. (1976) 'The City as a Growth Machine: Toward a Political Economy of Place' *American Journal of Sociology*, 82(2): 309–32.

Munasinghe, M. (1993) *Environmental Economics and Sustainable Development*, Washington, DC: International Bank for Reconstruction and Development/The World Bank.

Myrdal, G. (1956) *An International Economy, Problems and Prospects*, London: Routledge and Kegan Paul.

Naess, A. (1989) *Ecology, Community and Lifestyle* (tr and edited by David Rothenberg), Cambridge: Cambridge University Press

Newman, P. and Thornley, A. (1996) *Urban Planning in Europe. International Competition, National Systems and Planning Projects*, Routledge, London.

O'Connor, J. (1994) 'Is Sustainable Capitalism Possible?', in O'Connor, M. (ed.), *Is Capitalism Sustainable? Political Economy and Political Ecology*, New York: Guilford Press, 152–75.

O'Loughlin, J. and Friedrichs, J. (1996) 'Polarization in Post-Industrial Societies' in O'Loughlin, J. and Friedrichs, J. (eds), *Social Polarization in Post-Industrial Metropolises*, Berlin and New York: de Gruyter.

Pearce, D.W., Markandya, A. and Barbare, E.B. (1989) *Blueprint for a Green Economy. A Report for the UK Department of the Environment*, London: Earthscan.

Pepper, D. (1993) *Eco-socialism, From Deep Ecology to Social Justice*, London: Routledge.

Plumwood, V. (1993) *Feminism and the Mastery of Nature*, London: Routledge.

—— (1997) 'From Rights to Recognition, Ecojustice and non-humans', Paper given at the: Environmental Justice, Global Ethics for the 21st Century Conference, the University of Melbourne, Oct 1–3 (unpublished, obtainable from the author, Clyde Road, Braidwood NSW 2622, Australia).

Pulido, L. (1994) 'Restructuring and the Contraction and Expansion of Environmental Rights in the United States', *Environment and Planning A*, 26: 915–36.

Rawls, J. (1971) *A Theory of Justice*, Cambridge, Mass: Harvard University Press.

Rees, W.E. (1992) 'Ecological Footprints and Appropriated Carrying Capacity', *Environment and Urbanization*, 4(2): 121–30.

Rosenau, J. (1998) 'Governance and Democracy in a Globalising World', in Archibugi, D., Held, D. and Köhler, M. (eds), *Re-imagining Political Community: Studies in Cosmopolitan Democracy*, Cambridge: Polity Press.

Rowntree Foundation (1995) *Inquiry into Income and Wealth* (The Rowntree Report), York: Joseph Rowntree Foundation.

Safier, M. (1996) 'The Cosmopolitan Challenge in Cities on the Edge of the Millennium', *City*, 3/4: 12–29.

Sandercock, L. (1998) *Towards Cosmopolis: Planning for Multicultural Cities*, Chichester, Wiley.

Sassen, S. (1991) *The Global City: New York, London, Tokyo*, Princeton, N.J.: Princeton University Press.

Saul, J.R. (1997) *The Unconscious Civilization*, Harmondsworth (UK) and Ringwood (Australia): Penguin Books.

Self, P. (1993) *Government by the Market? The Politics of Public Choice*, Basingstoke: Macmillan.

Schwartzman, D.W. (1996) 'Introduction', *Science and Society*, 60(3): 261–5.

Shue, H. (1980) *Basic Rights, Subsistence, Affluence and U.S. Foreign Policy*, Princeton, N.J.: Princeton University Press.

Simon, J.L. and Kahn, H. (1984) *The Resourceful Earth: A Response to Global 2000*, Oxford: Blackwell.

Smith, D. and Blowers, A. (1992) 'Here Today, There Tomorrow: the Politics of Hazardous Waste Transfer and Disposal', in Clark, M., Smith, D. and Blowers, A. (eds), *Waste Location: Spatial Aspects of Waste Management, Hazards, and Disposal*, London Routledge: 208–26.

Stilwell, F. (1993) *Reshaping Australia: Urban Problems and Policies*, Pluto, Sydney.

Stren, R., White, R. and Whitney, J. (eds) (1992) *Sustainable Cities: Urbanization and the Environment in International Perspective*, Boulder and San Francisco, USA: Westview Press.

Sudjic, D. (1996) 'Megalopolis now: Hong Kong, Shanghai, Jakarta', *City*, 1/2: 30–7.

United Nations Commission on Human Settlements (UNCHS) (1996) *The Habitat Agenda*: Chapter 4. 'Global Plan of Action: Strategies for Implementation', New York: United Nations.

—— (1997a) Follow-Up to the United Nations Conference on Human Settlements (Habitat II). The UNCHS (Habitat) Indicators Programme: Summary of the Final Report on Phase One (1994–1996) and Summary of the Programme Document for Phase Two (1997–2001), Note by the Secretariat, New York: United Nations.

—— (1997b) Cooperation with Agencies and Organizations within the United Nations System, Intergovernmental Organizations Outside the United Nation System and Non-Governmental Organizations: Implementation of Agenda 21, Summary of Report of the Executive Director, New York: United Nations.

United Nations Department for Policy Co-ordination and Sustainable Development (UNDPCSD) (1997a) *Overall Progress Achieved Since the United Nations Conference on Environment and Development. Report of the Secretary-General*. Addendum: 'Demographic Dynamics and Sustainability', New York: United Nations.

—— (1997b) *Global Change and Sustainable Development: Critical Trends, Report of the Secretary-General*, New York: United Nations.

—— (1997c) *Overall Progress Achieved Since the United Nations Conference on Environment and Development. Report of the Secretary-General*. Addendum: 'Combating Poverty', New York: United Nations.

—— (1997d) *Overall Progress Achieved Since the United Nations Conference on Environment and Development. Report of the Secretary-General*. Addendum: 'Promoting Sustainable Human Settlement Development', New York: United Nations.

United Nations Development Programme (UNDP) (1998a) *Overview of Human Development Report*, http://www.undp.org/undp/hdro/98.htm.

—— (1998b) *Human Poverty Profile and Index*, http://www.undp.org/undp/hdro/hpprof.htm.

United Nations, World Commission on Environment and Development (1987) *Our Common Future* (The Brundtland Report, Australian Edition), Melbourne and Oxford: Oxford University Press.

Urry, J. (1995) *Consuming Places*, London: Routledge.

Wackernagel, M. and Rees, W.E. (1996) *Our Ecological Footprint: Reducing Human Impact on the Earth*, Gabriola Island, British Columbia, Canada: New Society Publishers.

Warf, B. (1996) 'Global Cities in the Age of Hypermobile Capital', *City*, 1–2, 40–3.

Weale, A. (1992) *The New Politics of Pollution*, Manchester: Manchester University Press.

Wolch, J. and Dear, M. (1993) *Malign Neglect: Homelessness in an American City*, San Francisco: Jossey-Bass.

World Bank (1991) *Urban Policy and Economic Development: An Agenda for the 1980s*, Washington, DC: International Bank for Reconstruction and Development/World Bank.

—— (1992) *World Development Report 1992: Development and the Environment*, Washington, DC: International Bank for Reconstruction and Development/World Bank.

—— (1996) *Poverty Reduction and the World Bank*, Washington, DC: World Bank.

2 The Rio Declaration and subsequent global initiatives

Ingemar Elander and Rolf Lidskog

Although there has been progress since the Summit, it can hardly be said that we have reversed the major trends that threaten our common future. If we ask whether there are more poor people today than in 1992, the answer is yes. If we ask whether environmental deterioriation persists, the answer is also yes. And if we ask whether governments have forgotten the financial commitments they made at the Summit, the answer is again, sadly, yes.

(Speth 1997: xii)

Introduction

Like any other kind of policy, environmental policy can be addressed from at least three different angles, i.e. a descriptive, a normative and a constructive one. The focus of the first angle is on how to describe and explain environmental policies, their formulation, implementation and outcomes. From a normative point of view it is a question of developing a realistic vision of what an environmentally just society should look like and on what values and norms it should be based. Finally, the constructive dimension pertains to the strategies, measures and activities on the part of various actors that should be taken to reach an environmentally just society. Strategies for sustainability or any other environmentalist goal include descriptive, as well as normative and constructive elements. In other words they include a view of what the world looks like in terms of environmental qualities, a vision of what it should look like and an idea of how to realise that vision.

So far the most comprehensive strategy for global action on ecologically sustainable development (ESD) is the Agenda 21 endorsed by the 178 government delegations that attended the Rio Summit in 1992. Agenda 21 consists of 40 chapters that cover almost everything about the planet and how humans interact with it. Although it is not legally binding, Agenda 21 constitutes a moral and political commitment, 'a blueprint for sustainable development', comprising economic, social and environmental dimensions (Lindner 1997: 4). Haughton and Hunter (1996: 296) argue that although 'most of the individual measures are in themselves fairly unexceptional, what makes the Rio agenda different is that the policy elements are brought together in a single package which has the backing of virtually the whole international community'. Achterberg (1996: 173) cautiously concludes that the safest judgement of the Earth Summit is that 'the conference was not a complete failure'. However, we find this conclusion a little too pessimistic and rather agree with Flavin (1997: 19): 'though the pace of change has been frustratingly slow so far, and disappointments abound, a purely negative verdict would be too harsh, and certainly premature. It is already clear that the Earth Summit set in motion historical processes that will bear fruit for decades to come.'

The aim of this chapter is to locate Agenda 21 within the framework of global environmental governance. Although the chapter is mainly descriptive it cannot and should not try to evade the normative and constructive aspects, as they are part and parcel of Agenda 21. Thus the rhetorical dimension of the Rio Declaration and Agenda 21 are critically discussed as well as the assumptions and perceptions behind them – their perspective of the scope, forms and causes of environmental problems, the possible solutions to them as well as the prescribed political action necessary to achieve these solutions. Following this introduction, the chapter is divided into five parts. The second part defines and discusses the concept of environmental governance with special reference to Agenda 21. The third surveys the post-war development of global environmentalism up to the Earth Summit in Rio 1992. In the fourth part the focus is directed to the post-Rio agenda, analysing international initiatives promoting ESD. Special attention is devoted to the City Summit in Istanbul in 1996 (Habitat II). The growing interest in broad urban policy initiatives towards ESD is the topic of the fifth part, whereas in the concluding section there is an assessment of the role of Agenda 21 within the framework of global environmental governance, highlighting the Local Agenda 21 achievements that can so far be observed.

Global environmental governance

During the last two decades the global environment has developed into the third major issue in world politics, comparable only to international security and international economy (Porter and Brown 1996: 1). The third UN Conference on the Law of the Sea (UNCLOS III) negotiations 1973–82, the 1972 Stockholm Conference on the Human Environment and the instigation of the UN Environment Programme (UNEP) in 1973 mark the beginning of a new era of global environmental governance (Keohane and Nye 1977: 35). At the beginning of the 1980s the depletion of the ozone layer, global warming and worries over the depletion of the world's fisheries were issues to be negotiated at a global level. Whereas the 1985 Vienna Convention on Depletion of the Ozone Layer developed from a toothless regime to a more efficient one with the revisions of the Montreal Protocol in London (1990) and Copenhagen (1992), negotiations in other areas made little progress (French 1997). Although today there are more than 170 negotiated regimes, their implementation leaves much to be desired. A number of good things have been agreed upon, but unfortunately they mostly lack efficient instruments for implementation. Indeed, one may even argue that the international environmental community today suffers from 'treaty congestion' (Vogler 1995: 147; Porter and Brown 1996: 147).

Different approaches to environmental governance

Environmental governance should be looked upon as one instance of a broader trend commonly referred to as 'governing without government' (Rhodes 1997: 47) or governance by 'self-organizing, interorganizational networks' (ibid.: 53; cf. Kooiman 1993). Hempel defines global environmental governance as 'the people, political institutions, regimes, and nongovernmental organisations (NGOs) at all levels of public and private policy making that are collectively responsible for managing world affairs' (Hempel 1996: 5; cf. similar definitions in Young (1996: 2) and Lipschutz with Mayer (1996: 249)).

Current trends of global environmental governance are closely linked to the normative discussion of how governance should be conducted. Hempel (1996), for example, in his book *Environmental Governance. The Global Challenge,* identifies three major approaches of global environmental governance: (1) development of a limited world federalist system, (2) confederal reform of the United Nations and its affiliated agencies, and (3) 'some mixed form of nationalism and nascent supranationalism' (Hempel 1996: 159–78). All approaches are represented in the real world system, and they all have their supporters and adversaries. Whereas the first two approaches are expressions of 'Globalization-from-Above', the third approach is mainly a case of 'Globalization-from-Below' (Brecher *et al.* 1993; Falk 1997, 1998a), sometimes related to the concept of global environmental citizenship (Irwin 1995; Christoff 1996).

The first approach, mainly relying upon negotiations between nation-states, is sometimes referred to as 'incrementalism' and comes close to the status quo. It is very state-centred and pins its faith very much on the development of international regimes:

> The most important determinant of progress in regime formation, strengthening, and implementation will continue to be the active leadership of a major state or groups of states in each case. An agreement is likely to be strong and effective only if a strong coalition of lead states is actively engaged. Lead states must provide some combination of diplomatic influence, technical expertise, financial commitments, and pace-setting unilateral initiatives to induce swing states and potential veto states to commit to the regime or its strengthening.
>
> (Porter and Brown 1996: 172)

However, as argued by one of the most prominent regime analysts, Oran Young, international regimes are issue-specific, and 'do not make good vehicles for addressing the basic problems of the overarching world order' (Young and Demko 1996: 238).

The second approach ultimately aims at some sort of world government, and its supporters hope that nation-states will somehow 'voluntarily surrender their sovereignty to a central world authority in exchange for untested assurances of collective security' (Hempel 1996: 160). This approach seems to be more or less identical with the model of 'cosmopolitan pacifism', negatively pictured by Zolo (1997: 166) as the basis of a world government that would 'of necessity be a despotic and totalitarian Leviathan, condemned to resort to the use of crushing military measures in response to the inevitable proliferation of violence'. Other critics of this approach talk about a New World Order not aiming at the reduction of the 'domination and exploitation of the Old World Order but, under new circumstances, to perpetuate them' (Brecher 1993: 6).

The third approach, sometimes labelled 'global partnership' or 'cooperative governance', advocates that national sovereignty be relaxed at least in two policy domains, environment and security, and the term 'glocalism' has been introduced to 'highlight the message that global changes in ecology and political economy . . . are beginning to foster a devolution of power and authority away from the nation-state and towards greater reliance on supranational, regional and local levels of governance' (Hempel 1996: xiii). This approach also appears under other labels such as 'one world community', 'Earth Democracy', 'transborder participatory democracy', 'global citizenship' and 'cosmopolitan democracy' (Brecher *et al.* 1993: Introduction; Archibugi and Held 1995). Arguably, this approach comes close to Zolo's proposal of a 'weak pacifism' exerted through 'a network of international and above all regional and national institutions specifically directed towards enhancement of ethnic-cultural identities' (Zolo 1997: 154). Although primarily developed within the context of military conflict resolution, Zolo's model has clear links to other areas, not least to global environmental governance. Crucial to this model is the insight that the popular environmentalist slogan 'think globally, act locally' is no longer appropriate, but has to be transcended by an approach, which also considers the need to change the global context of local action (Low and Gleeson 1998: 189).

Agenda 21 fits very well into the third approach, i.e. a 'glocalist' or 'weak pacifist' perspective. Firstly, although inaugurated at a supranational level, and signed by no less than 178 national governments, it has no strong means of implementation. Thus there are no legal sanctions and no financial guarantees. Secondly, implementation is largely delegated to a national and sub-national level, explicitly comprising nine major 'stakeholders' or 'partners': women, children and youth, indigenous people, non-governmental organisations, local authorities, trade unions, business and industry, the scientific and technical community, and farmers (Lindner 1997: 11). One of the chapters (no. 28) urges that local authorities should produce a local Agenda 21 in co-operation with local residents and institutions. Thus global commitment goes hand in hand with local implementation, and Agenda 21 can therefore be regarded as a test case of the third model of global environmental governance mentioned above. It is too broad and void of legal status to qualify as a regime. National financial support that was announced at the Rio conference has not been realised, and supranational institutional commitment is not very strong. Rather, it is a case of informal, 'glocalist' governance, at least having the potential to become an element of a strategy for cosmopolitan democracy. More will be said about the Rio Earth Summit and its shortcomings in Chapter 3 from the crucial perspective of the USA – the latter being not just another nation, but in many ways the nation which, while not determining global affairs, willy-nilly shapes the direction of development.

Agenda 21, Habitat II and the partnership approach

The spirit of the Earth Summit in Rio in 1992 has been characterised in terms of a global partnership for sustainability and peace. This partnership does not only include the nation-states as 'stakeholders', it also comprises governments and NGOs at the local level. As argued by Chip Lindner, who was Executive Director of the Centre for Our Common Future, International Co-ordinator responsible for organising the 1992 Global Forum in Rio de Janeiro during the Earth Summit, and Secretary of the Brundtland Commission:

> A new form of governance is emerging – that of "stakeholders". Local stakeholders in communities are linking together, whether they are local business or local authorities, non-governmental organizations or community-based organizations, women's groups or residents' associations. Groups that have an identifiable "stake" in the future of the community are making these links to create a vision for the future which has a set of good and measurable criteria or indicators.
>
> (Lindner 1997: 13)

The local partnership approach was further developed by the Habitat II conference in Istanbul in 1996. Over 3,000 delegates from 171 countries attended, as well as some 300 parliamentarians, 579 local authorities, 89 special agency representatives, 341 people from intergovernmental organisations, and 2,400 NGO representatives. The parallel NGO Forum '96 attracted 8,000 registered NGO representatives (Carlson 1996: 4).

Notably, also, the World Bank had a presence in Istanbul 'which went far beyond its mere number of participants' (Leaf 1997: vi). Responding to those who questioned the degree of commitment to the Habitat II principles on the part of the World Bank, Ismail Serageldin, Vice-President for Environmentally Sustainable Development, answered: 'Do I really want to change the World Bank? No, I want to change the world, and changing the World Bank is just one step along the way' (as quoted in Leaf, 1997: vi).

Indeed, the World Bank during the 1990s has changed its policy, at least rhetorically, and now adheres to the triangular model of sustainable development – comprising economic, social and ecological dimensions – that can easily be deduced already from the Brundtland Report. It has proclaimed the principle of 'sustainability as opportunity'; namely, that the opportunities of future generations should not be limited. To be able to calculate the progress – or otherwise – of society in terms of sustainability, the World Bank identifies four kinds of capital: economic, natural, human and social. When summarising action to preserve the four categories of capital, the Bank asserts the principle that one generation should not leave less capital to the next generation than it has itself received (Serageldin 1996). To implement this goal, of course, raises a number of questions that we will not enter into here. Thus although the World Bank has expressed its ambition to walk the road towards sustainability, it remains to be seen whether that road will also be followed in practice. So far, the Bank 'has failed even to develop an adequate environmental screening process for their loans' (Flavin 1997: 6).

In sum, taken together the Earth Summit and Habitat II strongly propagate a broad partnership approach to meet the challenges raised by a global commitment to sustainability in a very broad sense. In brief, the message of the Istanbul conference is that everyone has something to win (the synergy effect) and no one has anything to lose from a partnership approach. Of course, such a general concept, adhered to in rhetoric by a heterogeneous mass of actors and interests, begs the question of implementation in widely varying national contexts.

The rationale of partnership

The slogan 'partnership' is an ideological term indicating a view that we should all strive for a common goal – for example, 'peace', 'democracy', or 'sustainability' (Elander in press). Partnership in the Rio and Habitat II contexts has a very wide scope, encompassing international support programmes as well as capacity-building programmes at the national and sub-national levels. It demands 'increasing participation of men and women, creating effective partnerships, promoting a sense of public service, and removing barriers to mobilizing all kinds of resources'. Indeed, Habitat II 'intends to use a mechanism of agreements with all sectors of society' (Introduction to Habitat II 1996).

Partnership can be an attractive concept to government because it 'diffuses responsibility for success or failure' and 'could become a system for co-opting institutions into an extended system of repressive control'. Therefore, crucial questions to be addressed in research are: 'Which interests, and which players, will be included in partnerships, and which will be left outside? Who will be the leaders within partnerships? Whose agendas will prevail?' (Jewson and MacGregor 1997: 9). Although stated in a narrower, nation-bound context, the arguments and questions raised by Jewson and MacGregor are obviously relevant also to the Agenda 21 and Habitat II partnership approach. With so many actors and interests involved in a particular partnership it is most unlikely that all will come out as winners. For example, it is all very well for the World Bank to proclaim its adherence to sustainability, but it does not necessarily follow that the global corporate institutions which are supported by World Bank money will act in accordance with the principles of environmental and ecological justice. Partnerships that might look 'happy' at the formulation stage may well appear more conflict-ridden and controversial at the stage of implementation. In other words, what is said is not always what is done. This could also be discussed in terms of the rhetoric/practice dichotomy. Thus looking at the implementation of partnership initiatives taken by a central government, the EU or the UN, one has to be careful not to draw too much from the explicit intentions when it comes to policy evaluation. Words may sometimes function as triggers for efficient action, but they may in other cases mask failing policies, i.e. 'words that succeed and policies that fail' (Edelman 1977)

From environmental threat to sustainable development

International agreements are nothing new in history. As McGrew (1995: 30ff.) has shown, international mechanisms to 'govern' aspects of global affairs have existed since the mid-nineteenth century. These agreements and co-operation – for example concerning education (1864), international associations of unions (1865), international industrial standards and intellectual property (1875), human rights (1890) – laid the foundations of the more comprehensive, albeit fragmented, system of global regulation that exists in the late twentieth century. However, in the last three decades a new issue for international negotiation has emerged: the global environment. Today there are more than 170 international conventions concerning the global environment. More than two-thirds of these conventions have been established after 1972 (Wandén 1996), a year often regarded as the starting point of the development of international environmental politics. The UN conference on the environment in Stockholm and the publication of the Club of Rome report *Limits to Growth* and its 'counter-report' *Blueprints for Survival* are considered the first steps. Since then the environmental issue has remained on the international agenda (Hajer 1995: 24). However, it is important to note that the UN conference was as much a response to the environmental debate of the 1960s as it was an initiator of the transnational environmental politics of the 1970s and 1980s.

Between the Second World War and the beginning of the 1960s there was, in the industrialised world, a widespread belief in the virtues of economic growth. In the 1960s the situation changed rapidly, and 'nature' was increasingly seen as threatened to the extent that the whole of humankind itself might be in danger. Carson (1962) warned that continued use of chemicals (especially pesticides) would lead to a 'silent spring' with the death of the birds, Hardin (1968) and the Ehrlichs (1969) argued that population growth would lead to overexploitation of natural resources and shortage of food, and Commoner (1971) predicted that the new productive technology with its use of new raw material (plastics, heavy metal, artifical fertilisers, etc.) would lead to more, not less, environmental pollution. Even if they made different diagnoses, and to some extent also proposed different cures, these critics all shared the view that not only specific regions were threatened, but the whole planet. Problems such as overpopulation, resource depletion, decreasing water supply, air pollution, and the use and spread of chemicals and heavy metals in nature came into focus. The new critical message was that something must be done to save planet Earth and humankind, and that an adequate response must not be limited to technological improvements but must also include changes in fundamental values, ways of life and the design of society.

The Club of Rome study *Limits to Growth* (Meadows *et al.* 1972) is the most influential work of this period, and perhaps of the last two decades (Nelissen *et al.*, 1997: 179). In this study it was stated that if present growth trends in population, industrialisation, pollution, food production and resource use continue, the planet's carrying capacity would be exhausted within 100 years. In this report the environmental issue was presented as a global crisis and that, if relevant action was not undertaken, the present trends would lead to a collapse. Even if this report had no official status, the respectability of the Club of Rome made the report a key text in the following environmental debate.

The same year as this report was presented, The UN Conference on the Human Environment, took place in Stockholm. It was the biggest UN conference ever held, with 112 nations represented, and the first international platform for discussions and agreements on environmental problems. At this conference, it was made clear that there is a conflict between national interests and international co-operation and that the existing international order was relatively powerless in attacking global problems (Nelissen *et al.* 1997). The conference resulted in a declaration comprising 26 principles, and was an important step in the development of national as well as international policy programmes aimed at reducing environmental problems, especially acid rain and other kinds of air pollution. It led to the creation of the United Nations Environmental Programme (UNEP), which became an important institutional base for later international environmental negotiations.

During the years before and after the conference, several governments made environment protection a policy area in its own right. Between 1967 and 1974 many industrialised countries established a national environmental protection agency and enacted comprehensive environmental legislation (Jamison 1997). A 'new politics of pollution' emerged, where the aim was not only to react to pollution but to prevent it (Weale 1992). The institutionalisation of the environmental issue was very much a response to demands raised by the environmental movements. During the 1970s these movements

Table 2.1 Phases of post-war environmentalism

Period		Emphasis
Pre-1968	Awakening	Public education
1969–74	Organisation	Institution building
1975–80	Social movement	Political controversy
1981–86	Professionalisation	Environmental assessment
1987–	Internationalisation	Incorporation/integration

Source: Jamison (1997: 227).

were increasingly professionalised and integrated into the established political culture. Green parties emerged in several nations and became players in the games of the established political systems. At the end of the 1980s, collaboration between environmental organisations, business firms and the state became the norm. Table 2.1 summarises this development.

In 1987, the UN report *Our Common Future,* commonly labelled the Brundtland Report, was presented. The Stockholm conference had already demonstrated the impossibility of conducting global discussions on environmental problems without facing the development issues linked to them (Brenton 1994: 50). In line with this, the Brundtland Report states that economic development must be seen in close relationship to the natural environment. To deal successfully with the latter requires a broad perspective comprising the factors that underlie world poverty and global inequality. However, the proposed solution is not that of limits to growth, as the Club of Rome had stated 15 years earlier, but rather the opposite one. Through the improvement of both technology and social organisation, it was believed that economic growth could solve the problem of environmental destruction as well as that of poverty.

'Sustainable development' is the key slogan of the report, which summarises how the challenge of environmental problems and poverty should be met. Human needs are located both in a temporal and a spatial dimension, thus taking into account the needs of future generations as well as different regions of the world. Although the concept had been introduced earlier in the environmental debate (IUCNR, 1980), it was not until this UN report that it became well known, gained credibility and was rapidly integrated into the environmental policy language of supranational institutions, states, cities, business firms, etc. Thus ESD is not a matter of one single sector of society but a fundamental principle that has a relevance for most policy areas at all levels. The report stated that the ability to anticipate and prevent environmental damage requires that the ecological dimension of policy is considered as tightly intertwined with economy, trade, energy, agriculture and other dimensions.

One of the central controversies at Stockholm was the debate about whether economic growth and development are inherently destructive from an environmental point of view (Conca *et al.* 1995: 207). ESD holds out the promise of reconciling these different views, claiming that it is not only possible but also desirable to marry environmental concerns with economic growth.

The Brundtland Report did not give a detailed blueprint for action, but instead offered a pathway by which the peoples of the world may enlarge their scope of co-operation. This pathway has already become quite crowded by individuals and organisations in the spheres of civil society, the market and the state. ESD has enjoyed growing endorsement by many groups, including a wide spectrum of political and economic notables, and today it has become a sacred symbol which no one dares to question. However, the recommendation to combine ecologically sound forms of production and distribution with a call for renewed global growth to solve the problems of Third World poverty stands in sharp contrast to the recommendations given at the beginning of the 1970s.

Despite the public attention immediately following the publication of *Limits to Growth*, the notion was quickly rejected by academics, public officials, and business elites (Maddox 1972; Buttel *et al.* 1990). One reason for this was that the notion implied a threat to a wide range of interests, not least economic ones. However, environmental issues that were raised in the debates of the late 1960s and 1970s, and then rejected by the elites in policy and business communities, have now been reconsidered and integrated into an all-encompassing strategy for ecological, social and economic sustainability. Thus, one of the main achievements of the Brundtland Report was to present the environmental case in such a way that it could influence economic institutions like the World Bank and the International Monetary Fund (IMF) to proclaim their adherence to the path of sustainability.

On the other hand radical critics of the Brundtland Report claim that the whole idea of ESD is a rhetorical ploy which conceals a strategy for sustaining *development* rather than addressing the cause of the ecological crisis (Sachs 1992). ESD in this perspective will justify 'business as usual', putting a green face on current practices (Lohmann 1990: 82). Indeed, after the Earth Summit had made the World Bank one of the central agencies monitoring the greening of policies all over the world, radical critics argued that this did not mark the success of the environmentalists, but their total collapse (Finger 1993; Shiva 1993; Hajer 1995: 12). Thus, whether ESD and its inclusion of a wide diversity of interests is to be judged positively or negatively is contested.

From Earth Summit to City Summit and beyond

Whereas the Stockholm Conference in 1972 focused on relatively narrowly defined problems of air and water pollution, Rio embraced a far broader and more complex agenda, where the earth was viewed as one single, integrated system (Conca *et al.* 1995: 6). Rio was also extraordinary in its long-term commitment. Thus in his opening address to the UN Conference on Environment and Development (UNCED), the Secretary-General Maurice F. Strong said:

The Earth Summit is not an end in itself, but a new beginning. The measures you agree on here will be but the first steps on the new pathway to our common future. Thus, the results of this conference will ultimately depend on the credibility and effectiveness of its follow-up.

(quoted in Haas *et al.* 1992: 7)

During the 1990s, the United Nations arranged a series of major international summits – huge world conferences focusing on problems experienced by every inhabitant of the planet – aimed to constitute the basis for the UN's work for a better world in the twenty-first century (see Table 2.2). These conferences all expressed the view that the burden of responsibility should not just be placed on the international community and national governments, but on individuals, local governments, local communities, trade unions and business firms – that is, the whole range of individual and collective actors that make up societies.

Parallel to these six world conferences (Table 2.2), the UN has arranged other conferences on specific topics. One conference with high relevance to the post-Rio agenda was the ninth session of The UN Conference on Trade and Development in 1995 which pointed out the potential benefits of globalisation and trade liberalisation to developing countries, but warned that continued marginalisation of poor countries would make them unable to capitalise on new opportunities. Another conference was The World Food Summit in 1996 which called for a renewed effort to combat increasing hunger in the poorer regions of the world. Thus, since the Earth Summit, a number of conferences have been arranged by the United Nations, all with the aim of adopting plans of action that complement Agenda 21 or in some respects supersede it.

Human Settlements

In most nations, cities are major generators of economic activity, offering employment opportunities, education, health, and other social services. At the same time cities are the main consumers of natural resources and the main producers of pollution and waste. Furthermore, most of the world's population

Table 2.2 Major UN global conferences, 1992–6

Year	Theme	Place
1992	Environment and development	Rio
1993	Human rights	Vienna
1994	Population	Cairo
1994	Social development	Copenhagen
1995	Women	Beijing
1996	Human settlements	Istanbul

will soon live in cities. Urban issues are thus crucial to the environmental challenge of today. The environmental issue and the urban issue have one striking trait in common: they are both closely connected to the developed as well as to the developing world. Of course, the character and size of the problems, as well as the proposed solutions, vary considerably between different parts of the world. Poor cities, for example, do not only exist in the developing countries – the richest nation in the world has one million homeless, most of them living in cities (Tosics 1997: 368; and see the chapters on the USA in this volume). On the other hand, urbanisation is much more rapid in the developing world, where homelessness and other problems take on a different order of magnitude than in the developed world (see the chapters on India and China in this volume).

The first UN conference on human settlements, Habitat I in Vancouver 1976, was a product of the Stockholm conference in 1972. Whereas the Stockholm conference was about international environmental problems, Habitat I was convened to address local environmental problems such as housing, shelter, infrastructure, diminishing water supply, transport, etc. Astonishingly, human settlements had no place on the agenda in the preparatory meetings of the Rio Declaration (Carlson 1996: 4). However, a special chapter on human settlements was included in Agenda 21. As an overall objective the seventh chapter of Agenda 21 mentions the improvement of the social, economic and environmental quality of human settlements and of the living and working environments of all people, in particular the urban and rural poor. Human settlements, especially cities, were recognised both as a source of many global environmental problems and also as a key to their solution.

After Rio, the UN General Assembly decided that the cross-sectoral issue of human settlements was of crucial importance and decided to convene Habitat II. UNCED recognised the proper management of human settlements as a prerequisite to ESD. The United Nations Conference on Human Settlements took place in Istanbul from 3–14 June 1996. The overall theme of the conference was 'adequate shelter for all' and 'ESD in an urbanising world'. It can be viewed as the culmination of decades of efforts by the UN and other agencies to deal with the vast panorama of problems and sectors affecting the sustainability of planet Earth in supporting a rapidly increasing and urbanising human population (Carlson 1996: 1).

Habitat II is a reflection of the fact that national governments and international agencies, during the five years since Rio, have placed increasing emphasis on the critical role of cities and towns in global sustainable development. Thus fundamental to Habitat II is the notion that the future of the earth will be heavily determined by the quality of life in cities, i.e. urban policies become crucial. As the state and the market are not believed to solve urban problems by themselves, participation and partnership are put forth as necessary, i.e. actors at different levels and spheres in society must be mobilised and must co-operate. The emphasis is laid on the local level, with local authorities and NGOs having a much more important role to play than in previous UN conferences. Expert panels and representatives from NGOs, communities, trade and business firms gathered in various fora for dialogues. The 'enabling strategy' of Habitat II, including three principles – civic engagement, sustainability, and equity – is broader than comparable concepts in the prior, more narrowly focused conferences.

'Enablement' means changing the role of government intervention from being the sole responsible actor to becoming the creator of efficient partnerships between local and regional governments, the business sector, media groups, philanthropic organisations and other organisations. The 'civic engagement' element of the Habitat II strategy calls for broad participation in development processes by all people. It calls for participatory objectives that move beyond democratic pluralism, special-interest lobbying, civil rights for particular communities, and legitimate power in governance to further encompass the public interest, community inclusiveness, active mobilisation of vulnerable groups, and public service. It calls for consensus on ethical behaviour both in government and in society. However, no management indicators were given in the documents produced at the conference, and national governments are expected to set their own deadlines. In practice, few mechanisms have been established to monitor whether states meet their own established goals, and the credibility for effective progress has been rendered indeterminate.

Nevertheless, Habitat II provided a place in the UN system for other than national governments, as exemplified by the creation of the NGO Forum that ran as an official event during the conference, and by the inclusion of local government voices in the Dialogues (day-long events) around a variety of topics. It also emphasised the important role of individuals and market forces. Governments should serve as facilitators and not providers of housing, i.e. a reversal of the Vancouver recommendations 20 years before. The Habitat Agenda will be revised and updated periodically, and it will serve as an educational document for all constituencies concerned with the development and improvement of human settlements around the globe.

Although Habitat II had a very broad agenda concerning development issues (e.g. population growth and the right to housing), environmental issues (e.g. fossil fuel based transportation) and social issues (e.g. criminality and civil violence), it has been criticised for missing the link to the environment. Thus, Michael Cohen, senior adviser to the Vice-President of the World Bank's office for Environmentally Sustainable Development in a critical assessment, concludes:

> The biggest gap in Habitat II was the lack of progress in operationalizing the notion of environmentally sustainable development. At Vancouver, 'habitat' had been discussed with reference to human settlements but without an environmental context; at Rio 'habitat' had been used to mean ecosystem. The Istanbul conference did not succeed in

bringing the two definitions together; it neither demonstrated the interdependency of the two nor the risks to settlements associated with deterioration of natural resources.

(Cohen 1996: 8)

Beyond Habitat

The nineteenth special session of the United Nations General Assembly (UNGASS) – Earth Summit Plus Five – took place in New York, from 23–28 June 1997. Its aim was to review progress achieved over the five years that had passed since the Earth Summit (UNCED), and to re-energise the commitment to further action on goals and objectives set out by the Earth Summit. The meeting gathered 53 Heads of State, representatives of 165 nations spoke, and for the very first time in the UN history, NGOs were allowed to participate in the meeting together with the official representatives. The meeting – which again confirmed that ESD must include measures to combat poverty and change patterns of production and consumption – did not succeed in producing a political declaration, but it did establish a programme for continuing work with Agenda 21. However, much of the enthusiasm surrounding the Rio Conference had vanished, and assessments of the conference have mostly been negative, as will be illustrated in the concluding section of this chapter.

In December 1997, The Conference of the Parties (COP) for the Framework Convention on Climate Change was held at Kyoto. Preceeding the Kyoto meeting, the Cities for Climate Protection (CCP) held a world summit in Nagoya. Its declaration challenged the national delegations at the Kyoto Summit to follow local governments' lead by setting early and aggressive targets for reducing greenhouse gas emissions. The declaration called for a 20 percent reduction target by 2010. However, at Kyoto, national governments agreed to legally binding targets at a much lower level of ambition. Thus for the post-2000 period the developed countries must reduce their combined emissions of six key greenhouse gases by at least 5 percent by the period 2008–12, calculated as an average over these five years and with 1990 and 1995 as a baseline (depending on the particular kind of gas). Notably, three countries are even allowed to *increase* their emissions of greenhouse gases, i.e. Australia (by 8 percent), Iceland (10 percent) and Norway (1 percent). Nevertheless, by reducing the average level of greenhouse gas emissions to 5 percent below 1990 levels by the year 2010, the Protocol will result in 2010 emission levels that are approximately 29 percent below what they would have been in the absence of the Protocol (UN 1997b). Although this would at least mean a minor step forward towards achieving the Convention's ultimate objective of preventing 'dangerous anthropogenic [man-made] interference with the climate system' (UN 1998a) the Kyoto meeting has mostly been pictured in dark colours. As argued by an anonymous climate researcher just before the meeting:

The Kyoto Summit is a political ploy without effect. However, would it be the case that the Winter Olympic Games in Nagano cannot take place due to the fact that this winter has been the warmest one for 1,000 years in Japan as a consequence of El Niño, then the world will react.

(quoted in *Dagens Nyheter*, 1997)

As we all know, snow eventually fell, and the world could relax and follow the games on their television sets.

The follow-up climate change conference in Buenos Aires in November 1998 may become best remembered for having introduced the habit of 'emissions trading', i.e. the sanctioning of not decreasing carbon dioxide emissions on the part of the developed countries provided this is outweighed by decreasing amounts of emissions in the developing world (UN 1998).

Sustainable development and urban policy

During the twentieth century the focus of the environmental debate changed from preservation of specific natural areas, over environmental planning, to sustainability and ecocycles. It has been a process *from* scattered local problems due to various emissions from industries and communes and the preservation of some specific areas from human intrusion, *to* diffuse, large-scale threats like the greenhouse effect or the decreasing ozone layer. Adequate policy measures to tackle such big issues must include different levels and sectors of state and society. International agreements have to be negotiated, strong national policies must be developed, and local government has to take its responsibility as co-ordinator, facilitator and enabler of environmental strategies and policies. Central to this role is the development of strategies 'which bring together actors and agencies at local, national and international levels, across public, private and voluntary sectors' (Haughton and Hunter 1996: 300). The huge environmental problems are closely linked to the value-systems anchored in the daily lifestyle of ordinary people (concerning consumption, transportation, etc.), which is thus becoming a main arena for the implementation of green policies. Therefore, with regard to many environmental questions, both their causes and their remedies are deeply rooted in the kitchens, yards and streets, where most of the people spend their lives, i.e. in the cities and their neighbourhoods (cf. Elander *et al.* 1995).

The urgency of the problem depicted in 1972 by the Club of Rome has in no way diminished. Indeed in a return, after twenty years, to the issues they examined in *Limits to Growth* Donella and Dennis Meadows and Jürgen Randers argue that the world has simply approached closer to the limits (Meadows *et al.*, 1992). They say: 'the transition to a sustainable society requires ... an emphasis on sufficiency, equity, and quality of life rather than on quantity of output. It requires more than productivity and more than technology; it also requires maturity, compassion, and wisdom' (ibid.: xvi). The authors of Factor Four, while embracing the hope offered by technological progress, nevertheless argue that we may have just fifty years left to bring our levels of consumption in line with our capacity to provide for the world's population and maintain a satisfactory and sustainable environment (von Weizsäcker *et al.* 1997: ch. 11).

Although, in the vast and increasing literature on environmental issues, comparatively little attention was paid until recently to cities (McLaren 1992: 56; Stren 1992), sustainable city development today has become a commonplace in the set of policy goals given high priority by the political elites in various countries. The higher priority given to environmental issues in city politics, however, is not only an effect of the nature of such issues. It can also be seen in the light of a general trend towards decentralisation in many countries. From a top-down perspective it has been a conscious strategy to diversify responsibilities and 'export' crisis management, thereby hopefully reducing the demand overload and financial stress put on central state administration by the growth of the welfare state. From a bottom-up perspective, on the other hand, the decentralisation trend has been an expression of popular demands for local autonomy and self-government (Elander and Montin 1990).

As a reflection of the latter view many environmentalists have argued that society should be restructured by a radical decentralisation of the nation-state (Naess 1973: 98; O'Riordan 1981: 7–10, 307; Dobson 1990: 117–22; Eckersley 1992: 160–70). Eckersley uses the generic term 'ecocommunalism' to cover the diverse range of theories seeking the development of 'human scale, cooperative communities that enable the rounded and mutualistic development of humans while at the same time respecting the integrity of the nonhuman world'. Progress in these communities would generally be measured by the degree to which they have been able to adapt human communities to ecosystems and by the degree to which the full range of human needs are fulfilled (Eckersley 1992: 160).

As argued earlier in this chapter, to become efficient, local empowerment must be coupled with transnational governance structures. However, governance is a very broad and loose concept, which does not necessarily include open, democratic government set within the framework of a constitutional order. Thus there are today many ideas pointing in the direction of a reformed UN system, including such features as a global parliament, a new charter of rights and duties and a global legal system – including an International Court (Archibugi 1995; Held 1995; Bienen *et al.* 1998; Falk 1998b). This does not mean that local actors should just sit and wait for a World Government to solve all their problems. On the contrary, it is 'necessary today not only to think about the global consequences of local action, but to act to change the global context of local action: "Think and act, globally and locally"' (Low and Gleeson 1998: 189).

Practitioners who have turned to various eco-city paradigms or movements for guidance in applying the concept of 'sustainability' and its neighbours may have found 'much inspiration but relatively little guidance'. Nevertheless, there are by now 'numerous examples of citizen and community initiatives that demonstrate that creative, transferable solutions to seemingly intractable social and environmental challenges are being initiated by citizen organizations and municipal officials in cities and towns around the world' (Roseland 1997: 200–1).

In fact, the literature on sustainability and related themes has become so abundant that the problem is rather one of orientating oneself among the many more or less overlapping approaches. Haughton and Hunter (1996: 286–312), for example, identify four urban forms which they discuss in terms of 'balanced regional hierarchy, concentrated centre, concentrated decentralisation, and deconcentrated development'. They end their book by presenting a 'Sustainable City Manifesto' in eight points, largely reflecting the spirit of Agenda 21 and Habitat II (ibid.: 311).

At the supranational level, the European Union has been very active in uniting member states on environmental issues in general and on their urban manifestations in particular, issuing a *Green Paper on the Urban Environment* in 1990. In this document arguments were given in favour of compact cities, high-density cities 'which would see a renaissance of urban living and urban quality of life' (ibid.: 297–8). Four years later a European Commission report on the sustainable city argued that environmental sustainability cannot be perceived without social equity and economic sustainability. In this report, as in many other contexts, ESD 'is being defined more and more as a process and not as an endpoint, as a trip rather than a destination' (Mega 1996: 135).

At the opening of the Aalborg Conference on European Sustainable Cities and Towns in May 1994 it was stated that sustainability is 'equity extended into the future' (as quoted in Mega 1996: 135) This conference was the starting point for the European Campaign for Sustainable Cities and Towns, which was to be joined, in November 1995, by 177 cities. A common policy framework for the development of ESD performance indicators has been created in the Charter of European Cities and Towns: Towards Sustainability (ibid.: 140). In addition, an 'amazing number of conferences, books, articles and summer schools' have been devoted to urban ESD in Europe, for example several publications (including internet) on 'Good Practice'. 'All over Europe, cities are becoming laboratories for ecological innovation' (ibid.: 143), including a broad range of measures to improve collective transportation, create efficient recyling systems, and develop environmentally less harmful energy systems. However, cities also compete with each other to become major growth centres. These policies sometimes demand policies which conflict with the aims of ESD. The state of knowledge so far does not allow for any firm conclusions about the net result of the equation, i.e. whether the many plans and actions taken in favour of ESD in cities have had any substantial impact on the overall urban situation.

Agenda 21 and global environmental governance: concluding remarks

The Agenda 21 perspective is so general that it can serve as a source of inspiration for different strategies aiming at ecological sustainability. On the one hand this could be seen as a fundamental weakness. One could rephrase the point originally formulated by Aaron Wildavsky (1973): 'If sustainability is everything, maybe it's nothing.' On the other hand, one could also argue that the breadth and openness of interpretation characterising the Agenda 21 initiative contributes to its strength. Given its core commitment ('a blueprint for sustainable development') it leaves to the nine 'major groups', including local government, the task of operationalising the way forward. As argued in the first chapter of this volume, ESD is commonly considered within the context of a triangle, or three overlapping circles, where economic development, social justice and ecological sustainability are the three fundamental principles. Needless to say, this opens the way for different interpretations and applications at all levels and sectors of society.

An assessment of the outcomes of Agenda 21 is difficult to make for at least four reasons. Firstly, the scope of Agenda 21 is so broad that it is impossible to see whether the balance between success on one front is outweighed by failures on other fronts. Secondly, five years is too short a period for making a conclusive assessment of this broad programme. Thirdly, there is no simple yardstick of assessment. Let it suffice to recall the three-dimensional operationalisation of ESD, written into Agenda 21: economic growth, social development and environmental sustainability. Finally, there is always the problem of considering what would have happened if the programme had never existed.

With these reservations in mind a few general conclusions can still be drawn. Firstly, Agenda 21 and subsequent confirmations in supranational, national and local activities represent a moral commitment which at least provide a lever or yardstick for critique, although adequate resources for efficient action will only slowly and unevenly materialise. Secondly, to implement Agenda 21 many institutions have been built up at all levels of government and society, and these institutions, however efficient (or inefficient) they may be, at least have the potential to become vehicles for development towards ecological sustainability and environmental justice. Thirdly, although inaugurated as a top-down project, Agenda 21 strongly propagates a participatory approach, urging not only national governments but first and foremost local authorities, NGOs and other stakeholders to mobilise in favour of partnerships for saving the planet through prompt and concrete action. Fourthly, Agenda 21, in the spirit of the Brundtland Report, embodies a positive strategy for ecological modernisation which escapes the dead-end of the traditional environmental debate between the 'business-as-usual' approach and the 'fundamentalisms' of many deep ecologists and ecosocialists.

On the negative side, one can easily list a number of items. It is obvious that the relationship between the solemn declarations and the real actions taken so far is very tenuous. Considering the bleak record of the state of the world by the end of the millennium (Flavin 1997), Agenda 21+5 represents but a drop in the ocean. No new money to implement the programme has been offered. Business and industry have been very slow to change their behaviour in a more environmentally friendly direction. Issues which are crucial for ESD at the urban level such as transport, energy and tourism were not even given priority in the programme itself and neither are any other UN bodies dealing with them. Thus one should not be surprised by the negative assessment made by the UN General Assembly at its nineteenth special session in June 1997:

> Five years after the United Nations Conference on Environment and Development, *the state of the global environment has continued to deteriorate* ... and significant environmental problems remain deeply embedded in the socio-economic fabric of countries in all regions. Some progress has been made in terms of institutional development, international consensus-building, public participation and private sector actions and, as a result, a number of countries have succeeded in curbing pollution and slowing the rate of resource degradation. Overall, however, *trends are worsening.*
>
> (UN 1997a; our emphasis)

Even stronger words have been used by NGOs such as Friends of the Earth:

> This summit reveals a scandalous betrayal of the promise raised at Rio. After five years we have seen little progress in implementing Agenda 21. And now after two weeks of negotiations we have an utterly shameful outcome from the Earth Summit Two. The political will demonstrated here is entirely inadequate to meeting the challenges of sustainable development.
>
> (Friends of the Earth 1997)

However, although the overall assessment of Agenda 21 has a strong negative flavour, one may find positive examples in some national contexts, as will be demonstrated in the country-specific chapters of this volume. With regard to urban politics and the environment, one should be especially attentive to the implementation of the Local Agenda 21 (LA 21).

As of 30 November 1996, more than 1,800 local governments in 64 countries were involved in LA 21 activities. As many as 1,631 or 90 percent of the LA 21 activities were located in the developed countries. Local government associations and organisations from these countries were able to participate in the UNCED process, and were therefore able rapidly to disseminate information about LA 21 in their countries. Local governments in the developed countries, unlike their counterparts in developing countries, also tended to adapt existing environmental planning procedures. Municipal association LA 21 campaigns were under way in eight countries – Australia,

Denmark, Finland, Netherlands, Norway, Republic of Korea, Sweden, and the United Kingdom. In addition, national governments had established campaigns in Bolivia, China and Japan. These eleven campaigns involved 82 percent of the documented LA 21 planning efforts. Most local governments were still in the early stages of LA 21 planning and gave greater attention to participation and consensus-building in the preparation of action plans than to measures required for the implementation of these plans. As concluded by the International Council for Local Environmental Initiatives (ICLEI) on the basis of an evaluation of twenty-nine case studies: 'the greatest impact of Local Agenda 21 during its first years has been to reform the process of governance at the local level so that the key requirements of sustainable development can be factored into local planning and budgeting' (ICLEI 1997: 16; cf. Brugman 1997).

The ICLEI report ends with three recommendations. Firstly, to initiate LA 21 planning efforts national campaigns should be launched through a national municipal association or other local government associations, including a wide variety of 'stakeholders'. Secondly, national and international investment and development assistance programmes should be responsive to LA 21 plans. Considerable national and international assistance is needed to implement LA 21 plans successfully. Thirdly, the successful implementation of LA 21 action plans will require the establishment of a supportive national policy and fiscal framework.

The conclusions drawn from this study of LA 21 naturally remind us of the characteristics of global environmental governance discussed earlier in this chapter. Thus in whatever way we like to define ESD and envision the road towards environmental justice, the measures we now have at our disposal can be summarised as an 'eclectic mix of international agreements, sensible government policies, efficient use of private resources, and bold initiatives by grassroots organisations and local governments' (Flavin 1997: 4). Or, to put it briefly, we have to think and act, globally and locally.

References

Achterberg, W. (1996) Sustainability, community and democracy. Pp. 170–187 in Doherty, B. and de Geus, M. (eds) *Democracy and Green Political Thought.* London and New York: Routledge.

Agenda 21(1992) *United Nations Conference on Environment and Development,* Rio de Janeiro, 3–14 June 1992. New York: United Nations.

Archibugi, D. (1995) From the United Nations to cosmopolitan democracy. Pp. 23–40 in Archbugi, D. and Held, D. (eds) *Cosmopolitan Democracy. An Agenda for a New World Order.* Cambridge: Polity Press.

Archbugi, D. and Held, D. (eds) (1995) *Cosmopolitan Democracy. An Agenda for a New World Order.* Cambridge: Polity Press.

Bienen, D., Rittberger, V. and Wagner, W. (1998) Democracy in the United Nations System. Pp. 287–308 in Archbugi, D., Held, D. and Köhler, M. (eds), *Re-imagining Political Community.* Oxford: Polity Press.

Brecher, J. (1993) The Hierarch's New World Order – and Ours. Pp. 3–12 in Brecher, J., Brown Childs, J. and Cutler, J. (eds) *Global Visions. Beyond the New World Order*. Boston: South End Press.

Brecher, J., Brown Childs, J. and Cutler, J. (eds) (1993) *Global Visions. Beyond the New World Order*. Boston: South End Press.

Brenton, T. (1994) *The Greening of Machiavelli. The Evolution of International Environmental Politics*. London: Earthscan.

Brugman, J. (1997) Local authorities and Agenda 21. Pp. 101–12 in Dodds, F. (ed.) *The Way Forward. Beyond Agenda 21*. London: Earthscan.

Buttel, F.H., Hawkins, A.P. and Power, A.G. (1990) From limits to growth to global change. Constraints and contradictions in the evolution of environmental science and ideology. *Global Environmental Change* 1: 57–66.

Carlson, E. (1996) The legacy of Habitat II. *The Urban Age* 4 (2): 1, 4–6.

Carson, R. (1962) *Silent Spring*, Cambridge, Mass.: Riverside Press.

Christoff, P. (1996) Ecological citizens and ecologically guided democracy. Pp. 151–69 in Doherty, B. and de Geus, M. (eds) *Democracy and Green Politics*, London: Routledge.

Cohen, M. (1996) Habitat II: a critical assessment. *The Urban Age* 4(2): 8, 21.

Commoner, B. (1971) *The Closing Circle. Nature, Man and Technology*. New York: Knopf.

Conca, K., Alberty, M. and Dabelko, G.D. (eds) (1995) *Green Planet Blues. Environmental Politics from Stockholm to Rio*. Boulder, Colo.: Westview Press.

Dagens Nyheter (1997) Editorial, 30 November [Swedish Daily]

Dobson, A. (1990) *Green Political Thought*. London: Unwin Hyman.

Eckersley, R. (1992) *Environmentalism and Political Theory. Toward an Ecocentric Approach*. London: UCL Press.

Edelman, M. (1977) *Political Language: Words that Succeed and Policies that Fail*. New York: Academic Press.

Ehrlich, P. and Ehrlich, A. (1969) *The Population Bomb*. New York: Ballantine Books.

Elander, I. (in press) Partenariats et gouvernance urbaine. Un agenda pour la recherche comparative internationale. (Partnerships and urban governance: towards an agenda for cross-national comparative research), En V. Hoffman-Martinot et L. Koszinski (sous la direction de) *Partenariats urbaine*. Collection Vie Locale. Paris: Cedone.

Elander, I. and Montin, S. (1990) Decentralization and control: Central and local government relations in Sweden. *Policy and Politics* 18: 165–80.

Elander, I., Gustafsson, M., Sandell, K. and Lidskog, R. (1995) Environmentalism, sustainability and urban reality. Pp. 85–114 in Khakee, A., Elander, I. and Sunesson, S. (eds) *Remaking the Welfare State. Swedish Urban Planning and Policy-making in the 1990s*. Aldershot: Avebury.

Falk, R. (1997) Resisting 'globalisation from above' through 'globalisation from below'. *New Political Economy* 1: 13–26.

—— (1998a) Global civil society: perspectives, initiatives, movements. *Oxford Development Studies* 26 (1): 99–110.

—— (1998b) The United Nations and cosmopolitan democracy: bad dream, utopian fantasy, political project. Pp. 309–31 in Archibugi, D., Held, D. and Köhler, M. (eds) *Re-imagining Political Community*. Oxford: Polity Press.

Finger, M. (1993) Politics of the UNCED process. Pp. 36–48 in Sachs, W. (ed.) *Global Ecology: A New Arena of Political Conflict*. London: Zed Books.

Flavin, C. (1997) The legacy of Rio. Pp. 1–22 in Brown, L.R. *et al.*, *State of the World 1997. A Worldwatch Institute Report on Progress Toward a Sustainable Society.* New York/London: Norton and Company.

French, H. (1997) Learning from the ozone experience. Pp. 151–72 in Brown, L.R. *et al.*, *State of the World 1997. A Worldwatch Institute Report on Progress Toward a Sustainable Society.* New York/London: Norton and Company.

Friends of the Earth (1997) Press Release, 27 June.

Haas, P.M., Levy, M.A. and Parson, E.A. (1992) Appraising the Earth Summit: how should we judge UNCED's success? *Environment* 34 (8): 7–11, 26–32.

Habitat II (1996) Introduction. [http://www.unhabitat.org/]

Hajer, M.A. (1995) *The Politics of Environmental Discourse. Ecological Modernization and the Policy Process.* Oxford: Clarendon Press.

Hardin, G. (1968) The tragedy of the commons. *Science* 162: 1243–8.

Haughton, G. and Hunter, C. (1996) *Sustainable Cities.* London/Bristol: Jessica Kingsley.

Held, D. (1995) *Democracy and the Global Order.* Stanford, Calif.: Stanford University Press.

Hempel, L.C. (1996) *Environmental Governance. The Global Challenge.* Washington, DC: Island Press.

ICLEI (1997) *Local Agenda 21 Survey. A Study of Responses by Local Authorities and Their National and International Associations to Agenda 21.* International Council for Local Environmental Initiatives in co-operation with United Nations Department for Policy Co-ordination and Sustainable Development, February. [http://www.iclei.org/la21/la21rep.htm]

Irwin, A. (1995) *Citizen Science. A Study of People, Expertise and Sustainable Development.* London: Routledge.

IUCNR (1980) *World Conservation Strategy. Living Resource Conservation for Sustainable Development.* Gland: International Union for Conservation of Nature and Natural Resources.

Jamison, A. (1997) The shaping of the global environmental agenda: The role of non-governmental organisations. Pp. 224–45 in Lash, S. Szerszynski, B. and Wynne, B. (eds) *Risk, Environment and Modernity. Towards a New Ecology.* London: Sage.

Jewson, N. and MacGregor, S. (1997) Transforming cities: social exclusion and the reinvention of partnership. Pp. 1–18 in Jewson, N. and MacGregor, S.S. (eds) *Transforming Cities: Contested Governance and New Spatial Divisions.* London and New York: Routledge.

Keohane, R. and Nye, J. (1977) *Power and Interdependence: World Politics in Transition.* Boston/Toronto: Little, Brown and Company.

Kooiman, J. (ed.) (1993) *Modern Governance – New Government–Society Interactions.* London/Newbury Park/New Delhi: Sage.

Leaf, M. (1997) The many agendas of Habitat II. Guest Editorial. *Cities* 14 (1).

Lindner, C. (1997) Agenda 21. Pp. 3–14 in Dodds, F. (ed.) *The Way Forward. Beyond Agenda 21.* London: Earthscan.

Lipschutz, R.D. with Mayer, J. (1996) *Global Civil Society and Global Environmental Governance.* Albany: State University of New York Press.

Lohmann, L. (1990) Whose common future? *The Ecologist* 20(3): 82–4.

Low, N. and Gleeson, B. (1998) *Justice, Society and Nature: an Exploration of Political Ecology.* London and New York: Routledge.

McGrew, A. (1995) World order and political space. Pp. 11–64 in Anderson, J., Brook, C. and Cochrane, A. (eds) *A Global World? Re-ordering Political Space*. Milton Keynes: The Open University Press.

McLaren, D. (1992) London as ecosystem. Pp. 56–68 in Thornley, A. (ed.) *The Crisis of London*. London and New York: Routledge.

Maddox, J. (1972) *The Doomsday Syndrome*, London: Macmillan.

Meadows, D.H., Meadows, D.L., Behrens, W.W. and Randers, J. (1972) *The Limits to Growth – A Report to the Clube of Rome's Project on the Predicament of Mankind*. London: Pan Books.

Meadows, D. H., Meadows, D.L. and Randers, J. (1992) *Beyond the Limits. Global Collapse or Sustainable Society: Sequel to the Limits to Growth*. London: Earthscan.

Mega, V. (1996) Our city, our future: towards sustainable development in European cities. *Environment and Urbanization* 8 (1): 133–54.

Naess, A. (1973) The shallow and the deep, long-range ecology movement: a summary. *Inquiry* 6: 95–100.

Nelissen, N., van der Straaten, J. and Klinker, L. (eds) (1997) *Classics in Environmental Studies. An Overview of Classic Texts in Environmental Studies*. Utrecht: International Books.

O'Riordan, T. (1981) *Environmentalism*. London: Pion.

Porter, G. and Brown, J.W. (1996) *Global Environmental Politics*, Boulder, Colo.: Westview Press.

Rhodes, R.A.W. (1997) *Understanding Governance. Policy Networks, Governance, Reflexivity and Accountability*. Buckingham: Open University Press.

Roseland, M. (1997) Dimensions of the eco-city. *Cities* 14(4): 197–202.

Sachs, W. (1992) Environment. Pp. 26–37 in Sachs, W. (ed.) *The Development Dictionary – A Guide to Knowledge as Power*. London: Zed Books.

Serageldin, I. (1996) *Sustainability and the Wealth of Nations. First Steps in an Ongoing Journey*. Washington, DC: World Bank.

Shiva, V. (1993) Vetenskap, teknik och 500 år av kolonisationer (Science, technology and 500 years of colonisations) *VEST: Tidskrift för vetenskapsstudier* 6(2): 3–16. [Swedish Journal for Science Research]

Speth, J.G. (1997) Preface. Pp. xii-xiii in Dodds, F. (ed.) *The Way Forward. Beyond Agenda 21*. London: Earthscan.

Stren, R. (1992) Introduction. Pp. 1–7 in Stren, R., White, R. and Whitney, J. (eds) *Sustainable Cities: Urbanization and the Environment in International Perspective*. Boulder: Westview Press.

Tosics, I. (1997) 'Habitat II Conference on Human Settlements, Istanbul, June 1996', *International Journal of Urban and Regional Research*, 21 (2): 366–72.

UN (1997a) Resolution adopted by the General Assembly at its nineteenth special session, June 28 [http://www.un.org/esa/earthsummit/]

UN (1997b) *Kyoto Protocol to the United Nations Framework Convention on Climate Change*, 11 December.

UN (1998a) *UN Climate Change Convention*, Press Release, Bonn, 16 March.

UN (1998b) *UN Climate Change Convention*, Press Release, Buenos Aires, 14 November.

Vogler, J. (1995) *The Global Commons: A Regime Analysis*. Chichester: John Wiley and Sons.

Wandén, S. (1996) *Miljö och ansvar. En fråga om insikt, etik och effektivitet*. (Environment and responsibility. A matter of insight, ethics and efficiency) Report

46 29. Stockholm: Environmental Protection Agency.

Weale, A. (1992) *The New Politics of Pollution.* Manchester: Manchester University Press.

White, R. and Whitney, J. (1992) Cities and the environment: an overview. Pp. 8–51 in Stren, R., White, R. and Whitney, J. (eds) *Sustainable Cities: Urbanization and the Enviroment in International Perspective.* Boulder: Westview Press.

von Weizsäcker, E., Lovins, A.B. and Lovins, L.H. (1997) *Factor Four: Doubling Wealth, Halving Resource Use,* Sydney: Allen and Unwin.

Wildavsky, A. (1973) If planning is everything, maybe it's nothing. *Policy Sciences* 4: 127–153.

World Commission on Environment and Development (1987) *Our Common Future. The Brundtland Report.* Oxford: Oxford University Press.

Young, O.R. (1996) The effectiveness of international governmental systems. In Young, O., Demko, G.J. and Ramakrishna, K. (eds) *Global Environmental Change and International Governance.* Hanover and London: University Press of New England.

Young, O. and Demko, G.J. (1996) Improving the effectiveness of international environmental governance systems. Ch. 10 in Young, O., Demko, G.J. and Ramakrishna, K. (eds) *Global Environmental Change and International Governance.* Hanover and London: University Press of New England.

Zolo, D. (1997) *Cosmopolis. Prospects for World Government.* Cambridge: Polity Press.

3   A rough road out of Rio

The right-wing reaction in the United States against global environmentalism

Timothy W. Luke

Introduction

Most of the delegates to the 1992 Earth Summit in Rio de Janeiro came from countries which sincerely supported the meeting's professed values and ecological goals. On this count, however, the representatives of the United States stood apart from the rest. Because of President Bush's opposition to the meeting's climate change and biodiversity treaties, the 300 American delegates intransigently resisted the conference's basic agenda, which severely dampened the entire affair's overall diplomatic success. This difference is significant. While the other chapters in this collection examine the more positive intended effects of the Rio summit outside America, this chapter looks more closely at a few of its negative unintended effects on politics in the United States.

Since 1992, the meliorative tenor of Rio has been spun very perversely in the USA as a threat and menace to its national sovereignty, economic security, and domestic economy. While these anti-environmental interpretations of the Rio summit first came from the extreme right, they soon were endorsed, albeit obliquely, by the Bush administration as it felt the mounting pressures of the 1992 presidential primaries. Even though Bush lost the 1992 election, and the victorious Clinton–Gore ticket has espoused more pro-environment views, the new Democratic government holds views on free trade, engaged globalism, and American superpower that closely parallel those of President Bush. Yet, these Cold War internationalist positions have only inflamed the nationalistic right-wing backlash against transnational economics and ecology.

In fact, the split over international ecology agreements, like those propounded at the Rio Summit, express some of the most fundamental divisions now splitting the American body politic during the post-Cold War era. This chapter, then, will examine these political conflicts and ideological contradictions as they have surfaced all over the United States in the aftermath of the Earth Summit at Rio. It surveys the public discourse about the Rio conference – first, during the days leading into the 1992 summit and then afterwards in order to, second, illustrate how this global meeting has become reinterpreted in such a menacing manner to many American citizens.

The Earth Summit and its promise

After two years of planning and preparatory meetings, the United Nations Conference on Environment and Development was convened in Rio de Janeiro on June 3–14, 1992. The scope of the Earth Summit and its antecedents has already been outlined in Chapter 2. Before the Rio summit, many hoped that it might prove to be a watershed event that would change the direction of human history, like the treaties of Westphalia, the Congress of Vienna, or the San Francisco Charter. New nations from the mostly poor, industrializing South wanted the richer, industrialized North to limit their pollution and provide more development aid; older countries in the North hoped the nations of the South would make greater efforts to curb their population growth and protect biodiversity (Adler with Hager, 1992: 22).

Very little real progress has been made after the meeting because of the unwillingness of both sides to either make real unilateral concessions or engage in serious joint collaboration. Even though the United States has purported to be 'pro-environment' to all of its domestic and foreign audiences, the Bush administration persistently maintained throughout the proceedings that 'the American life-style is not up for negotiation' (Elmer-Dewitt, 1992: 58).

As a *Washington Post* editorial remarked, the Rio summit in the eyes of official Washington was not about global warming or species depletion. Instead, it was read as being about environment and development, or 'translating economic wealth into genuinely better living conditions over the next generation' (June 3, 1992: A18). Because just about every government in the world would be represented, many hoped the meeting also could make 'a contribution to the education of governments, particularly the one here in Washington' (June 3, 1992: A18). Unfortunately, as subsequent events during the 1990s have shown, this educational potential essentially was lost on both the American people and its government in Washington.

The Rio summit on the environment, for all the promise held by its impressive convocation of the world's countries, nongovernmental organizations, and environmentalists, did not produce many decisive outcomes due to a certain lack of decisive leadership (Babbit, 1992; Shabecoff, 1992; Begley *et al.*, 1992). Going into the summit, 139 countries had voted for a mandatory stabilization agreement on greenhouse gas emissions, which would have fixed year 2000 outputs at 1990 levels. From the beginning, the Bush administration disagreed with these targets (Rensberger, 1992: A22). Only the USA was standing in opposition to this accord in March 1992; yet, by May, most of the European nations, led by Germany, also were wavering.

The targets on emissions were moved, and the mandatory enforcement provisos of the agreement were revoked by the time the Rio conference was convened. Still, the European Community with Germany in the lead won plaudits for its firm commitment to work with the developing countries, regulate greenhouse gases, and provide $4 billion in environmental aid

(Weisskopf, 1992a: A1, 8). By and large, Japan played no special role in the conference, while Russia and the former Soviet republics only sought minimal recognition of their difficulties in making the transition to a market economy. Most developing nations placed their major emphasis, like President Bush and the United States, on promoting more economic growth (Greenhouse, 1992: 6). Mexico's President Carlos Salinas, for example, nodded with respect to 'nature's equilibriums,' but he also asserted that 'the cause of ecology must not be converted into the cause of protectionism' (Weisskopf and Devroy, 1992: A1, 12). Consequently, none of the Earth Summit's agreements were legally binding, and there was no effective means of insuring real compliance with any measures endorsed by the conferences in Rio (Lewis, 1992: 10).

By the end of the proceedings, very few real achievements could be attributed to the gathering. Of course, some general ideals were affirmed. The Rio Declaration, a six-page philosophical brief connecting poverty to environmental degradation, and Agenda 21, a lengthy blueprint for many environmental reforms, were approved by the assembled body, but these were totally non-binding declarations carrying only moral force to ensure compliance (Weisskopf and Devroy, 1992: A1, 12). A statement of principles on forest preservation was discussed; yet, poor, timber-producing countries resisted a tough treaty to protect the world's forests as wildlife habitats, carbon sinks, and biodiversity preserves. A binding treaty on biodiversity was drawn up. The United States, however, refused to sign it, arguing that it would cripple the nation's booming biotechnology industries (Robinson and Preston, 1992: A1, 26). The climate change convention on greenhouse gases was accepted, but without specific targets or timetables for reductions, even though most nations made a moral commitment to keep year 2000 emissions at 1990 levels (Elmer-Dewitt, 1992: 58).

In many ways, the Earth Summit in Rio was organised to mediate some new shared understandings between the highly industrial countries of the North and the newly industrializing nations of the South (Easterbrook, 1992: 33). The nations of the North were to concede their responsibility for polluting their own backyards, and everyone else's, while the countries in the South were to strike a fresh bargain with the North not to repeat all of the North's industrial missteps. The South was to pledge protection for biodiversity within their borders in exchange for aid, and the North was to provide aid to get real protections for the biosphere in the South. Regrettably, however, things on this account neither started out well nor ended positively. William K. Reilly, head of the United States Environmental Protection Agency, indicated on the eve of the conference that Washington was concerned about 'a certain amount of posturing by developing countries to try to get us to contribute more funds' when, in fact, 'those contributions are not in the cards' (Weisskopf and Robinson, 1992: A21). Similarly, Ting Wen Lian, Malaysia's ambassador to the UN Food and Agriculture Organization complained about such 'high-handed' diplomatic tactics, while suggesting that Third World nations would not be cast as 'scapegoats' for the world's environmental crises (Weisskopf and Robinson, 1992: A21).

When the conference ended, the United States had not shifted from its original intransigent positions. Reilly continued to claim, in keeping with the Bush administration, that signing all of the Rio treaties was out of the question. Indeed, it would be 'contrary to the interests of the United States' (Weisskopf and Robinson, 1992: A23). Maurice Strong, the multimillionaire architect of the Rio Earth Summit, estimated $6 or $7 billion was pledged by the assembled nations to aid the poor countries of the South; yet, the USA essentially baulked at a binding pledge to give only 0.7 percent of its GNP to pay for ecologically sustainable development in the Third World (Shabecoff, 1992: 101, 99). These penny-pinching tactics only demonstrated, once again, to Strong how few of the world's powerful nations saw the close connection between global economic inequalities and world-wide ecological disasters (Saul, 1992: 32–33). For a man in search of 'historic civilizational change' (Preston, 1992: B1), Rio proved to be very frustrating.

The official American opposition to Rio

The American response to the Rio conference cannot be comprehended without recognizing that the diplomatic positions of official Washington often reveal very little about where most of the state or civil society actually stand. On one level, environmentalism is treated as a 'Mom and apple pie' issue in everyday political life: everyone says they want clean air, clean water, clean land. On a second level, however, environmentalism often plays out as a very selfish type of localism: NIMBYism on either the neighborhood or national level always pushes for others to incur all the costs of otherwise narrowly distributed benefits. On a third level, environmentalism is increasingly being typecast as a real threat to 'the American way of life'. For many average consumers there is a new enemy. Dictatorial state regulators, who undoubtedly are socialists, want to take away backyard barbecues, fast cars, red meat, and air conditioners in the name of the ozone layer, global climate, international equity, and world environment. In other words, the various levels of the government bureaucracy and diverse quarters of civil society are becoming quite divided over the real meaning of the environmental crisis, which makes it easy for all Americans to be easily divided by mismanaged ecological concerns.

The Rio Declaration's sense of alarm, then, clearly is not at all shared by many neoliberals and nationalists, who are now the recruits for antiglobalization fundamentalist movements in the United States. While the Earth might have an 'integral and interdependent nature', this does not necessarily require everyone to conform to economic, political, and social directives from would-be transnational ecocrats intent upon protecting what they imagine is 'the global environmental and developmental system' (Grubb *et al.*, 1993: 87). On one level, neoliberals and nationalists rightly complain that the operational science that documents like the Rio Declaration rest upon is nothing but ordinary scientific research. As such, it is completely contestable, entirely subject to second opinion, and expressly mandates by itself no clear moral solutions (Rensberger, 1992: A1, 22). On a second level, antiglobalization advocates see the figure of globalization as a new strategy for simply redistributing the costs and benefits of unequal growth from one network of currently overprivileged localities to another collection of presently underprivileged localities. Both implicitly acknowledge there are great disparities within and between nations, but anti-transnational, anti-intergovernmental, anti-environmental resistances explicitly oppose any policies that will take something away from the United States in general, or impose new costs upon particular localities within the USA, in order to benefit some other unknown nations and localities elsewhere.

The American position at the Rio summit, which was defined by a Bush administration still flush with success after the Gulf War and the collapse of the Soviet Union, ironically, was not very yielding to international pressures during the June 1992 conference. During his official speech, Bush told the assembly of world leaders that 'America's position on environmental protection is second to none, so I did not come here to apologize' (Greenhouse, 1992: 4–1). Consequently, the USA compelled everyone to accept a very weakened version of the climate change treaty, and it refused to sign the biodiversity accord, suggesting that both initiatives would damage America's economic growth and industrial performance (Weisskopf, 1992b: A42). Three and a half years of very little real economic progress had left high percentages of Bush supporters during the 1992 primary season in New Hampshire, Pennsylvania, and Wisconsin saying they wanted any other GOP candidate besides President Bush (Belliveau *et al.*, 1992: A12). So, caught in a tough re-election race at home, Bush stressed the salient importance of 'the economy' over all environmental concerns.

In fact, the Bush administration was quite strident, attacking Japan and Germany as latecomers to the ecological cause, rebuking indigenous peoples for worrying excessively about local biodiversity, and dismissing larger efforts to cap year 2000 carbon dioxide emissions at 1990 levels (Begley *et al.*, 1992: 30–33). Germany's Minister of the Environment, Klaus Topfer, explicitly articulated his country's alarm over these hardnosed American positions. Fearing that the Cold War rivalry over ideology between East and West was being supplanted by a new North–South rivalry over the environment, Topfer said, 'I am afraid that conservatives in the United States are picking "ecologism" as their new enemy' (Greenhouse, 1992: 4–1).

Even though tough environmental regulations have sparked the creation of almost 70,000 environmental companies with nearly two million employees and $130 billion in sales (Schneider, 1992: 4–4), President Bush rebuffed the Rio conference's biodiversity negotiators. In a news conference prior to the Rio summit, Bush reaffirmed, on the one hand, that 'I will not sign a treaty that in my view throws too many Americans out of work', and, on the other hand, he refused 'criticism from what I consider some of the extremes in the environmental movement, internationally and domestically' (Schneider, 1991: 4–4). In this open assault on all environmentalists, foreign and domestic, who opposed his allegedly strong ecological record, Bush echoed antiecological claims made by the network of timber, coal mining, agriculture, and land developing interests allied together in the 'wise use' movement across the American West (Harvey, 1996: 384–385). Bush's take on the Earth Summit was captured best in an interview with *Jornal do Brasil* on the eve of his departure to Rio: 'I am president of the United States, not president of the world, and I'll do what is best to defend USA interests' (cited in Shabecoff, 1992: 89).

This official White House approach simply stuck to a nationalistic neoliberal understanding of the world's economy and ecology. Partly a response to global economic competitions, and partly a response to global ecological scarcities, today's neoliberal and antiglobal readings of the earth's political economy construct the attainment of national economic growth, security, and prosperity as a zero-sum game. Having more material wealth or economic growth in *one* place, such as the USA or any given locality within its borders, means not having it in *other* places – namely, rival foreign nations and all of their many local communities. These positions also assume that material scarcity is an inflexible constraint; hence, all resources, everywhere and at any time, should be treated as private property whose productive potentials must be subjected ultimately to economic exploitation and not obstructed by ecological regulation.

Many anti-transnational and anti-environmental popular groups in the USA accept the prevailing form of mass market consumerism as it presently exists, because it defines many material private benefits as the public ends that advanced economies ought to seek (Harvey, 1996: 383–385). This, then, affirms the need for hard discipline in elaborate programs of local productivism, only now couched within rhetorics of highly politicized national competition, as the means for sustaining mass market consumer lifestyles in advanced nations like the United States. Creating economic growth, and producing more of it than other equally aggressive developed and developing countries, is the *sine qua non* of 'national security' in the 1990s. As Richard Darman, President Bush's chief of the Office of Management and Budget declared after Earth Day in 1990, 'Americans did not fight and win the wars of the twentieth century to make the world safe for green vegetables' (cited in Sale, 1993: 77). Not everyone in the USA, then, sees environmentalism as tantamount to moral salvation by leaving behind an entire way of life tied to using increased levels of natural resources to accelerate economic growth.

These nationalistic readings of the environment have sparked into life some new discourses about collective social responsibility. Even the green geopolitics of the Clinton administration carries a very nativistic reading in its codes of ecological reflexivity. This became obvious when President Clinton made green geopolitics an integral part of his global doctrine of 'engagement' in 1995. 'To reassert America's leadership in the post-Cold War world', and in moving 'from the industrial to the information age, from the Cold War world to the global village', President Clinton opened up to both antiglobalization localists and neoliberal nationalists when he claimed:

> we know that abroad we have the responsibility to advance freedom and democracy – to advance prosperity and the preservation of our planet . . . in a world where the dividing line between domestic and foreign policy is increasingly blurred . . . Our personal, family, and national future is affected by our policies on the environment at home and abroad. The common good at home is simply not separate from our efforts to advance the common good around the world. They must be one in the same if we are to be truly secure in the world of the 21st century.
>
> (*Foreign Policy Bulletin*, 1995: 43)

The Rio summit simply turns into one more piece in an emergent mosaic of international accords and transnational understandings that are disturbing for many Americans as they contemplate the world system, and their nation's place within it, after the Cold War. Along with the North American Free Trade Agreement (NAFTA) and the Uruguay Round of the General Agreement on Trade and Tariffs (GATT) negotiations in 1993, the Rio summit in 1992 and the Kyoto climate conference in 1997 all now appear to be, as Henry Kissinger celebrates, the emerging architecture of 'a new international system', which will permit the USA to take the final steps toward 'the new world order' (1993: 1C). Advocates of 'free trade' and 'sustainable development' like Henry Kissinger, George Bush, Warren Christopher, and Bill Clinton, support the growing openness of the American economy and environment to global competition and regulation. For others, however, like Ross Perot, Jerry Brown, Jesse Jackson, and Patrick Buchanan, the dictates of Agenda 21 or NAFTA represent more than restrictions of greenhouse gas emissions or industrial jobs going to the South. Instead, they are now all 'about American sovereignty going south' (Buchanan, 1998: 264).

Antiglobalism and ultranationalism in the USA

Many Americans will resist intrusions of any sort into their material way of life, but the most active and militant opposition to transnational environmentalism comes from loosely organized but quite widespread conservative groups, like the self-identified Patriot movement. Spanning a very narrow band of the ideological spectrum from the Christian Identity, Counties' Rights, Wise Use and National Taxpayers' Union to the John Birch Society, Ku Klux Klan, Posse Comitatus, and Christian Coalition, this growing band of fellow travelers also counts many right-to-life, neo-nazi, gun advocates, and anti-Semitic groups among its ranks. United by their distaste for what they imagine as the New World Order, these ultranationalist groups all see themselves as legitimate expressions of the popular sovereignty underpinning the American republic. Like other militant, self-organized, and well-armed associations of citizens throughout the history of the United States, the Confederate States, or the original colonies of North America, these nationalists now dispute the decision-making authority and legitimacy of the state behind today's incumbent bipartisan regime of free trade globalists.

After the siege at Waco and the passage of NAFTA in 1993, many Patriot groups turned up the rhetorical heat in their interpretations of the present moment. For many, the activities of the Clinton administration, in particular, soon,

> led to the conviction that the government was proceeding to disarm citizens, to subdue them later, submitting Americans to surveillance from hidden cameras, and black helicopters, and implanting biochips in the newborn. To this global threat, on jobs, on privacy, on liberty, on the American way of life, they oppose the Bible and the original American Constitution, expunged of its Amendments. In accordance with these texts, both received from God, they affirm the sovereignty of citizens and its direct expression in county governments, not acknowledging the authority of the federal government, its laws, its courts, as well as the validity of the Federal Reserve Bank.
>
> (Castells, 1997: 86–87)

Not surprisingly, the Patriot movement has little use for any international resolutions to preserve the environment that the current federal government has chosen to back with diplomatic, economic, and scientific support.

For antiglobalization advocates, there is, in fact, no better example of the New World Order than the workings of transnational environmental conferences, groups, and institutions as they have emerged out of the Montreal ozone protocols or the Rio environmental conference. All of them appear clearly poised to extinguish American economic and political sovereignty. When positioned alongside other more bread-and-butter decisions imposed by the World Trade Organization, the International Monetary Fund, and the United Nations, even feel-good issues, like protecting the environment or preserving biodiversity, assume a more insidious quality for those who question the liberal meliorism of these international institutions. Such organizations are interpreted as a very real threat to the American way of life, particularly to the well-paying jobs, privacy rights, personal freedoms, and political powers of individual American citizens.

The Preamble to Agenda 21 from the Earth Summit on the global environment at Rio is full of technocratic talk that highlights the policy imperatives, not for individuals, localities or even nations, but rather for all human beings, planetary ecosystems or especially global partnerships:

> Humanity stands at a defining moment in history. We are confronted with a perpetuation of disparities between and within nations, a worsening of

poverty, hunger, ill health and illiteracy, and the continuing deterioration of the ecosystems on which we depend for our well-being. However, integration of environment and development concerns and greater attention to them will lead to the fulfillment of basic needs, improved living standards for all, better protected and managed ecosystems and a safer, more prosperous future. No nation can achieve this on its own; but together we can – in a global partnership for sustainable development.

(Grubb *et al.*, 1993: 83)

Because many communities in the USA already enjoy a safe, prosperous present, antiglobalization advocates see these sorts of pious liberal pledges expressing a set of transnational tactics to reduce their local community's security and prosperity. Moreover, the vague designs of a 'global partnership for sustainable development' upsets neoliberals and nationalists in the USA who believe existing markets are working well enough to serve American national interests.

Transnational environmentalism cannot strike a worthwhile bargain for the United States because it apparently stands for an unknown, unfixed and untrustworthy 'environmentalist transnation' whose ecocratic overseers will be certain to put postnational ecosystemic interests over and before the national economic interests of Americans. Even though they would be extremely wary of some antiglobalization localists, elite groups of neoliberal nationalists also share this deep suspicion of well-meaning global conferences, like the 1992 Rio conference on the environment, that try to renegotiate the terms of global political economy by using ecological values to change who gets what, where, when, and how.

Antiglobalization localists fear their already marginal economic security will be abridged in the hope of enhancing life somewhere in China, India or Brazil where poverty, hunger, ill health and illiteracy are all quite severe. Yet, making this move also will certainly bring most of these negative factors very quickly down to many Americans on a local level. Neoliberals and nationalists, in turn, fear too many of these global partnerships will not sustain America's continuing economic development in the hopes of jogging new development elsewhere outside the USA. Almost every American industry feels this heat. Many ordinary workers know Japan produces light, fuel-efficient cars in new highly robotized factories. So when the USA produces full-sized vans in Ford automobile plants that date from 1919 and require twice as much energy per vehicle produced as comparable Japanese auto plants (Weisskopf, 1992a: A1), the ecological costs and economic benefits of America's existing sovereignty gets put into a more revealing global perspective.

Anti-statism as anti-environmentalism

These dynamics have been building in the USA since 1968. Amidst a losing war in Vietnam, an inability to halt the spread of Soviet-style communism, a stagnation in average family income, and an apparent favoritism for racial minorities over the white majority, a widespread backlash began to build against the cosmopolitan values of New Deal and Great Society liberalism in the early 1970s. The events of Waco, Ruby Ridge, and Oklahoma City in the 1990s are not entirely the work of a lunatic minority. They do express the rage of many rural, white, working class Americans, who now distrust almost all government officials, corporate middlemen, and scientific experts. In many cases, these men and women have been struggling for decades to establish and/or maintain their place upon America's vast middle classes. As the economy and environment of the United States were opened up to foreign competition and pressure during the 1970s by corporate liberals with free trade ideals and transnational environmentalist values, 'a growing number of white men in rural America had come to believe that this kind of liberalism had little or nothing to offer to them' (Stock, 1997: 150).

In partisan terms, these citizens were angry. Even so, when they vote, most are not always racist, sexist or conservative in their politics. During 1968, the vast majority of those who voted for George Wallace of Alabama named Robert F. Kennedy as their second choice for President (Bennett, 1988: 337). It was the anti-elitism, populism, and anti-corporate tone of both Wallace and Kennedy that excited these people. Richard Nixon, of course, named them 'the silent majority', but mostly they were, as this odd preference ordering for George Wallace *or* Bobby Kennedy suggests, those Americans 'for whom New Deal and Great Society liberalism had not delivered on its promise, if it had made a promise at all' (Stock, 1997: 152). In 1980, many of these voters turned out for Ronald Reagan; and in 1992, they voted for Bill Clinton. Each time, they basically were seeking something new, something different, something not unlike what was expressed in the words of Bill Clinton at the launch of his 1996 campaign when he exclaimed that 'the age of big government is over' (*New York Times*, January 24, 1996: A14).

Endorsing environmental efforts, like Agenda 21 or the Rio Declaration, however, is not a sign of big government going away. In fact, this sort of transnational environmental treaty seems far more ominous to far too many ordinary Americans, because it looks as though the era of big American government will end only to be replaced by an era of even bigger foreign governance: the New World Order. This fracture down the center of the Cold War consensus is what defines much of America's politics in the 1990s, and there is every indication that it will continue into the next century as new environmental imperatives are pushed more and more on to center stage in Washington and every state capital. 'Within both parties', as Buchanan asserts, 'nationalists are in rancorous conflict with the globalists . . . this is the new conflict of the age that succeeds the Cold War' (Buchanan, 1998: 265).

Repudiating even weak environmental regulations, such as those from the Rio Summit, represents America's new populist 'economic nationalism' very well. For many average Americans, the work of the EPA within the USA is living proof that 'environmental policy is out of control, costing jobs, depressing living standards and being run by politicians, scheming business people and social extremists' (Brimelow and Spencer, 1992: 59). Nationalistic populists believe in 'wise use' philosophies of market-driven conservationism, but they do not endorse stronger campaigns to regulate consumer choice or producer prerogative directly by intrusive state intervention in the name of all humanity. Buchanan, for example, is quite explicitly anti-global in his version of economic nationalism, which means,

> tax and trade policies that put America before the Global Economy, and the well-being of our own people before what is best for 'mankind' [*sic*]. Trade is not an end in itself; it is the means to an end, to a more just society and more self-reliant nation. Our trade and tax policies should be designed to strengthen USA sovereignty and independence and should manifest a bias toward domestic, rather than foreign, commerce. For as von Mises said, peaceful commerce binds people together, and Americans should rely more on one another.
>
> (Buchanan, 1998: 228)

The efforts to reduce greenhouse gases, losses of biodiversity, and ozone-destroying compounds can be dismissed as being based upon shoddy science and/or devious diplomacy, both of which certainly seem aimed at curtailing American sovereignty. Consequently, any additional attempts to impose unwanted environmental regulations must be, according to America's new anti-globalists, held before the demanding bar of an enlightened nationalism. The purposes of economic and environmental policy in the United States are not 'to prosper mankind – but Americans first: our workers, farmers, businessmen, manufacturers. And what is good for the Global Economy is not automatically good for America' (Buchanan, 1998: 284).

Grassroots opposition such as this to major international agreements on the environment also affects many higher level policy deliberations. Before the USA delegation departed to the conference on the global climate in Kyoto during 1997, the Senate unanimously passed the Byrd–Hagel resolution, which states that the United States must not sign any agreement on greenhouse gas emissions unless it stipulates specific commensurate reductions for developing nations. Sponsored by Senators Robert Byrd (Democrat) of West Virginia and Chuck Hagel (Republican) of Nebraska, this bipartisan resolution has influenced the debates and negotiations over the December 1997 treaty at home and abroad (Passacantando, 1998: C5). On one level, this resolution marks an intense level of lobbying by coal, gas, and oil interests in the United States, who do not want their markets to shrink until Mexico, India, China, and Brazil also agree to reduce their consumption of dirty fossil fuels. On another level, however, these moves also express the anxieties of ordinary voters who do not want their own high paying jobs or everyone's national security to be negotiated away in the name of environmental regulation, only to have the jobs reappear at unregulated sites somewhere in the Third World because Mexico, India, China or Brazil were exempted by the treaty.

For many Americans, even very conventional forms of environmentalism, including the initiatives advanced at the Montreal, Rio, Cairo or Kyoto international conferences on CFCs, climate, the environment, or population, can be cast as a serious threat to private property rights and individual free enterprise. United States Representative and Nevada Republican John Ensign, for example, attacks the leaders of the environmental movement, domestic and foreign, as 'socialists' or 'collectivists' who want to use big government, ecological regulations, and international treaties to take away people's privative property. In a recent campaign speech in his race for a USA Senate seat in Nevada, he repeated some very widespread beliefs: 'If you look at what modern environmentalists have become, they have become not about protecting the environment, but they have become about big government and regulations and putting things out of the hands of private citizens' (Vogel, 1998: 8B).

Because the environment has become a mainstream value for most Americans, such conservative voices cannot dismiss it. They instead endorse a particular type of environmentalism that emphasizes human stewardship of the environment against the environmental movement's apparent anti-humanism. As Michael S. Berliner, the executive director of the Ayn Rand Institute, suggests, Earth Day should actually be called 'Anti-Human Day', because the environmentalists behind such events believe nature ought to be revered 'for its own sake, irrespective of any benefit to man' (Berliner, 1998: 1–E). Conservative stewardship, on the other hand, argues in political debates and policy deliberations 'the Earth is here for us. We are to be good stewards; we are to take care of it' (Vogel, 1998: 8B). In fact, the preservation of private property rights for many conservatives will give all humans a very real stake in the process of environmental protection, and should show everyone how 'to manage in a way that is good for people and the environment' (Vogel, 1998: 8B).

While this conservative reaction to environmentalism supposedly resists the demolition of technological/industrial civilization, it implicitly also stands up in defense of the United States' uniquely important place in the world's economic system. Virtually no distinctions are drawn between international environmental accords and national ecology groups by their conservative opponents. Environmentalists are all cast as anti-human and pro-nature. Indeed, as Berliner asserts, in the United States, which remains

a nation founded on the pioneer spirit, they have made 'development' an evil word. They inhibit or prohibit the development of Alaskan oil, offshore drilling, nuclear power and every other form or energy. Housing, commerce and jobs are sacrificed to spotted owls and snail darters. Medical research is sacrified to the 'rights' of mice. Logging is sacrificed to the 'rights' of trees. No instance of the progress which brought man out of the cave is safe from the onslaught of those 'protecting' the environment from man, whom they consider a rapist and despoiler by his very essence.

(Berliner, 1998: 1–E)

Sincere efforts to protect the environment, then, are deflected immediately by such critics into the register of national sovereignty, economic freedom, and personal liberty rather than remaining on the level of global crisis, ecological collapse, and imperiled biodiversity.

Once the work of environmentalism like that expressed at the Rio conference is put on this plane, anything that is done to protect biodiversity can be trivialized, overstated, or distorted as another sorry example of anti-human extremism. Plainly, there are some environmental extremists at work today, and the attention-getting rhetoric continually comes back to haunt them. Most environmentalists, however, are not extremists, and their efforts are pitched at guaranteeing the survival of human life by ensuring the survival of all the nonhuman life that humans depend upon in their environment. Unfortunately, many conservatives continue to reduce all forms of environmentalism to their most extreme expression: 'Such is the naked essence of environmentalism: It mourns the death of one whale or tree, but actually welcomes the death of billions of people. A more malevolent, man-hating philosophy is unimaginable' (Berliner, 1998: 1–E).

Conclusions: rough road ahead

The United States does not always carry its responsibilities as a world superpower easily or effectively. On the one hand, it must accept, because it professes to believe in democracy, consensus, and law, the contradictory dictates embedded in clusters of environmental agreements. They have been negotiated, when all is said and done, by the world's governments as meaningful and valuable understandings. On the other hand, it can also ignore, because it possesses the wealth, power, and technology, the weak constraints created by these agreements inasmuch as they limit America's sovereign authority and economic growth. Indeed, during the triumphalist 1990s, the citizens of the United States virtually see this as the special prerogative of American superpower.

Ultimately, Buchanan and other patriotic Americans who buy into his sort of economic nationalism explicitly repudiate the position espoused by Maurice Strong at the Rio summit: 'No one place can remain an island of affluence in a sea of misery. We're either going to save the whole world or no one will be saved' (cited in Weisskopf and Preston, 1992: A20). Instead, the right-wing reaction to the Earth Summit pushes Principle 1 of the Rio Declaration all the way to its nationalist conclusion: 'Human beings are at the center of concerns for sustainable development. They are entitled to a healthy and productive life' ('Draft of Environmental Rules: "Global Partnership"', 1992: 10). Human beings must be front and center, but they should also be first and foremost *American* human beings. Thus, environmental regulations are great as long as they apply to everyone else, but not to Americans.

These intense nationalist reactions to the Earth Summit in the United States add a distressing quality to post-Cold War politics. They are also unlikely to fade anytime in the near future. Accepting costly structures of national economic disadvantage in order to support larger geopolitical strategic goals, which was quite common during the Cold War, is no longer an automatic feature of bipartisan politics in the USA. What were tactics for preventing communist expansion via a very biased system of privileged international exchange are now seen by antiglobal nationalists as examples of parasitical free-riding at America's expense. Because globalized environmental treaties will cost Americans their jobs, they must be rejected by the ruling elites in Washington as well as by many mass publics beyond the beltway out in the country.

More liberal observers may discount these right-wing reactions as the passing signs of a temporary fringe movement of extremists which really poses no serious threat to the emergent transnational regime on the environment. This analysis is wishful thinking. The sources of this right-wing reaction have been building for a generation, and the greater geopolitical forces that once kept them at bay now have changed decisively. Private property based conservation has a large growing constituency in the USA, but public regulatory intervention in the name of abstract transnational ecologies is increasingly regarded as an economic fifth column dedicated to destroying the United States from within. Public spirited environmental activists no longer dominate political discourses about ecology in the way they once did thirty years ago. Unless and until, those discursive battles are refought and won in the USA, economic nationalists, conservative populists, and xenophobic isolationists will block the effectiveness of any new ecological initiatives from international conferences like the Earth Summit in Rio de Janeiro.

References

Adler, J., with Hager, M. (1992) 'Earth at the Summit', *Newsweek*, CXIX, no. 22 (June 1), 20–23.

Babbitt, B. (1992) 'The World After Rio', *World Monitor*, 5, no. 6 (June), 28–33.

Begley, S. *et al.* (1992) 'The Grinch of Rio', *Newsweek*, CXIX, no. 24 (June 15), 30–33.

Belliveau, J. *et al.* (1992) 'The Race for President: Act 1', *The Washington Post* (June 5), A12.

Bennett, D.H. (1988) *Party of Fear: From Nativist Movements to the New Right in American History.* Chapel Hill: University of North Carolina Press.

Berliner, M. (1998) 'Earth Day? It Should Actually Be Called Anti-Human Day', *Las Vegas Review-Journal* April 19, 1–E.

Brimelow, P. and Spencer, L. (1992) 'How Politics is Strangling Environmentalism: You Can't Get There From Here', *Forbes*, 150, no. 1 (July 6), 59–64.

Buchanan, P.J. (1998) *The Great Betrayal: How American Sovereignty and Social Justice Are Being Sacrificed to the Gods of the Global Economy* Boston: Little Brown.

Castells, M. (1997) *The Information Age: Economy, Society and Culture, Vol. III: The Power of Identity*, Oxford: Blackwell.

President Bill Clinton, 'Address at Freedom House, October 6, 1995 [A White House Press Release],' *Foreign Policy Bulletin*, November/December, p. 43.

'Draft of Environmental Rules: "Global Partnership"' (1992) *The New York Times* (April 5), 1–10. (no author)

Easterbrook, G. (1992) 'How Rio Will Make History', *Newsweek*, CXIX, no. 24 (June 15), 33.

Elmer-Dewitt, P. (1992) 'Summit to Save the Earth: Rich vs. Poor', *Time*, 139, no. 2 (June), pp 42–58.

Greenhouse, S. (1992) 'Ecology, the Economy, and Bush', *The New York Times* (June 14), 4–1, 6.

Grubb, M. *et al.* (1993) *The Earth Summit Agreements: A Guide and Assessment*, London: Earthscan Publications

Harvey, D. (1996) *Justice, Nature & the Geography of Difference*, Oxford: Blackwell.

Kissinger, H. (1993) 'The Trade Route: NAFTA a Step Toward a Prosperous World Order', *The Cleveland Plain Dealer* (July 18), 1C.

Lewis, P. (1992) 'Environment Aid for Poor Nations Agreed at the UN', *The New York Times* (April 5), A1, 10.

New York Times (1996) January 24, A14. (no title or author given)

Passacantando, J. (1998) 'A Pothole in the Ozone Layer', *The Washington Post* (March 15), C5.

Preston, J. (1992) 'The Man with the Rio Plan', *The Washington Post* (June 3), B1, 8.

Rensberger, B. (1992) 'Where Global Warming Consensus Breaks Down: Human Activity as a Cause', *The Washington Post* (May 31), A1, 22.

Robinson, E. and Preston, J. (1992) 'Earth Summit Close to Accord', *The Washington Post* (June 14), A1, 26.

Sale, K. (1993) *The Green Revolution: The American Environmental Movement*, New York: Hill and Wang.

Saul, J. (1992) 'A Power Broker Who Doesn't Worship Power', *World Monitor*, 5, no. 6, 32–33.

Schneider, K. (1992) 'Environmental Policy: It's a Jungle in There', *The New York Times* (June 7), 4–4.

Shabecoff, P. (1992) 'Real Rio', *Buzzworm; The Environmental Journal*, IV, no. 5 (September/October), 39–43, 89, 99, 101.

Stock, C.M. (1997) *Rural Radicals: From Bacon's Rebellion to the Oklahoma City Bombing*, New York: Penguin.

Vogel, E. (1998) 'Ensign: Environmental Groups Steered by Socialist Leadership', *Las Vegas Review-Journal* (April), 8B.

Washington Post (1992) 'On the Way to Rio' (June 3), A18. (no author given)

Weisskopf, M. (1992a) 'Rust Belt Emissions Cloud Earth Summit', *The Washington Post* (June 2), A1, 8.

—— (1992b) 'USA Balks Anew Over Species Pact', *The Washington Post* (June 5), A42.

Weisskopf, M. and Robinson E. (1992) 'Rio Summit Highlights North–South Schism', *The Washington Post* (June 3), A21, 23.

Weisskopf, M. and Devroy, A. (1992) 'Global Leaders Set Course for Protecting the Earth', *The Washington Post* (June 13), A1, 12.

Weisskopf, M. and Preston, J. (1992) 'U.N. Earth Summit Opens with Calls to Save Planet', *The Washington Post* (June 4), A20, 30.

4 Contradictions at the local scale

Local implementation of Agenda 21 in the USA[1]

Robert W. Lake

Introduction

The large ambitions of Agenda 21, the global manifesto adopted at the 1992 United Nations Conference on Environment and Development (UNCED) in Rio de Janiero, are announced in the first words of the Preamble. 'Humanity,' it declares, 'stands at a defining moment in history.' As Tim Luke points out in Chapter 3, the document's framers clearly hoped to turn the tide of human history with an internationally agreed-upon blueprint for attaining sustainable development in the twenty-first century.

According to an UNCED report prepared by the Government of Australia, Agenda 21 'represents the current international consensus on actions necessary to move the world toward the goal of truly sustainable development' (Australia, Department of the Environment, 1996: 1). This language is symptomatic of Agenda 21's extraordinarily large pretensions: it represents an 'international consensus,' (a momentous claim in any substantive arena) that seeks to 'move the world' (an awesome accomplishment) towards 'truly sustainable development' (a glorious but undefined ideal lacking in either conceptual or substantive clarity). 'The objective,' continues Australia's *Guide to Agenda 21*, 'is no less than the alleviation of poverty, hunger, sickness and illiteracy worldwide while at the same time arresting the deterioration of the ecosystems on which humankind depends to sustain life' (ibid.).

The blueprint to achieve this objective is laid out in Agenda 21's forty chapters and hundreds of pages of text. Separate chapters address an array of issues covering consumption patterns, demographic dynamics, human settlements, the atmosphere, agriculture, desertification, biological diversity, biotechnology, women, children, indigenous peoples, non-governmental organizations, local authorities, trade unions, financial arrangements, education, legal issues, and much, much more.

Chapter 28 of Agenda 21 addresses the role of local authorities in achieving sustainable development. Local authorities are important, according to the document, because 'many of the problems and solutions being addressed by Agenda 21 have their roots in local activities' and, as a consequence, 'the participation and cooperation of local authorities will be a determining factor

in fulfilling [Agenda 21's] objectives ... As the level of governance closest
to the people, [local authorities] play a vital role in educating, mobilizing
and responding to the public to promote sustainable development.' In a
direct extension of this logic, Chapter 28 instructs each local authority world-
wide 'to enter into a dialogue with its citizens, local organizations and private
enterprises' to adopt a 'Local Agenda 21' to provide a strategy and a process
for attaining sustainability within the local sphere.

In what follows I examine the implementation of Local Agenda 21 programs
by sub-national governments in the United States, viewed against the backdrop of the brave goals embraced by the authors of the Rio Declaration and
Agenda 21. The success or failure of local government authorities in achieving
socio-ecological sustainability in the largest consuming nation on earth may
be the single most critical determinant of world-wide progress towards this
elusive goal. To what extent does the implementation of Local Agenda 21
in the United States contribute to 'the alleviation of poverty, hunger, sickness and illiteracy worldwide – while at the same time arresting the
deterioration of ... ecosystems?'[2] To answer this question, this chapter surveys
the number and regional distribution of Local Agenda 21 programs in the
United States, summarizes their principal features, and evaluates their contribution to the broad goal of ecological sustainability articulated in Rio. The
chapter concludes with an assessment of the contradictions inherent in the
attempt to achieve global ecological sustainability through programs situated
at the local scale.

The experience to date of local implementation of Agenda 21 in the United
States provides little grounds for optimism. Local Agenda 21 programs in
the USA are few in number and widely scattered in relatively non-central
locations. The handful of programs in place are necessarily driven more by
the imperatives of local political and economic dynamics than by the lofty
but generally unspecified goal of global sustainability. Attempts to promote
wider adoption of Agenda 21 at the local scale encounter fundamental contradictions between local and global priorities. Local governments caught in a
competitive struggle with other localities are more concerned with the
demands of a global economy (or with the need to fend off its unwanted
consequences) than with the goal of global ecological sustainability. Demands
for local control and self-determination often diverge in contradictory respects
from global priorities and agendas. Finally, programs that appear to contribute
to sustainability at the local scale may not contribute to global sustainability
if they merely entail shifting social or environmental problems elsewhere.

Adoption of local Agenda 21 programs in the USA

Subsequent to UNCED's promulgation of Agenda 21, the International
Council for Local Environmental Initiatives (ICLEI), a non-profit organization
with headquarters in Toronto, Canada, has adopted the role of co-ordinating
and reporting on implementation of Local Agenda 21 programs around the

world. ICLEI, which bills itself as 'the international environmental agency for local governments' (*http://www.iclei.org*), has produced a *Local Agenda 21 Planning Guide* (ICLEI, 1996) to aid local government authorities in preparing and implementing programs, and the organization serves an important clearing-house function disseminating information world-wide.

An ICLEI report released in August 1997 (ICLEI, 1997) evaluates the progress of Local Agenda 21 in the United States. The report identifies twenty-two local governments in the United States that, in ICLEI's view, 'have embarked on a comprehensive, local government-led, long-term effort to attain sustainability, which integrates planning and action in environmental, economic, and community spheres' (ICLEI, 1997: 1). Seventeen of the programs identified by ICLEI explicitly use the term 'sustainability' in their title and were expressly created to encompass Agenda 21's objectives, although few of these programs refer to Agenda 21 by name (Table 4.1). The five remaining cases are characterized by ICLEI as local government programs that have evolved over time to become substantially equivalent to explicitly created sustainability programs even though that term does not appear in their title. An assessment of these twenty-two programs provides a means for evaluating the implementation of Local Agenda 21 in the United States and is the focus of discussion in this chapter.[3]

The spatial distribution of Local Agenda 21 programs in the United States reveals several important characteristics (Figure 4.1). The map, of course, is most notably characterized by its sparseness. The twenty-two local jurisdictions

Table 4.1 Local Agenda 21 programs in the United States

Explicitly created local sustainability programs	*Substantially equivalent local programs*
Albuquerque, New Mexico	Berkeley, California
Austin, Texas	Burlington, Vermont
Boston, Massachusetts	Olympia, Washington
Boulder, Colorado	Portland, Oregon
Chattanooga, Tennessee	Seattle, Washington
Grantsville, Utah	
Metropolitan Dade County, Florida	
Mount Washington Valley, New Hampshire	
Pattonsburg, Missouri	
San Francisco, California	
San Jose, California	
Santa Cruz, California	
Santa Monica, California	
Sherwood, Oregon	
Thomas Jefferson Regional Planning District, Virginia	
Tucson, Arizona	
Wayne County, New York	

Source: ICLEI (1997).

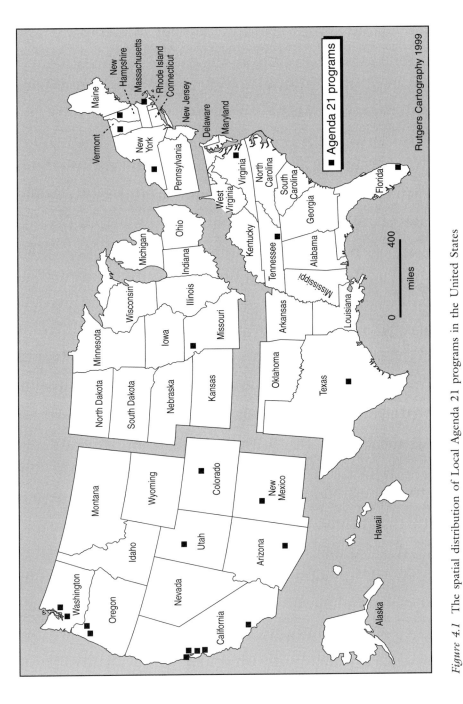

Figure 4.1 The spatial distribution of Local Agenda 21 programs in the United States

Agenda 21 programs

Rutgers Cartography 1999

0   400

miles

include only 2.3 percent of the total U.S. population in 1990 (US Bureau of the Census, 1990). As will be documented in further detail below, this woefully insignificant figure results both from the very small number of local authorities that have adopted Local Agenda 21 programs and from the circumstance that adopting communities are themselves relatively small in size.

The regional clustering of Local Agenda 21 programs is closely correlated with recent dynamics of economic growth and decline in the United States. As indicated in Figure 4.1, four of the twenty-two programs are located in the nation's north-east, four are in the south, only one is in the mid-west, and thirteen are in the west. The single mid-western case, Pattonsburg, Missouri, is itself an anomaly: a small town of 414 people that opted to relocate to a new site after experiencing two 500–year floods in a single year, 1993, and decided to rebuild in a 'sustainable' manner at its new location.

That not a single Local Agenda 21 program, aside from the anomaly of Pattonsburg, is located in the industrial mid-west of the nation strongly suggests that the population loss and economic decline characterizing this region are not conducive to discussions of sustainability. A vista of shuttered factories, deserted shopping malls, deteriorating infrastructure, depopulated cities, and abandoned toxic waste sites does not describe a landscape to be sustained. Four decades of post-war population and job loss at a massive scale have constructed a regional culture that does not welcome discussion of limits, carrying capacity, ecological footprints, or environmental constraints. The discourse of sustainability is politically untenable in this regional context, as indicated by the blank spaces on the U.S. map.

By contrast, thirteen of the twenty-two Local Agenda 21 programs are located in the rapidly growing western region of the country (Figure 4.1). The rhetoric of sustainability provides an acceptable language to legitimate local municipal controls on growth in a region where economic and population growth are imposing a severe strain on environmental resources and valuable scenic amenities. Landscapes in which great scenic beauty, still-abundant open space, and environmental amenities guarantee both environmental quality and economic value – but where these very qualities are susceptible to the onslaught of growth – lend themselves easily to discussions of local sustainability and the adoption of Local Agenda 21 programs as a means to achieve it.

Finally, an important characteristic of these places that is not apparent from the map (Figure 4.1) is that at least twelve of the twenty-two localities contain major universities that tend to have a strong influence on local affairs.[4] It is reasonable to expect that the presence of a large university in a relatively small community has the effect of raising median education levels within the general population, as well as increasing the likelihood of awareness of global sustainability issues in general and of the Rio Earth Summit and Agenda 21 in particular. It is also possible that individuals associated with these universities personally participated in the extensive governmental and nongovernmental organizational activities at the Rio Summit and brought their personal knowledge back to their home communities where they were able to influence adoption of Local Agenda 21 programs.

Further information on the characteristics of localities with Local Agenda 21 programs is derived from considering their position within the national (and international) urban system. The potential for widespread diffusion of such programs may vary considerably depending on whether initial adopters are major urban centers or relatively isolated peripheral locations not closely integrated into the urban system.

An assessment of the 'connectivity' of municipalities with Local Agenda 21 programs is obtained by overlaying their location on a map of the primary urban system in the United States (Figure 4.2). This urban system map, depicting transport and communication connections among the 270 largest metropolitan areas, identifies the primary urban nodes of New York, Atlanta, Chicago, Dallas, Los Angeles, and San Francisco and the secondary metropolitan areas most closely connected to these nodes (Abler and Adams, 1976). Mapping the twenty-two Local Agenda 21 programs on this urban system base map reveals that none is located in a primary urban center. Half of the programs, eleven of the twenty-two, are located in secondary metropolitan centers linked to one of the primary nodes, while the other half are only indirectly connected to the nation's urban system: six are suburbs of secondary urban centers and five are located in peripheral rural areas.

As would be expected, the socio-demographic characteristics of Local Agenda 21 municipalities are correlated with their position within the urban system hierarchy (Table 4.2). The eleven urban centers with Local Agenda 21 programs are on average in the half-million population range, placing them in the third or fourth tier of U.S. metropolitan areas, well below the million-plus centers at the top of the national urban hierarchy. Only one, Metropolitan Dade County, Florida, a consolidated area including the city of Miami and its surrounding densely populated county, exceeds a million in population. The smallest center in this category, Olympia, the capital of Washington State, had a 1990 total population of only 33,840. The six suburban jurisdictions have an average total population of less than 75,000 and two of these, Grantsville, Utah and Sherwood, Oregon, have populations of less than 5,000. The five unattached rural places are below 50,000 in population on average. These data support the observation made above that the minuscule proportion of the U.S. population residing in localities with Local Agenda 21 programs is attributable to both the small number as well as the small population size of such places.

Data on racial and ethnic characteristics reveal that most of the twenty-two Local Agenda 21 places are relatively homogeneous in population compared to the nation as a whole (Table 4.2). The eleven urban centers have the highest average proportion of ethnic and racial minority population, as expected.[5] This figure ranges from a high of 46.4 percent in San Francisco to a low of 15.3 percent in Portland, Oregon, and 7.9 percent in Olympia, Washington. The six suburban jurisdictions contain average minority

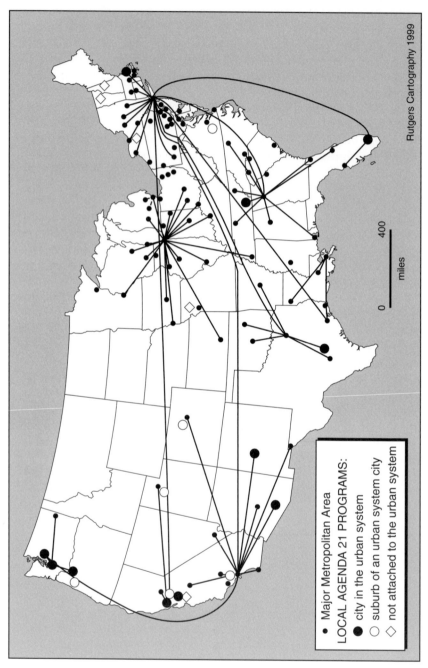

Legend (within figure):

• Major Metropolitan Area

LOCAL AGENDA 21 PROGRAMS:
● city in the urban system
● suburb of an urban system city
◇ not attached to the urban system

Rutgers Cartography 1999

0   400
   miles

Figure 4.2 Local Agenda 21 places in the urban system

population concentrations below the national average. This rate, however, is pushed upward by Berkeley, California, which contains nearly 40 percent minority population. The remaining five suburban areas excluding Berkeley have minority populations of only 9.9 percent, about half of the national average. The five unattached places contain populations that are only 5.7 percent minority on average, confirming the relatively homogeneous population composition of municipalities with Local Agenda 21 programs.

Municipalities with Local Agenda 21 programs are not only homogeneous in ethnic and racial composition but also economically middle-class. Median household income in most of the twenty-two municipalities is close to the 1990 national average of just over $30,000 (Table 4.2). Among the eleven urban centers, median household income reported in 1990 ranges from a low of $21,748 in Tucson, Arizona to a high of $46,206 in San Jose, located in California's high-tech Silicon Valley. Median household income is close to the national figure in the six suburban jurisdictions and slightly lower in the unattached rural places (the latter pulled downward primarily by the very low figure of $15,500 in tiny Pattonsburg, Missouri). The income data, in short, suggest that places with Local Agenda 21 programs include neither the wealthiest nor the poorest municipalities but are relatively middle-income communities seeking to maintain their way of life.

Perhaps the most anomalous characteristic of these twenty-two communities is their unusually high level of educational attainment (Table 4.2). This finding is in line with the observation made above concerning the high proportion of university communities among this self-selected set of places. The proportion of the population in these communities with a college degree or higher is half again above the national average. The proportion nears 60 percent in university communities such as Berkeley (University of California) and Boulder (University of Colorado) and even exceeds the national average in the five unattached rural municipalities.

In summary, data on geographic location, socio-economic characteristics, and urban system dynamics provide a consistent profile of the twenty-two

Table 4.2 Socio-demographic characteristics of Local Agenda 21 places, by urban system category

	Total population	*Ethnic/racial minority (%)*	*Median household income ($)*	*College degree or higher (%)*
City in the urban system (N = 11)	583,201	27.9	28,668	33.7
Suburb in the urban system (N = 6)	74,259	14.5	30,429	34.5
Unattached place (N = 5)	44,426	5.7	26,337	23.8
US total	248,709,873	19.7	30,056	20.3

Source: U.S. Bureau of the Census (1990).
Notes: All data are for 1990. Figures in table are unweighted means for each group.

sub-national jurisdictions with Local Agenda 21 programs in the United States. These programs are not found in the nation's first-tier urban centers nor, indeed, anywhere in the declining industrial mid-western section of the country. Only half of the twenty-two places are metropolitan areas closely linked to the national urban system. The remaining eleven municipalities are suburban communities or small rural places relatively unattached to the system of cities. All are relatively small, middle-income communities with relatively homogeneous populations and high levels of educational attainment. These characteristics strongly influence the motivations for adopting Local Agenda 21 programs and the approaches to sustainability evidenced in those programs, as we discuss in the following section.

Motivations for initiating Local Agenda 21 programs

The definition of sustainability emanating from the Brundtland Report (United Nations, 1987) emphasizes the rights (or needs) of future generations. This definition promotes taking actions today (e.g., resource conservation, promotion of clean industry, etc.) that safeguard future generations or, alternatively, avoiding actions today (e.g., excessive resource extraction, contamination of air, soil or water) that would compromise the ability of future generations to meet their needs.

Such forward-looking definitions of sustainability pervade the planning documents produced by the twenty-two sub-national governments in the U.S. with Local Agenda 21 programs. Borrowing from the Brundtland Report, Pattonsburg, Portland, San Francisco, and Santa Monica, among others, define sustainability as 'the ability to meet current needs without compromising (or sacrificing) the ability of future generations to meet their own needs.' Several municipal documents express an even more explicit future orientation. Burlington, Vermont's Municipal Development Plan asserts that, 'Sustainable development is nothing more than "future-oriented" common sense. Decisions and choices made today should not limit the choices and opportunities of future generations' (City of Burlington, Vermont, 1996: 4). The Thomas Jefferson Sustainability Council (Virginia) refers to 'our responsibility to proceed in a way that ... will allow our children, grandchildren, and great-grandchildren to live comfortably in a friendly, clean, and healthy world' (Thomas Jefferson Sustainability Council, 1996: 2). The General Plan for a Sustainable Community prepared for the city of Grantsville, Utah, asserts a responsibility to protect future generations: 'If we fail to convert our self-destructive economy into one that is environmentally sustainable, future generations will be overwhelmed by environmental degradation and social disintegration. Simply stated, if our generation does not turn things around, our children may not have the option of doing so' (*State of the World Report 1993*, quoted in University of Utah, 1994: I–1). The rhetoric of concern for future generations legitimates local government's regulatory foray into economic and environmental arenas through the mechanism of Local Agenda 21.

However, this forward-looking rhetoric notwithstanding, adoption of Local Agenda 21 programs in the United States is predominantly a response to a perceived threat to the quality of life of *existing*, rather than future, residents. The origin and the character of these perceived threats, furthermore, differ according to the specificities of the social, political, and economic contexts within which the adopting municipalities are embedded (Table 4.3).

The perception of negative consequences of impending or actual regional economic growth is the most pervasive motivation for adopting Local Agenda 21 programs in the United States. This concern, voiced by twelve of the twenty-two programs considered here, reflects the regional distribution of Local Agenda 21 programs as mapped in Figure 4.1 above. With only one exception (Wayne County, New York), the municipalities citing impending growth pressures as their motivation for adopting Local Agenda 21 programs are all located in the rapidly growing southern and western rim of the nation.

Table 4.3 Motivations for adoption of Local Agenda 21 programs in the United States

---

Impending growth pressure (N = 12)
Albuquerque, New Mexico
Austin, Texas
Grantsville, Utah
Metropolitan Dade County, Florida
Olympia, Washington
Portland, Oregon
San Jose, California
Seattle, Washington
Sherwood, Oregon
Thomas Jefferson Regional Planning District, Virginia
Tucson, Arizona
Wayne County, New York

Environmental concern (N = 5)
Berkeley, California
Boulder, Colorado
San Francisco, California
Santa Cruz, California
Santa Monica, California

Lagging economic growth (N = 3)
Boston, Massachusetts
Chattanooga, Tennessee
Mount Washington Valley, New Hampshire

Natural disaster (N = 1)
Pattonsburg, Missouri

Commitment to multifaceted sustainability (N = 1)
Burlington, Vermont

Source: Compiled by author.

References to the negative effects of impending growth are prevalent in the planning documents describing Local Agenda 21 programs. The 'Sustainable City Initiative' of Olympia, Washington, observes: 'Like many rapidly urbanizing cities across the country, Olympia . . . must contend with increased suburban sprawl, pollution, and traffic jams' (University of Washington, 1995: 1). Olympia's 'Sustainable City Initiative' constitutes that city's attempt to contend with those problems of economic and population growth affecting the region. A similar observation prompted neighboring Seattle's Local Agenda 21 program:

> The [Washington] state legislature in 1990 passed a Growth Management Act . . . out of a widespread perception in the state that growth . . . had been sprawling out of control for a decade or so, wreaking [*sic*] havoc on environmental quality, housing affordability, and other facets of society.
> (Lawrence, 1997: 5)

Across the country, identical concerns motivate sustainability planning in the Thomas Jefferson Planning District of Virginia:

> In the Thomas Jefferson Planning District, growth pressure from Washington, D.C., Richmond, and Northern Virginia has started to influence population growth [equal] to a doubling of the population every twenty years . . . These growth extremes with the commensurate environmental and economic problems they bring are set in a backdrop of pastoral piedmont countryside and pristine Appalachian mountains.
> (Thomas Jefferson Planning District Commission, 1994: 10)

The use of Local Agenda 21 as a means to rationalize growth and modulate its effects is explicitly stated in 'A Pathway to Sustainability' prepared for the city of Sherwood, Oregon:

> The city [of Sherwood] was growing rapidly, doubling its population within a span of a few years, and elected officials, city staff, and residents were concerned about the future. Would Sherwood continue to be the kind of community that its citizens desired or would it become a place that no one would recognize? . . . Faced with unprecedented growth, this community . . . set out to develop a strategy that would maintain and enhance the livability or sustainability of their community.
> (Institute of Portland Metropolitan Studies, 1995: iii)

The concern for future generations expressed in the generic definition of sustainability is often lost in the rhetoric of sustainability as an antidote to growth. The immediate concern expressed by the authors of Sherwood's sustainability plan is whether current residents will continue to recognize the city's charm in the face of rapid growth.

Concern over deteriorating environmental quality provided a motivation for adoption of Local Agenda 21 programs in five of the twenty-two municipalities, a distant second behind fears of growth (Table 4.3). Four of these communities – Berkeley, San Francisco, Santa Cruz, and Santa Monica – are in California and, together with Boulder, Colorado, share a long-standing history of environmental activism.

The Berkeley, California General Plan explicitly identifies concern over environmental quality as a motivation for sustainability planning:

> Environmental quality is fundamental to a community's livability . . .
> Environmental quality is continuing to deteriorate, both globally and
> locally . . . The City [of Berkeley] intends to take a leadership role in
> the use and conservation of resources by integrating basic environmental
> principles into public and private decision-making processes to promote
> pollution prevention and reduce environmental hazards.
>
> (City of Berkeley, California, 1997: 1)

The document 'Sustainable San Francisco' sounds a similar warning: 'The environmental practices of people in the City of San Francisco are currently such that the quality of human life and the ecological health and biodiversity of the region cannot be sustained for future generations' (Sustainable San Francisco, 1996). Santa Monica's 'Sustainable City Program' asserts that 'Santa Monica is committed to protecting, preserving, and restoring the natural environment' (Santa Monica City Council, 1994: 1).

The motivation for adopting Local Agenda 21 programs in three additional communities reflects a quite different economic setting. In Boston, Massachusetts, Chattanooga, Tennessee, and Mount Washington Valley, New Hampshire, the motivation for adoption was a desire to stimulate lagging economic growth and (in Boston and Chattanooga) to support urban redevelopment. Both Boston and Chattanooga experienced substantial deindustrialization, job loss, and economic decline in the post-war era. In both places, sustainability is defined in terms of a reinvention of economic functions, and Local Agenda 21 is a program for municipal government intervention in support of renewed economic growth. In Mount Washington Valley, New Hampshire, a tourist and agricultural region, residents and local officials sought a means to cushion the local economy from seasonal fluctuations of tourism and to promote economic diversity. In Pattonsburg, Missouri, as noted above, Local Agenda 21 provided a template for design of an entirely new community at a new location after the town's original location on a tributary of the Missouri River was inundated by two unprecedented floods in a single year. Adoption of Local Agenda 21 in only one community, Burlington, Vermont, approaches the kind of integrated, comprehensive, and multifaceted program envisioned in Agenda 21 and the Rio Declaration. Because Burlington's program is virtually unique in this respect, it will be described in further detail below.

Local Agenda 21 program characteristics

According to ICLEI's survey of Local Agenda 21 programs in the US, to be included in ICLEI's report, a municipal program had to be 'comprehensive, encompassing environmental, economic and social issues' (ICLEI, 1997: 4). This criterion notwithstanding, most of the programs reviewed here and by ICLEI are extremely limited in both scope and ambition. While several programs are in early stages of development and may become more comprehensive over time, most are characterized by a narrow rather than a comprehensive focus, emphasize rhetoric and public relations rather than meaningful action, and reflect the limited authority and jurisdiction available to municipal governments in the United States.

The most frequent and predominant emphasis among these programs is a narrow focus on issues of environmental quality (Table 4.4). For example, the *Sustainability Plan for the City of San Francisco*, adopted by the city's Board of Supervisors in 1997, unabashedly asserts: 'The primary focus of this version of San Francisco's sustainability plan is the environmental component ... This plan addresses primarily the physical systems of the planet that often get short shrift from planners, and the social systems that have a direct impact on them' (City and County of San Francisco, 1997: vi).

The resulting plan addresses a panoply of environmental matters including air quality, biodiversity, energy, climate change, ozone depletion, hazardous materials, parks and open space, solid waste, water and wastewater, and more. While the authors of the San Francisco plan indicate that they expect to turn to the 'economic and community aspects of the plan' in the future, the present single-minded emphasis on repairing environmental damage and improving environmental quality is symptomatic.

This program emphasis in part is consistent with the importance of environmental concerns as the announced motivation for developing Local Agenda 21 programs in the first place. It also reflects a tendency in practice to reduce the multifaceted concept of sustainability to a program of environmental remediation and the correction or elimination of the negative environmental consequences of private consumption and production. The 'Sustainable City Program' adopted by the Santa Monica City Council makes this claim explicitly:

> The City of Santa Monica recognizes that we live in a period of great environmental crisis. As a community, we need to create the basis for a more sustainable way of life both locally and globally through the safeguarding and enhancing of our resources and by preventing harm to the natural environment and human health. We are resolved that our impact on the natural environment must not jeopardize the prospects of future generations.
>
> (Santa Monica City Council, 1994)

Table 4.4 Local Agenda 21 program emphasis

Environmental quality (N = 14)
Albuquerque, New Mexico
Berkeley, California
Boulder, Colorado
Boston, Massachusetts
Chattanooga, Tennessee
Metropolitan Dade County, Florida
Pattonsburg, Missouri
Portland, Oregon
San Francisco, California
San Jose, California
Santa Cruz, California
Santa Monica, California
Seattle, Washington
Wayne County, New York

Quality-of-life/livability (N = 7)
Grantsville, Utah
Mt. Washington Valley, New Hampshire
Pattonsburg, Missouri
Sherwood, Oregon
Thomas Jefferson Regional Planning District, Virginia
Tucson, Arizona
Wayne County, New York

City government operations (N = 6)
Austin, Texas
Berkeley, California
Boulder, Colorado
Chattanooga, Tennessee
Olympia, Washington
Tucson, Arizona

Indicators projects (N = 7)
Albuquerque, New Mexico
Austin, Texas
Boulder, Colorado
Boston, Massachusetts
Olympia, Washington
Seattle, Washington
Thomas Jefferson Regional Planning District, Virginia

Multifaceted sustainability (N = 1)
Burlington, Vermont

Source: Compiled by author.
Note: Several programs appear under multiple categories.

Based on this principle, three of the four major policy areas in Santa Monica's Sustainable City Program address environmental quality issues: resource conservation, transportation, pollution prevention and public health protection.

Within the focus on environmental quality, specific local programs reflect particularities of site and situation. In Albuquerque, New Mexico, regional population growth in a desert ecology is causing severe depletion of water supplies. In response, a Water Resources Management Strategy encouraging water conservation is the centerpiece of the city's sustainability program (City of Albuquerque, 1997). Sustainability planning in Metropolitan Dade County, Florida is driven by the necessity to direct population and economic growth away from the ecologically sensitive Everglades National Park (Governor's Commission for a Sustainable South Florida, 1995). The Local Agenda 21 programs in Berkeley, Boulder, Boston, Portland, San Jose, and Santa Cruz similarly address particular local matters, including brownfields clean-up, waterfront reclamation, watershed management, energy conservation, recycling and waste management, and the like. Sustainability in all of these programs has been reduced to the task of cleaning up past environmental problems and, to a lesser extent, preventing new ones, while maintaining the integrity of market-based production and consumption.

A secondary focus of Local Agenda 21 programs in the United States emphasizes issues of quality-of-life, livability, and community identity (Table 4.4). This program emphasis is most apparent in the smaller and more rural communities where sustainability is equated with preservation and reproduction of an idealized small-town character.

The General Plan for a Sustainable Community prepared for Grantsville, Utah by a Community Planning Workshop at the University of Utah exemplifies this approach: 'The essence of this plan is to help create an enjoyable community with a sense of place' (University of Utah, 1994: I–2). Towards this end, the plan proposes a series of design solutions to improve the town's physical appearance, mark its boundaries, and contribute to historic preservation. 'The goals of urban design should include maintaining a small town environment and rural lifestyle' (University of Utah, 1994, III–43). Proposed design solutions include construction of a median island for the town's Main Street and signs identifying the town limits: 'The sign should be repainted on a regular basis . . . Flowers should be planted in the spring' (University of Utah, 1994: III–27).

The segue from sustainability to livability has been made explicitly in Tucson, Arizona. According to the ICLEI report, 'Tucson [is] preparing to modify its terminology, replacing *sustainability* with *livable community* because of a perception that the meaning of sustainability is unclear to practitioners and the community (ICLEI, 1997: 17: emphasis in original). Sustainability planning in Sherwood, Oregon and the Thomas Jefferson Regional Planning District, Virginia seeks to safeguard amenities and maintain livability by keeping impending growth at bay. Planners in Mt. Washington Valley, New Hampshire hope to encourage economic diversity as a means to maintain community character by preventing decline. Sustainability in Wayne County, New York is essentially synonymous with a program of farmland preservation.

Sustainability planning in at least six of the twenty-two communities surveyed focuses primarily on making internal city government operations more environmentally friendly (Table 4.4). In Boulder, Colorado, for example, 'The Sustainability 2000 Project is designed to reduce the environmental impacts of the City government's operations . . . By "getting its own house in order," the City hopes to create a model for other organizations' (City of Boulder, Colorado, 1997: 1). Berkeley, California has adopted the Sierra Club's 'Valdez Principles' to guide city government operations. These include the use of recycled materials and water saving devices, retrofitting city buildings for energy conservation, the purchase of smaller vehicles for the city-owned fleet and the conversion of some vehicles to alternative fuels, and reduction in the use of toxic products and pesticides (City of Berkeley, California, 1997). Austin, Texas has adopted 'sustainable purchasing guidelines' for city offices. Chattanooga's electric bus fleet is now one of the largest in the nation.

While such improvements in city government operations contribute important environmental benefits, they are a far reach from a comprehensive approach to sustainability that integrates environmental, economic, and political components. Instead, such programs are an indication of the extremely limited reach and authority of municipal administrations that address internal operations because they are unable to initiate meaningful change within the sphere of the private market.

A further indication of the relative impotence of municipal authorities to achieve sustainability is the number of Local Agenda 21 programs that are limited to development and reporting of sustainability indicators (Table 4.4). The rationale supporting indicators projects is that the public sector can marshal data and monitor trends towards or away from sustainability and that the evidence so produced will motivate others to action. Bracketing, for the moment, the methodological problems inherent in the choice of indicators and the availability of data, limiting the public role to such data gathering is a tacit admission of the inability of local government directly to effect fundamental change.

Across the set of twenty-two Local Agenda 21 programs, only Burlington, Vermont can be said to approach both the spirit and the reality of a comprehensive approach linking economic, political, and environmental change implied in the concept of sustainability. Burlington is a small city of 40,000 people in north-western Vermont, the home of the University of Vermont and one of the 'progressive cities' analyzed by Pierre Clavel (1986) more than ten years ago.

Burlington's innovative program is based on six 'principles of sustainable development' that together seek change in economic, environmental, social, and political spheres (City of Burlington, 1997: 4–6):

1   Encouraging economic self-sufficiency through local ownership and maximum use of local resources.
2   Equalizing the benefits and burdens of growth.
3   Leveraging and recycling scarce public funds.
4   Protecting and preserving fragile environmental resources.
5   Ensuring full participation by populations normally excluded from the political and economic mainstream.
6   Nurturing a robust 'third sector' of private, non-profit organizations capable of working in concert with government to deliver essential goods and services.

What is notable about Burlington's approach to sustainability is that it implicates the municipal government in seeking change in fundamental economic and political institutions. Within the economic sphere, the municipal government has developed an eco-industrial park and funds a microenterprise loan program and a small business incubator to stimulate development of local small business. The commitment to 'equalizing the benefits and burdens of growth' engages the city in a redistributive role that it has pursued through inclusionary zoning regulations that require construction of affordable housing units within market-rate developments. The city commits its own resources through the small business revolving loan fund and the use of city revenues to leverage additional sources of public and private capital. Environmental improvement projects are nestled within the context of these economic and social programs. Burlington contributes to the goal of full participation through a program of employment training for women in non-traditional occupations. Perhaps most importantly, Burlington's sustainability program includes an attempt to restructure processes of political decision-making through development of 'neighborhood planning assemblies' that meet monthly in each of the city's six wards, providing a means for direct and widespread public participation in the design and implementation of the city's programs.

Discussion

This review suggests that Local Agenda 21 programs in the United States encompass at least three different approaches to the idea of sustainability. One approach, exemplified by the plans developed for Grantsville, Utah and Sherwood, Oregon, defines sustainability as livability. These plans rely on architectural, landscape, and design solutions to augment the supply of physical and environmental amenities and maintain or improve the 'quality of life' for community residents. In most if not all cases where this strategy is employed, community residents and officials perceive themselves at risk of the negative effects of surrounding regional growth in the form of increasing population density, traffic congestion, and declining environmental quality. Here sustainability is synonymous with stasis, the prevention of change, and the maintenance and replication of a desired landscape achieved through the exclusion of undesirable effects.

A different approach among these twenty-two municipal programs defines the public role in sustainability in terms of facilitating private development and/or cushioning its negative effects. This definition in large measure underlies programs that emphasize environmental improvements such as the clean-up of abandoned toxic waste sites, waterfront reclamation, and air, soil, and water purification, as well as programs aimed at reducing negative environmental effects of internal city operations. In the former case, environmental clean-up programs serve to offset the negative externalities of private production and consumption. In the latter case, improvements in internal municipal operations, however admirable they may be, allow the negative externalities of market processes to proceed unabated.

A third approach to sustainability challenges local government to instigate fundamental change in socio-economic institutions and political decision-making processes. Such change is necessary to move beyond cushioning the negative externalities of market processes or simply shifting them to other locations and, instead, developing new systems of production and consumption that reduce negative externalities in absolute terms. This in turn requires changes in the production of negative externalities and not simply their redistribution. As we have seen, there is little evidence that this approach to sustainability is represented among Local Agenda 21 programs in the United States.

The central position of environmental initiatives in local sustainability programs raises some difficult and contentious questions. The emphasis on environmental quality improvements in Local Agenda 21 programs in Berkeley, Boulder, San Francisco, Santa Cruz, Santa Monica, and elsewhere continues a long record of environmental activism in these localities and may, indeed, produce substantial quantifiable improvements in air and water quality and similar environmental indicators. Such programs may suggest a coming-of-age of environmentalism indexed by the integration of environmental action into local government practices on a day-to-day basis.

An alternative and more worrisome interpretation may also be plausible, however, especially in the absence of evidence of fundamental transformation in private market production and consumption relations. Where market relations remain unchanged but local government absorbs the associated negative environmental externalities, the growing acceptance of environmentalism by municipal governments may best be described as the greening of capitalism (Faber, 1998). In this interpretation, the adoption of environmentalism by municipal governments may herald not the coming-of-age of ecological consciousness but rather its continuing co-optation and the end of environmentalism as a movement capable of bringing about fundamental change in economic and political relations.

The role of environmental actions in Local Agenda 21 programs problematizes the traditional relationship between economic growth and environmental protection. Under the cloak of sustainability, a few municipal governments have adopted programs that simultaneously improve local environmental quality, support economic development through the attraction of clean jobs and industry, and export environmental disamenities beyond their borders. For these communities, the high level of environmental amenity that contributes to economic viability does not depend on changes in the production of environmental problems but rather on their spatial redistribution.

The search for sustainability through the spatial exclusion of environmental problems raises a reversal of the traditional charge of environmental justice advocates. The conventional indictment on the part of environmental justice concerns the disproportionate concentration of environmental risks in low-income communities. The evidence reviewed here may document the concentration of environmental benefits in middle-class communities. An equity issue arises if the improvement in local environmental quality is conditioned not on a reduction in the production of environmental risk but simply its exclusion from the locality to distant sites.

Conclusion

Reasoned assessment suggests that ICLEI's (1997) characterization of the sustainability plans of these twenty-two municipalities as Local Agenda 21 programs may have been either unwarranted or premature. Negating the expectation that Local Agenda 21 programs should entail a comprehensive integration of economic, environmental, and social spheres, the programs reviewed here are decidedly limited in objective, scope, and effectiveness.

But reasoned assessment also suggests that it is both unfair and unrealistic to expect that government at the local scale can accomplish more than is encompassed in these programs. Local government in the United States lacks the authority, the resources, and, most importantly, the power to initiate and accomplish the fundamental transformations in systems of production and consumption that are required to, indeed, 'move the world toward the goal of truly sustainable development' (Australia Department of the Environment, 1996: 1). While U.S. municipal government may well fit the characterization, in Chapter 28 of Agenda 21, as the level of government closest to the people, it also is the institution that is least able to effect fundamental change in the private market relations that stand in the path of truly sustainable development.

Notes

1   Preparation of this chapter would not have been possible without the research assistance provided by Anne Leavitt-Gruberger and Liesje DiDonato. Helpful comments on earlier versions were provided by participants at the International Conference on Environmental Justice: Global Ethics for the 21st Century, held at the University of Melbourne, the Urban Affairs Colloquium at the University of Delaware, and the CUPR faculty seminar at Rutgers University.

2   While it may seem unfair to judge local municipal programs on their ability to alleviate problems 'world-wide', it is also the case that local decisions may have profound effects at distant locations.
3   The twenty-two local government programs identified in the ICLEI report (ICLEI, 1997) comprise the focus of this chapter. The many sub-national program initiatives (perhaps numbering in the hundreds) seeking some form of sustainability through nongovernmental organization (NGO) sponsorship are beyond the purview of this discussion. In line with the directive in Agenda 21's Chapter 28 regarding local government authorities, this chapter focuses only on Local Agenda 21 programs or their equivalents that have been adopted through official statutory authority of a sub-national unit of government at either a municipal, county, or regional level.
4   Communities with major universities include: Austin (University of Texas); Berkeley (University of California); Boston (Boston University); Boulder (University of Colorado); Burlington (University of Vermont); Thomas Jefferson Regional Planning District, centered on Charlottesville (University of Virginia); Portland (Portland State University); San Francisco (San Francisco State University); San Jose (San Jose State University); Santa Cruz (University of California); Seattle (University of Washington); Tucson (University of Arizona).
5   Ethnic and racial minority includes population identified in the Census as African American, Native American, Pacific Islander, or 'Other Race.'

References

Abler, R. and Adams, J. (1976) *A Comparative Atlas of America's Great Cities.* Washington, DC: Association of American Geographers and the University of Minnesota Press.

Australia. Department of the Environment (1996) *A Guide to Agenda 21.* Online from Environmental Resources Information Network at http://www.erin.gov.au/portfolio/esd/nsesd/a21summ.html

City and County of San Francisco (1997) *The Sustainability Plan for the City of San Francisco.* City of San Francisco Board of Supervisors Resolution No. 692–97, adopted 21 July 1997.

City of Albuquerque (1997) *City of Albuquerque Water Resources Management Strategy.* Albuquerque Public Works Department.

City of Berkeley, California (1997) *City of Berkeley General Plan, Environmental Management Element.* Berkeley: Planning and Development Department.

City of Boulder, Colorado (1997) *Sustainability 2000 Project.* City Government Operations Progress Report 1997. Boulder: Office of Environmental Affairs.

City of Burlington, Vermont (1996) *Municipal Development Plan.* Burlington: Department of Planning and Zoning.

City of Burlington, Vermont (1997) *Creating a Sustainable City: The Case of Burlington, Vermont.* Burlington: Office of the Mayor.

Clavel, P. (1986) *The Progressive City: Planning and Participation, 1969–1984.* New Brunswick, N.J.: Rutgers University Press.

Faber, D. (ed.) (1998) *The Struggle for Ecological Democracy: Environmental Justice Movements in the United States.* New York: Guilford Press.

Governor's Commission for a Sustainable South Florida (1995) *Initial Report.* Coral Gables: Florida Department of Community Affairs.

Institute of Portland Metropolitan Studies (1995) *A Pathway to Sustainability.* Portland, Oreg.: Portland State University, Institute of Portland Metropolitan Studies.

International Council for Local Environmental Initiatives (ICLEI) (1996) *The Local Agenda 21 Planning Guide: An Introduction to Sustainable Development Planning.* Toronto: ICLEI.

International Council for Local Environmental Initiatives (ICLEI) (1997) *Local Agenda 21 in the United States: Municipal Sustainability Efforts.* Berkeley, Calif.: ICLEI-US.

Lawrence, G. (1997) 'Sustainability in Seattle: Turning Best Practices into Common Practice,' *City*, 3, 4.

Santa Monica, City Council (1994) *Santa Monica Sustainable City Program.* Santa Monica, Calif.: Environmental Programs Division.

Sustainable San Francisco (1996) *Sustainable San Francisco: How We're Trying to Do It.* online at http://www.igc.org/sustainable/structure.html

Thomas Jefferson Planning District Commission (1994) *The Thomas Jefferson Study to Preserve and Assess the Regional Environment: Economic Growth Within the Capacity of the Environment to Support It.* Charlottesville, Va.: Thomas Jefferson Planning District Commission.

Thomas Jefferson Sustainability Council (1996) *Indicators of Sustainability: Interim Report.* Charlottesville, Va.: Thomas Jefferson Planning District Commission.

United Nations. World Commission on Environment and Development (1987) *Our Common Future.* Oxford: Oxford University Press.

United States. Bureau of the Census (1990) *Census of Population and Housing.*

University of Utah (1994) *Grantsville General Plan for a Sustainable Community.* Salt Lake City: University of Utah, Center for Public Policy and Administration.

University of Washington (1995) *Olympia Sustainable City Initiative and Sustainable Community Roundtable.* Seattle: University of Washington, Center for Sustainable Communities.

5 Britain: unsustainable cities

Andrew Blowers and Stephen Young

Britain in international context

Britain was the first country to undergo the process of urbanisation based on the industrialisation of the modern age. This process was completed in the early part of the twentieth century with about four-fifths of the population living in cities and towns. At the end of the twentieth century this proportion was similar, with urban uses occupying about 11 per cent of the land surface (of England and Wales). However, the economic and social context of urbanisation has vastly changed during this period.

The nineteenth-century capitalist economy, with its accent on the private sector, individualism and the concept of the enabling state, has certain echoes today. But late twentieth-century Britain is a post-colonial country, somewhat reluctantly enmeshed in the European Union and inextricably linked into the processes of global capitalism. Britain has proceeded further than most western states in the direction of deregulation but has been constrained by wider processes. These include the growth of service industries and the informal economy together with the huge expansion of technological innovation, especially information technology. In ecological terms, British cities, like their counterparts elsewhere in the west, leave a footprint that is felt far afield, throughout the non-urban parts of the country and throughout the world. In their consumption of resources and production of pollution British cities are, literally, environmentally unsustainable.

All this has added its imprint to urban development in the late twentieth century. There has been a vast expansion of office development both in the heart of the big cities and in suburban locations. London has accumulated a concentration of corporate headquarters, financial institutions and information control centres. Castells (1990) speaks of the rise of the 'information city' with both concentration and decentralisation. He considers that the new relations between capital and labour that have emerged amount to a change in class formation, with a dominant and expanding elite engaged in the control of information flows and a subordinate class engaged in routine decentralised processing operations, or working in the informal economy. This has created a 'dual city', with the information city being grafted on to, rather than superseding, the urban forms and class structure of the preceding age.

Three underlying themes are explored in this chapter. First, the theme of *sustainability* is concerned with the impact of British cities on the environment. Secondly, cities are not simply physical forms, they reflect the social processes of *development*. Spatial social segregation and fragmentation is a structuring characteristic of the city. Together the physical and social aspects of the city constitute a third theme, that of *ecologically sustainable development*. This brings into focus the problems of the management and governance of cities and particularly the role of planning as a means of promoting sustainability. The chapter will first explore the evolution of British cities since the Second World War to establish the physical and social context of the contemporary problem. It will go on to examine the role of town planning in conditions of the liberal market economy. This will lead to an evaluation of the constraints and opportunities presented by the Rio process for achieving ecologically sustainable cities in Britain. Finally, the concluding section will examine the prospects for change.

Urbanisation and the construction of social inequality

The post-Second World War period in Britain saw the apotheosis of state intervention in the creation of the welfare state inaugurated by the 1945 Labour government. The political change ushered in during the war marked a major transformation. The sense of common cause, even of 'community', forged in the compulsory collectivism of the war was translated into a generous social reform and reconstruction programme in health, housing and welfare focusing on the needs of the whole population (Hennessy, 1992). However, the post-war programme created a low density, contained but open landscape of urbanisation which extended the social segregation process that had become firmly established between the wars.

By the 1960s, as attention turned to the redevelopment of the cities themselves, a combination of central government policy, local authority implementation and the growing influence of private sector construction companies had wrought a transformation that had both physical and social consequences. Over the period 1955–70 half a million high rise flats (the tallest reaching 33 storeys) were built, housing a million people. The builders and their architects exerted influence over receptive council officials and, through the nexus of power which developed, promoted what Dunleavy (1981: 124) has called a 'technological short-cut to social change'. For a time, this revolution in urban development went largely unchallenged (though see Young and Willmott, 1962) despite the high costs: 37 per cent more conventional housing units could have been built at higher standards, the destruction of physically sound houses to make way for vast construction sites, and the absorption of resources into maintenance. Socially, the high rise housing boom represented 'the reproduction of a sanitised status quo' (Dunleavy, 1981: 72) replicating spatial inequalities by confining the working class to deteriorating, soulless blocks, lacking in facilities and accessible open space.

Although this phase has left its imprint on contemporary British cities, it was an aberration. By the mid-1970s the post-war consensus was drawing to a close. Economic growth was checked, especially by the oil price rise of 1974, and the combination of labour costs and welfare expenditure created budget deficits and a resulting 'fiscal crisis' (O'Connor, 1973). Soon enough, public expenditure cuts indicated that 'the party was over'. By the 1980s, the ideological transformation was completed as the Thatcher governments, first elected in 1979, espousing the doctrines of the New Right set about an accelerating programme of deregulation, privatisation and retrenchment in the public sector. The power and influence of business grew while trade union power diminished. Enterprise zones, urban development corporations and simplified planning zones were introduced as deregulated areas to encourage investment. They represented a 'shift away from local democratic processes, away from public involvement and an increase in decision-making by central government accompanied by more freedom of action for developers' (Thornley, 1986: 7).

By the early 1990s, the rampant capitalism and its attendant ideology of the New Right of the 1980s had become less strident. The notion of 'partnership' had become the new dispensation. It was at the heart of the Major government's City Challenge and Single Regeneration Budget programmes. These had been designed partly to get away from the physical regeneration of small areas of cities and to give emphasis to the economic potential for tackling social regeneration in more deprived areas. There was now a rhetorical emphasis on the combination of the market's flexibility and responsiveness, with the state adopting an 'enabling' role facilitating favourable conditions for market operations and providing a regulatory framework to ensure standards of performance and a level playing field. In addition, the voluntary sector has come to be seen as a key element in partnerships, partly, no doubt, to legitimate the process but also reflecting its increasing significance– just as local government has diminished in importance.

The powers of local government had been severely reduced in three ways: by financial squeeze starting from the IMF crisis of 1976; by the loss of functions to the central state, the private sector and the proliferating quangos; and by the abolition of those councils governing the great conurbations, notably the Greater London Council (Cochrane, 1993). The processes of centralisation and fragmentation had serious consequences for the post-Rio era. The Conservatives' aim in the 1980s and early 1990s had been to shift local authorities from being providers of services, to a model of the council working through a mix of public, private and voluntary bodies, enabling them to deliver services; and to provide planning frameworks for private investment based on a system of central government subsidies.

In terms of social relations, then, an enduring feature of urbanisation during the period of modernisation has been the process of spatial social inequality, created and structured by the combination of market/state relationships. This patterning reflects the different balances of private investment and state intervention in the provision of housing, business, industry and supporting facilities which together, developing over time, make up the urban form. This social inequality is also physically reflected in the contrasting environmental quality in British cities.

The social impact of urban planning

Although the social purpose of planning has been emphasised, notably in the visions of its early advocates such as Howard and Geddes, in practice it has been primarily concerned with land use and urban form. Planning is a local government activity with central government exercising overall control through policy guidance. The intention has been to promote the 'public interest' in land-use decisions through development plans and the instrument of development control. The social conditions of the cities have been addressed mainly through 'urban policy', which has been especially directed at providing solutions for the combination of poverty, poor education and health, deteriorating environments, unemployment and crime concentrated in the inner cities and huge overspill estates outside cities like Liverpool and Glasgow.

The problem of social inequality has scarcely been tackled by town planning, which has a basically spatial remit. In physical terms, the Greater London Plan of 1944 had initiated a policy of urban containment and planned dispersal. This was given impetus through post-war planning, town expansion and new towns legislation. Green Belts (eventually covering 12 per cent of England and Wales) were established to check urban sprawl, safeguard the countryside, to preserve the character of towns and assist in urban regeneration. Within the conurbations 'twilight areas' were redeveloped, and a series of new towns and expanding towns were designated to encourage planned dispersal from London or economic regeneration in the declining regions.

While British town planning has contributed to the spatial patterning of British cities, it has been far less influential in achieving the social purposes envisaged by early town planners and by the first post-war government. In its formative period, planning was imbued with a clear social vision, the improvement of living conditions. Howard's 1898 blueprint for the garden city remains, a century later, the touchstone of British attitudes and the source of a particular set of solutions to its urban problems. His concept of social balance identifies social inequality as a major constraint on co-operation. His ideas find expression today in such things as the provision of neighbourhood facilities, and continue to influence contemporary debates about sustainable cities. The Town and Country Planning Association, which continues the tradition of Howard, Geddes and other early planning visionaries, still proclaims the principles of town planning to be social equality and the promotion of 'a better world for the enjoyment of present and future

generations'. The sense of social purpose found its expression in the development of new towns, and planning was a key element in the post-war
programme of housing provision, slum clearance and economic regeneration.

Thereafter it lost its social vision and became more and more a regulatory
activity. Planning was professionalised and bureaucratised, a technical process
shorn of 'political' interest. This was reflected in its 'procedural planning
theory' (Faludi, 1973; Healey *et al.*, 1982) which was not a theory so much
as a set of technical procedures incorporating the paraphernalia of systems
theory, modelling, 'rationality', organisational decision-making and the like.
Although these ideas have long been abandoned, planning has not fully
restored the link between purpose and policy which characterised its early
existence.

A key reason for this is that planning lacks effective powers; indeed, it
failed to secure the powers that might have made a difference in the balance
between public and private. Although planning supposedly provides for development in the right place at the right time, its powers are negative. Planning
can prevent (though refusal may be overturned on appeal), but not promote
(allocation of land for a purpose does not guarantee that it will occur). While
planning can influence the use and form of development (in terms of density
and layout) it cannot ensure that needs are met (such as local employment,
social housing, types of retailing). While planning influences the value of
land, the unearned increment or 'betterment' value is a windfall gain to the
owner or developer instead of to the community, as was originally intended.
The failure to link land use and land value has weakened the ability of the
community to secure benefits from the planning process. 'Planning gain' in
the form of financial compensation or infrastructure provision in return for
planning permissions amounts to a form of bribery in which the public interest
at best is subordinate to the claims of the highest bidder.

Added to its lack of financial powers, planning has a weak institutional
base. It is a local government activity, and has been seriously affected by the
Thatcherite programme to reduce the role of local government referred to
above. Unlike most other western European countries the regional tier of
government in Britain is advisory and thus planning lacks a regional strategic
dimension. The 1990s restructuring of local government has promoted urban
unitary authorities and increased the potential for conflict between rural and
urban areas. This conflict is likely to increase the problems of inequality as
cities try to cope with problems of urban capacity and rural areas try to resist
further development emanating from the cities.

Furthermore, planning is weak in its reach. It only deals with land use –
it does not deal with the other aspects of environmental sustainability. The
control of pollution, water quality and waste management, formerly under
separate bodies, is the responsibility of the Environment Agency in England
and Wales, and the Scottish Environmental Protection Agency. There is
greater emphasis on the environmental impacts of land use, including the
use of environmental impact assessment procedures, but spatial planning and

environmental control remain separate functions. With the urge towards an integrated approach to sustainability, this divorce may be construed as a critical weakness.

As a consequence of these weaknesses, planning has tended to facilitate the market. Indeed, for a time during the 1980s, this was seen as its prime purpose. In the climate of deregulation, privatisation and promotion of private development, planning was seen as an obstacle to progress (HMSO, 1986). It was accused of imposing a 'cost on the economy and constraints on enterprise that are not always justified by any real public benefit in the individual case' (HMSO, 1985: 3.1). In the ideological thrust of the time there was a perceived need to 'simplify the system and improve its efficiency and to accept a presumption in favour of development' (ibid.: 3.4).

The period of high capitalism eventually softened and the idea of planning-led development was encouraged (Stoker and Young, 1993: ch. 3). But, by this time, much damage had been done with the spread of private housing around towns and the countryside unrelated to any principles of sustainability. The proliferation of out of town shopping centres encouraged motorised shopping trips and drained the life blood out of town centres and local neighbourhood centres across the country. By the time sustainability became a planning priority the problem of unsustainability had intensified. Planning became the key government function in the development of the Rio process.

British cities and the Rio process

Action at the National Level

Between the publication of the Brundtland Report in 1987 and the Rio Earth Summit in 1992, there was a considerable growth of interest in sustainable development in British local government (Ward, 1993). This was the period of Friends of the Earth's Environment Charter developed at Kirklees (Huddersfield); the launch of the Environment Cities programme sponsored by British Telecom (BT) at Peterborough and Leicester; and the appointment of the first environmental co-ordinators. By the time of Rio, a small number of pioneering authorities were doing State of the Environment reports and addressing the issue of how to apply ecologically sustainable development ideas within their areas (see Figure 5.1 for locations discussed).

Meanwhile there had been important developments at the national level. Although the 1990 White Paper, *This Common Inheritance* (DoE, 1990), had been widely criticised, it was nevertheless an important step. It led to a new planning framework emerging in 1991/2, and to the local authority associations taking the concept of sustainable development more seriously (Stoker and Young, 1993: ch. 4).

After Rio, the Conservative government was generally positive, but limited in its approach. The UK's Agenda 21 Report – *Sustainable Development: The*

Figure 5.1 Locations in the UK with plans for ecologically sustainable
        development

UK Strategy – encouraged councils to respond to the challenge (DoE, 1994: para 30.4). However, this was not supported by programmes with resources and clear targets. A pattern emerged of discussing options, but only taking tentative steps and developing programmes with voluntary targets (e.g. recycling or composting 25 per cent of all household waste by 2000). But, with few exceptions (some of the London boroughs), councils made little progress. Only £66.4m of local council borrowing was approved during 1991–6, for investment in recycling and composting (DoE, 1996: 35 and 81, Ref 233). The new landfill tax gave firms an incentive to reduce their waste but it did not become operational until September 1996.

During the period after Rio, the Local Government Management Board (LGMB) became the dominating organisation in terms of promoting LA21. It built on the pre-Rio interest and involvement of central government, and on the work of the pioneering councils in the early 1990s. It also drew increasingly from its involvement in European and global networks on LA21. After Rio the local authority associations established the LA21 Steering Group, which included representatives from other sectors, to oversee and develop what was now called the LA21 Initiative (DoE, 1994: paras 30.6, 30.10). The LGMB was very active in publishing guidance and good practice (LGMB, 1993), promoting training and developing the network of more than 400 LA21 officers that emerged.

One of Labour's first actions in office in 1997 was to take the potentially significant step of amalgamating the Departments of Transport and Environment to form a new Department of the Environment, Transport and the Regions (DETR). This opened up the possibility of giving environmental criteria precedence over economic criteria when reviewing road proposals. Some big road schemes were axed, but the initial emphasis of Labour in office was on rhetoric and reviews, rather than clear action. There were strong speeches on climate change at the United Nations General Assembly Special Session (UNGASS), and at the Kyoto conference later in 1997. Discussion documents were published. Policies on topics like biodiversity were developed incrementally.

One of the key environmental issues facing the government was where to locate the 4.4 million new homes forecast as needed by 2016. The Conservative government set a target of half of these homes to be built on brownfield sites within cities. Labour increased the target to 60 per cent. Quite aside from whether such a percentage was achievable on expensive, often polluted and difficult urban sites, was the issue of the social implications. By levering more housing on to vacant sites or on to infill plots, precious open space would be lost, services put under greater pressure and environmental conditions in danger of deteriorating still further. Some observers concluded that the result would be to leave the 'poorest stranded in social housing ghettoes', replicating the pattern of disadvantage from one generation to another (Breheny and Hall, 1996: 5). At the same time policies to limit development in small towns and rural areas would create a rise in house prices that would disadvantage the less well-off, leaving a danger that 'sustainable development may become merely a device for preserving rural amenity at the expense of everything and everyone else' (ibid.: 51). Overall, Labour's initial approach did not promise to tackle the social consequences of environmental change.

Action at the level of the city

At this point an earlier, contextual, pre-Rio point needs to be borne in mind: the Thatcher era had greatly extended the post-war weakening of local government. However, despite the loss of powers and funds, the responses of many cities to the Rio challenge was enthusiastic. These are discussed here. The ways in which central government undermined this energy are analysed later.

A central concern was the search for new tools. Policy-makers were becoming increasingly aware that their traditional policy instruments could not effectively tackle the multidimensional problems posed by LA21. Experiments with new instruments became an important part of policy-makers puzzling out what applying sustainable development meant in practice. Environmental Impact Assessment became widespread in the early 1990s following an EU directive. There were experiments with green housekeeping schemes; environmental audits and appraisals (DoE, 1993); and sustainability indicators (LGMB, 1995). In particular, interest in Environmental Management Audit Systems (EMAS) grew. EMAS aimed to identify the detrimental effects on the environment of the routine application of council policies and to adopt a systemic approach to managing local services in environmentally friendly ways (Morris and Hams, 1997: 18–19).

On the participation front, there was a ferment of activity. In some respects this became an end in itself (Young, 1997; Morris and Hams, 1997: 41–50). Councils like Leicester and Reading tried to move away from top-down, consultation strategies towards bottom-up strategies that aimed to empower local communities. The imaginative range of approaches reflected a wider concern in Britain about the need to regenerate local democracy (DETR, 1998: ch. 4). Together with some of the estate regeneration and urban renewal programmes, LA21 has been in the forefront of a frenzy of experimentation. The redevelopment of the Hulme crescents in Manchester, that symbol of 1960s inner city comprehensive redevelopment, involved an ambitious participation programme that had a considerable impact on detailed policy-making (Harding and Garside, 1996).

Progress on LA21

The LGMB surveys of LA21 activity during the 1994–6 period made it clear that the aim of completing all LA21 strategy documents by the end of 1996 would not be achieved.[1] By then, only 194 LA21 strategy documents had been completed out of a potential UK total of 475. About eighty of the 194 were from urban councils (Morris and Hams, 1997: 5, fig. 15 and Appendix 1). Even in those cases where LA21 strategy documents have been produced, they vary enormously in scope – as Church (1995) and Whittaker (1995) had earlier argued. Some are genuine action plans, but others have little claim to be serious LA21s. The total includes Manchester's, which the City Council had virtually disowned because it was out of step with council policy on expanding the airport.

If an LA21 is to reflect the detail and the spirit of the Rio Agenda 21 document, and to be a real action plan with the potential to make a serious impact, then it needs a number of key features. It needs to analyse issues; establish priorities and implementation targets; set out adequately resourced programmes; and create effective monitoring techniques. In terms of these features it is clear that only a handful of Britain's 80 urban LA21 strategy documents – together with some of those completed after the November 1996 survey – can really claim to be serious action plans.

Even where an LA21 has the features of a serious action plan, its significance will only emerge slowly. In the British case this will depend on its impact on two further issues. First, the budgets of the council concerned, and the organisations with which it deals, will have to be dismantled and rebuilt around sustainable development priorities. Second, the status of LA21s has yet to be determined in the context of Britain's planning system. The system of structure and local plans, and Unitary Development Plans, provides the legal basis of the planning system (Rydin, 1993). These plans set out policies on land release, minerals, housing investment, transport infrastructure, tourism and a whole range of other issues. They reflect major central government policies, as over airports and motorways; as well as setting out the council's conclusions from its own surveys and consultation programmes. Once they have been formally approved, these are the statutory plans against which planning applications are judged. Where the LA21 is at variance with the existing statutory plan, it will be the latter which will almost always determine the outcome.

Other impacts at the level of the city

To understand the British situation it is necessary to distinguish between producing LA21 strategy documents, and cities being involved in the LA21 process (Morris and Hams, 1997: figs 1 and 15). Being involved in the process has been interpreted much more widely to include not just the LA21 strategy document, but trying to apply the principles agreed at Rio across the whole range of urban problems. At this broader level, much more progress has undoubtedly been made. It needs to be examined in the context of the postwar developments analysed earlier. During the 1970s and 1980s, policies on issues like housing and industry were developed in a narrow, self-contained way. Their side-effects and ecological consequences were not fully appreciated.

Transport policies were predominately roads-based for example. They spread development outwards, generating yet more traffic, making cycling more difficult, adding to air pollution, increasing incidences of asthma, threatening wildlife sites, and contributing further to the greenhouse effect.

Together, Brundtland and Rio highlighted the links between policy domains. Policy-makers began to appreciate the need to think laterally, across the artificial boundaries between issues imposed by departmental and committee structures. In the 1970s and 1980s attempts to co-ordinate policies on issues like planning, transport, wildlife, crime and health had been limited.

The initial 1990s response was to examine policy through an environmental lens. Sustainable development was seen not in terms of its social and economic dimensions, but in terms of planting trees, putting in a few cycle lanes and tidying up the environment. This can only be understood in terms of the inherited approach outlined earlier exerting a continuing influence. But as the 1990s advanced, more holistic approaches were developed – as in health (Crombie, 1995) – so as to reduce the side-effects of conventional programmes.

There were two ways in which cities adapted their decision-making processes to help them think more holistically. The more ambitious focused on structures, on giving those involved a strong position within the council's hierarchy. This approach is most in evidence in councils like Wrexham in North Wales, where the environment unit was put into the Chief Executive's Department; or, as with Lancashire, where a strong and influential Environment Department was established. Committed political leadership, as in Kirklees, has also driven change. From 1996 onwards, the establishment of LA21 groups to co-ordinate departmental inputs became more common (Marston, 1996; Morris and Hams, 1997: 17).

On the policy side, thinking across issues was sometimes straightforward. Greater numbers of Combined Heat and Power projects provided cheap heat. Increased recycling reduced the volume of waste going to landfill. The most common response, though, seems to have been to get into more detail within one issue – as on biodiversity (Young, 1995). However, there was little sign of more ambitious ideas like using access to public transport as a criterion by which to assess major planning applications. Such holistic, lateral thinking is routinely used in other countries in western Europe (Barton *et al.*, 1995).

However, in British cities sustainable development has mainly been linked to environmental issues like land-use planning and waste management (Wood, 1994, 1995; Morris and Hams, 1997: fig. 9). Much less progress has been made in integrating sustainable development with policies on social and economic issues like housing, anti-poverty strategies and economic development. A strong feature of Britain's approach to sustainable development has thus been to focus on the environmental dimensions rather than the economic or social ones.

There is a much greater understanding than before Brundtland of how the local affects the global; of how, for example, all the forms of air pollution come together to contribute to global warming and ozone depletion. But the extent to which there is a conscious attempt to tackle this at the local level is limited to isolated cases like Edinburgh's and Oxford's transport initiatives. Some environmental conditions have deteriorated. Urban smog and traffic congestion are examples. But some have improved. Salmon have returned to cleaned-up rivers. Improvements have happened for reasons that have nothing to do with Rio. CO_2 emissions have been reduced to 1990 levels ahead of the target date of 2000. This was mostly due to the Conservatives' decision to run down the coal industry, and to encourage the replacement of coal-fired power stations with gas-fired ones (DoE, 1996: paras 7–14).

The significance of Rio

Some positive steps had thus been taken in Britain's cities. But they were very tentative when compared to what had been envisaged at Rio. Moreover, they were isolated, tending to affect a minority of cities, and usually only some neighbourhoods or some programmes. State urban policies in Britain continued to develop in a slow, incremental way. Overall, the impact of Rio was very limited when compared to the scale of the problems. The issue of inequality – identified above as crucial to sustainable development – received scant attention. Very little progress was made in producing LA21 strategy documents that were serious action plans.

However, Rio's lack of impact does not mean that the post-Rio process has been insignificant in Britain. It is important to draw out the way reacting to Rio has made those involved analyse their policy-making processes. This has helped in two senses. First, producing the LA21 strategy documents was initially seen as a goal in itself. But by the time of UNGASS in June 1997 more and more urban authorities were following LGMB's advice and seeing LA21 as just one stage in a longer process. Labour's 1997 commitment to get all councils to complete LA21s by 2000 was another positive step.

In addition, policy-makers had approached LA21 and sustainable development partly via changed policy-making processes. They used EMAS and other tools to help them think across issues and develop holistic approaches (Audit Commission, 1997). This has started to change the policy-making processes themselves. Applying EMAS to challenging issues like transport and central resource allocation processes, as in Hereford, represents a significant change. Rio has thus released pressures and ideas that are beginning to have a long-term impact. But holistic approaches will only produce significant change via slow, cumulative pressures beginning to generate critical mass over the next decade. This could start to happen, though it is not inevitable. If it is to happen, urban policy-makers need to transform the way they routinely think about policy. This would require nothing less than a culture change in local government. Having analysed the urban level, it is necessary to refocus on the national picture.

Factors affecting progress on LA21

In the mid-1990s in Scotland, Wales and some English counties, many councils were preoccupied with the Conservative government's local government reorganisation agenda. This diverted attention away from serious consideration of big emerging policy issues. Beyond this – temporary – issue, there were three other sets of constraints on progress towards ecologically sustainable urban development.

First, there has been the role of central government. The impact of Thatcherism had been to turn Britain into an even more centralised unitary state, especially when compared with countries like Germany and Sweden. Rio demands local responses to local needs. Part of the impact of the Thatcher and Major governments had been to promote the fragmentation of power and the concept of the enabling authority. This meant that councils often have to work through partnerships – as in crime, health, poverty, and transport. These partnerships are time-consuming to set up, and often fail for lack of public sector funding. Councils lack experience of how to operate them, except in urban renewal (Healey *et al.*, 1992).

Transport provides a good example of the complexities of implementation for councils in the contexts of both centralisation and fragmentation. Councils drew up proposals to tackle local problems – park and ride schemes, road improvements, traffic-calming measures, cycle paths, and so on. But decisions as to the amount of available funding were largely taken in Whitehall. Any action on introducing road pricing depended on a central government lead. The Department of Transport determined the road building programme. Buses and trains were now largely privatised; and decisions about public transport infrastructure projects were even more dispersed across private and public sector interests. The implementation of LA21 proposals on transport thus depended on councils working through others.

The other aspect of central government acting as a constraint concerns the limited and confusing national policy frameworks on so many issues. Attempts to respond to Rio were undermined by policies on out-of-town superstores; the roads programme; and development on greenfield sites, spreading cities outwards (Owens, 1997). These contradictions were symbolised by the increasingly open disagreements over the roads programme and car-orientated policies between the environment and transport ministries from 1989 through to the 1997 election (Young, 1994). A continuing problem was the lack of clear regional strategic guidance on the location of the 4.4 million new households and major infrastructure projects. The routine references to sustainable development in the regional policy documents and Labour's proposal in 1997 to make sustainable development a responsibility of the new Regional Development Agencies did not add up to a coherent approach.

The second factor constraining progress on LA21 was the weakness of political support in both state and civil society for a strong assertive promotion of sustainable development by government at all levels. The neo-pluralist model illuminates the problems (Dunleavy and O'Leary, 1987: ch. 6). The central state encouraged debates over transport and sustainable development. It responded to environmental interests where it could without creating conflicts. Examples include exhorting people to adapt their lifestyles by using public transport and recycling waste. It tried to buy off opposition with token projects – like the planting of trees along motorways. It also responded to pressures for environmental initiatives on uncontroversial topics as with traffic-calming schemes, community forests and urban wildlife projects. However, where there were conflicts, the neo-pluralist state invariably sided with business. The debate on transport after the Royal Commission on Environment and Pollution report (RCEP, 1994) led to the weakening of support for a roads-based transport policy. But ministers were unable to generate an alternative. During the 1992–7 period Conservative ministers were fearful of anything that would make the loss of support in the polls and at by-elections even worse. LA21 was more amenable to democracy at the neighbourhood level than at the sub-regional or higher levels where strategic issues were discussed. Economic interests largely retained the inside track, elbowing out the environmental lobbyists and marginalising the new social movements. In the conflicts within Whitehall, the economic departments largely won out.

However, in the urban areas where councils have actively promoted LA21, a different situation has arisen. The main departments and committees have started to take the non-economic dimensions more seriously, and have been more prepared to limit the influence of economic interests, as in Leeds over restricting traffic in the city centre. They have also tried, as in Leicester, to build coalitions of support for LA21 among groups interested in environmental quality; public transport users, cyclists, wildlife, and other environmental groups; and voluntary sector bodies. Wherever councils took the lead on LA21 they largely failed to persuade industrialists to take the environment more seriously and to generate business support for LA21 (Wood, 1995). In spatial terms support for LA21 came more from the better-off suburbs than from inner city areas.

The third constraint has been the organisational structures and processes within local government. The promotion of ecologically sustainable development requires integrated, multi-disciplinary, cross-departmental processes. Only then is it possible to pursue holistic approaches that build from the environmental ideas that are becoming more and more detailed, and draw in the economic and social dimensions that have been so neglected. However, it is clear that in the mid-1990s only a tiny proportion of councils fully appreciated this. Compartmentalism remains a strong feature of British local government (Stoker, 1991). Manchester, Leeds and Bradford, three large cities each aspiring to be leaders in planning for sustainability illustrate how attempts to promote holistic approaches broke down in a tangle of departmental and committee rivalries, and the withdrawal of political support for

LA21 (Littlewood and Whitney, 1998). They each reveal how ecologically sustainable development is a concept 'which has captured the intellectual and philosophical high ground but failed to be defined in such a way as to subvert the continuation of the old practices which have signally failed to adequately address sustainability in the past' (ibid.: 17).

The conditions for sustainable urban development

Sustainable development involves both physical and social adjustment in response to environmental change. At a theoretical level there are both pessimistic and optimistic perspectives of future prospects. The pessimistic suggests that nothing short of an environmental catastrophe would reveal the need for change and, by then, the damage could be irreversible. As Beck, peering at the abyss, pronounces, modern societies are Risk Societies 'confronted by the challenges of the self-created possibility, hidden at first, then increasingly apparent, of the self-destruction of all life on this earth' (1995: 67). He argues that societies have the capacity to avert catastrophe provided they are engaged in what he calls 'reflexive modernisation'. By this he means that recognition of the dangers may open the way for self-criticism and self-transformation on the part of individuals and society at large, so that new technologies, new ways of living and new institutions are developed which enable society to adjust to the limits imposed by the natural environment. But his move from analysis to prescription is vague and speculative. Also the tentative steps taken so far in Britain on LA21 have been so limited and weak that they do not encourage the prospect of reflexive modernisation.

A more comforting picture is portrayed by those who believe there is the capacity within present society for a transition towards sustainable development based on the idea of 'ecological modernisation' (Hajer, 1995; Mol, 1995, 1996; Christoff, 1996; Blowers, 1997; Spaargaren, 1997). Essentially this means giving greater priority to ecological needs in the production process. It regards the free market operating within the regulatory framework provided by the state as the most efficient way to reduce demands on resources and to minimise pollution. Economic growth and environmental conservation are thereby compatible objectives. This politically convenient greening of 'business as usual' approach, underlies the policies put forward by business and governments as they respond to the call for sustainable development. Here too there are problems. Ecological modernisation is a rather vague concept. It is used both as a broad sociological concept and as a prescriptive programme. The features of progress at the sub-national level have not been worked out. It cannot be claimed that Britain's faltering steps on LA21 add up to a serious application of this model. Ecological modernisation encourages the seductive belief that sustainability can be achieved with little sacrifice. Also in the British case, manufacturing industry – the focus of almost all the writing on ecological modernisation – only accounted by the mid-1990s for 20 per cent of economic activity.

From the British perspective, sustainable urban development requires certain conditions to be fulfilled. First there is a need for policies to promote greater social equality, an essential precondition for the social cohesion necessary to secure a transformation towards sustainability. Second, sustainable urban development needs to be planned over the long term. It requires a system of integrated and strategic planning linking social and physical criteria at the global, national and local levels. This is not the narrowly based British land-use planning system. Nor is it the discredited centralised bureaucratic planning which was operated by the former Soviet Union and its satellites. Third, there is a need for leadership at the national level promoting policies which are relevant to the scale of the issues, and adequately resourced.

Fourth, with regard to changed values, it is evident that the shift from the unsustainable cities of today to the sustainable cities of tomorrow will not take place unless there are fundamental changes in society itself. But social change is not simply an autonomous process. It both influences and responds to political change. The question is whether political purpose and social values can become so aligned as to make possible a shift in the direction of ecologically sustainable urban development. Clearly sustainable development can only be achieved through a social transformation. It is difficult to see how this might come about because the values that have led to unsustainable cities are so deeply implanted in the structures that drive modern society.

Finally, planning for sustainable development requires a holistic approach that transcends economic policy sectors, and involves social criteria as well as environmental ones (Blowers, 1993). But the land-use planning system remains too narrowly focused for holistic approaches to develop. Broadening its scope will require not just new legislation, but a culture change that places environmental sustainability and social equality at the heart of the process of urban development.

Note

1   The LGMB commissioned the University of Westminster to do surveys of progress on LA21 by all councils in England, Wales, Scotland and Northern Ireland. These were carried out in December 1994, February 1996, and November 1996. (The findings are summarised in Morris and Hams 1997: ch. 2.) There is a lot of useful material, but there are several reasons why researchers need to be wary in quoting from it. First, the surveys coincided with a period of local government reorganisation, with 542 councils being surveyed in December 1994, and 475 in November 1996. Some were abolished while others were taking on enhanced responsibilities. Second, the response rate varied. It averaged about 60 per cent. Third, from the point of view of this chapter, there is the complication that the figures are totals, and do not distinguish between urban and rural. Last, there is a difficulty with quoting from the data as 14 of the 16 figures are given as percentages of respondent authorities. So, for example, Figure 15 says that in the November 1996 survey 24 per cent of the respondents would produce an LA21 strategy document in 1996; and 44 per cent would later on. Twenty-four per cent of the 297 respondents is 71, but the total of all councils surveyed was 475.

So 24 per cent translates into 71 out of 475 – which is about 15 per cent. Similarly, 44 per cent of the 297 respondents translates into 131 out of 475 – which is about 28 per cent. Little is known about the 178 non-respondents to the November 1996 survey, although 70 of them had returned the February 1996 survey. So while the surveys are valuable they need to be used with a little caution.

References

Audit Commission (1997) *It's A Small World: Local Government's Role as a Steward of the Environment*, London: Audit Commission.

Barton, H., Davis, G. and Guise, R. (1995) *Sustainable Settlements: A Guide for Planners, Designers and Developers*, Luton: LGMB.

Beck, U. (1995) *Ecological Politics in an Age of Risk*, Cambridge: Polity Press.

Blowers, A. (1993) 'Environmental policy: the quest for sustainable development', *Urban Studies* 30, 4–5, May: 775–96.

—— (1997) 'Environmental politics: ecological modernisation or the risk society?', *Urban Studies* 34, 5–6, May: 845–71.

Breheny, M. and Hall, P. (eds.) (1996) *The People – Where Will They Go?*, London: Town and Country Planning Association.

Castells, M. (1990) *The Informational City: Economic Restructuring and Urban Regional Development*, Oxford: Blackwell.

Christoff, P. (1996) 'Ecological modernisation, ecological modernities', *Environmental Politics*, 5, 3: 476–500.

Church, C. (1995), *Towards Local Sustainability: A Review of Current Activity on Local Agenda 21 in the UK*, London: United Nations Association.

Cochrane, A. (1993) *Whatever Happened to Local Government?*, Buckingham, The Open University Press.

Crombie, H. (1995) *Sustainable Development and Health*, Birmingham: Public Health Trust.

Department of the Environment (DoE) (1990) *This Common Inheritance*, Cm 1200, London: HMSO.

—— (1993) *Environmental Appraisal of Development Plans: A Good Practice Guide*, London: HMSO.

—— (1994) *Sustainable Development: The UK Strategy*, Cm 2426, London: HMSO.

—— (1996) *This Common Inheritance: UK Annual Report 1996*, Cm 3188, London: HMSO.

Department of the Environment, Transport and the Regions (DETR) (1998) *Modernising Local Government: Local Democracy and Community Leadership Consultation Paper*, London: DETR.

Dudley Report (1944) *The Design of Dwellings*, London: HMSO.

Dunleavy, P. (1981) *The Politics of Mass Housing, 1945–75*, Oxford: Clarendon Press.

Dunleavy, P. and O'Leary, B. (1987) *Theories of the State: The Politics of Liberal Democracy*, Basingstoke, Macmillan.

Faludi, A. (1973) *Planning Theory*, Oxford: Pergamon.

Hajer, M. (1995) *The Politics of Environmental Discourse*, Oxford: Clarendon Press.

Harding, A. and Garside, P. (1996) *Hulme City Challenge: First Residents Survey*, Liverpool: European Institute for Urban Affairs, Liverpool John Moores University.

Healey, P., McDougall, G. and Thomas, M. (1982) 'Theoretical debates in planning: towards a coherent dialogue', in P. Healey *et al.* (eds.) *Planning Theory: Prospects for the 1980s*, Oxford: Pergamon.

Healey, P., Davoudi, S., Tavsanoglu, S., O'Toole, M. and Usher, D. (1992) *Rebuilding the City: Property-led Urban Regeneration*, London: Spon.

Hennessy, P. (1992) *Never Again: Britain 1945–51*, London: Vintage.

HMSO (1985) *Lifting the Burden*, Cmnd 9571, London: HMSO.

—— (1986) *Building Business not Barriers*, Cmnd 9571, London: HMSO.

Littlewood, S. and Whitney, D. (1998) 'Sharing the vision? Triumphs and tribulations planning a sustainable future in Manchester, Bradford and Leeds?, Paper presented at the Conference on the Twentieth-century Planning Experience, University of New South Wales, Sydney, 15–18 July.

Local Government Management Board (LGMB) (1993) *Agenda 21: A Guide for Local Authorities in the UK*, Luton: LGMB.

—— (1995) *Sustainability Indicators Research Project: Consultants Report of the Pilot Phase*, Luton: LGMB.

Marston, A. (1996) 'Local Agenda 21 progress: a survey of the new councils', *Scotland's 21 Today*, Autumn, Issue 10: 6–7.

Mol, A. (1995) *The Refinement of Production: Ecological Modernisation Theory and the Chemical Industry*, Utrecht: van Arkel.

—— (1996) 'Ecological modernisation and institutional reflexivity: environmental reform in the late modern age', *Environmental Politics*, 5, 2: 302–23.

Morris, J. and Hams, T. (1997) *Local Agenda 21 in the UK: The First 5 Years*, London: LGMB.

O'Connor, J. (1973) *The Fiscal Crisis of the State*, New York: St Martin's Press.

Owens, S. (1997) 'Interpreting sustainable development: the case of land-use planning', in M. Jacobs (ed.) *Greening the Millennium? The New Politics of the Environment*, Oxford: Blackwell.

Royal Commission on Environmental Pollution (RCEP) (1994) *Transport and the Environment*, London: HMSO.

Rydin, Y. (1993) *The British Planning System*, Basingstoke: Macmillan.

Spaargaren, G. (1997) 'The Ecological Modernisation of Production and Consumption', Thesis, Landbouw Universiteit, Wageningen, The Netherlands.

Stoker, G. (1991) *The Politics of Local Government*, (2nd edn), Basingstoke: Macmillan.

Stoker, G. and Young, S.C. (1993) *Cities in the 1990s: Local Choice for a Balanced Strategy*, Harlow: Longman.

Thornley, A. (1986) *Thatcherism and Town Planning*, Planning Studies No. 12, Polytechnic of Central London.

Ward, S. (1993) 'Thinking global, acting local? British local authorities and their environmental plans', *Environmental Politics*, 2, 3: 453–78.

Whittaker, S. (ed.) (1995) *Local Government Policy Making: Special Issue on LA21*, 22, 2, October.

Wood, C. (1994) *Painting By Numbers*, Lincoln: Royal Society for Nature Conservation.

—— (1995) *Stepping Stones II*, Lincoln: Wildlife Trusts.

Young, M. and Willmott, P. (1962) *Family and Kinship in East London*, Harmondsworth: Penguin.

Young, S.C. (1994) 'The Environment', in P. Allan, J. Benyon and B. McCormick (eds), *Focus On Britain 1994: A Review of 1993*, Deddington, Oxon: P. Allan Publishers.

—— (1995) 'Running up the down escalator: developments in British wildlife policies after Mrs Thatcher's 1988 speeches', in T. Gray (ed.) *UK Environmental Policy in the 1990s*, Basingstoke: Macmillan.

—— (1997) 'Local Agenda 21: The renewal of local democracy?', in M. Jacobs (ed.) *Greening the Millennium? The New Politics of the Environment*, Oxford: Blackwell.

6 Sustainability and urban policy in Germany

Retrospect and prospect[1]

Anke Valentin, Martin Gürtler
and Joachim H. Spangenberg

Introduction

Germany, as a federal state with three levels of decision-making (federal – *Bundesrepublik*, states – *Länder*, and local government – *Kommune*), has political and decision-making structures quite different from many other European countries. German cities also have much more administrative autonomy and financial resources than cities of other countries (with municipal budgets twice the size per capita, for example, of those of many US cities) and therefore have more options to pursue their individual urban policies. The underlying political structure has its foundation in modern history, so understanding present urban challenges in Germany and current policy responses means that it is necessary to be aware of a number of factors.

First one should recall the long urban history of the country. Two millennia ago, the Romans founded a dense network of colonial cities in their province of 'Germania'. A thousand years later, the Middle Ages saw the emergence of independent cities and the powerful networks of 'Hanse' cities. In feudal times, new cities of residence were established by feudal sovereigns to demonstrate their power and their dedication to the arts. The evolution of industrial cities and the subsequent foundation of industrial new towns during the period of the Third Reich was the last effort to complement the already dense urban network in Germany. The rights of local self-administration enshrined in the constitution give the cities considerable power, in political as well as in financial terms.

Secondly, it is important to understand the established decentralised federal system which gives considerable power to the states. These states were equipped with a set of powerful tools, including independent parliaments, a (federal) senate representing the states and controlling the national government and parliament, a policy monopoly over education and culture, and the power to formulate and implement their own urban development policies and strategies.

Thirdly, we must take account of the practical repercussions of the reunification of Germany. For more than forty years, the urban systems in East and West Germany developed in different directions until the reunification in 1989 caused their full integration into the established West German system. The required transformation process in East Germany has been financed by the western part because the indigenous economic base of the east had been eroded within a short period of time and left behind a deindustrialised geography extremely vulnerable and quite unprepared to handle the mechanisms of the Western market economy. Germany today, therefore, includes both the richest (Hamburg) and the poorest (Mecklenburg) regions of the European Union.

Germany has a population of about 81 million and a gross population density of 228 inhabitants per km^2. Of these 81 million about 63 million (77.8 per cent) live in West Germany and 18 million (22.2 per cent) in East Germany, including Berlin. Compared with other cities in Europe, the political and administrative urban system in Germany is well balanced, in that large cities are scattered more or less evenly throughout the country, resulting in a dense functional network of rural, medium and large cities.

Because of the historical development described above, the administrative structure in Germany distinguishes between states and cities as autonomous governing organs. Consequently, in the debate about sustainability one also has to distinguish between processes on national and processes on local levels.

In this chapter, the national discourse of sustainability is first described, taking into consideration governmental as well as non-governmental activities. Due to the 1998 change of government after sixteen years of conservative rule, major changes seem to be ahead. We here mainly describe the policy approach of the governments up to 1998 since that regime brought about the status quo, but we also discuss the changes in perspectives based on the recent coalition agreement as accepted by the parties forming the new government on 24 and 25 October 1998. Following this, the Agenda processes taking place on local levels will be elaborated.

Sustainable development at national level

Federal government policy

At national level 'sustainability' had become a slogan but has not (yet) been elaborated into a coherent and substantial political programme. The conservative German government perceived sustainability as consisting of two components: environmental conservation (a target that is to be promoted at national and international level), and the rest of Agenda 21 which was considered to be an aspect of (foreign) 'development', to be brought about by development aid or by direct investment. Sustainability has hardly been regarded as an obligation to promote a lifestyle in Germany that is compatible with the social and environmental needs of future generations world-wide, nor have equity and gender issues been addressed. The new 'red–green' government, by contrast, regards ecologically sustainable development (ESD) as its basic orientation, with a strong focus on the integration (or at least parallel pursuing) of economic growth, social security, CO_2 reduction and nuclear phase-out, as well as strengthened public participation. However, drastic behavioural changes, like a reduction of mobility or economic interventions involving, for example, reducing transport volumes, are not on the government's agenda.

The prime responsibility for the realisation of sustainable development currently rests with three ministries: the Federal Ministry of Environment (BMU – Bundesministerium für Umwelt, Naturschutz und Reaktorsicherheit), the Federal Ministry for Economic Co-operation and Development (BMZ – Bundesministerium für wirtschaftliche Zusammenarbeit und Entwicklung) which will be strengthened and committed to sustainable development to some degree under the new government, and the Federal Ministry for Construction (BMBau – Bundesministerium für Raumordnung, Bauwesen und Städtebau).

This division of responsibilities has been problematic because the three ministries did not in the past work together sufficiently, and they have pursued entirely different goals and concepts. Other relevant ministries exhibit only token participation, and have in fact been boycotting any move towards sustainable development, for example in transport and agriculture (see Fues, 1997: 10). Whilst the BMU seeks to provide back-up for the national and international debate about environmental protection, the BMZ concentrates on its project of foreign aid in the South, although it is now working more within the guidelines of sustainable development than before. The third ministry mentioned, the BMBau, is responsible for urban development and planning. Together with the municipalities, this ministry gives effect to the goals of the Habitat conferences in seeking to achieve urban sustainable development. Thus there are several different political tasks to which we now turn.

Development co-operation: the BMZ

Working together with the South to enhance the social, political and economic security of poorer countries is seen as the most important responsibility within the development policy of the BMZ, taking into account gender issues and environmental concerns. In the context of sustainable development the policy concentrates on setting new criteria (or relabelling old ones) for foreign aid in developing countries. In particular the following are emphasised (see Forum Umwelt und Entwicklung, 1997: 44–46):

1   Conversion and liquidation of debts through 'debt for nature' swaps, and – new – debt relief for poor countries.
2   International agreements about environmental conservation, like the conventions on desertification, biodiversity, and ozone layer protection.
3   The programme to support the sustainable development of small islands (AOSIS).
4   Education of people in developing countries by the Deutsche Stiftung

für internationale Entwicklungszusammenarbeit (DSE – German Fund for Co-operation on Development).

5 Sponsoring of non-governmental organisations (NGOs) and projects in the South – to be increased with the new government.
6 The transfer of non-polluting technology.
7 The transfer of know-how in environmentally sound energy use.
8 Support for family planning programmes.

With the change of government, steps towards the goal of 0.7 percent of GNP for development aid are now promised.

All of these are measures whose impact lies outside Germany's frontiers. At national level, the non-'environmental' goals of sustainable development have been mainly understood as urban planning issues, as formulated at the Habitat conferences. Now, however, the social dimension and institutional aspects such as extended citizen participation are also on the agenda; however, these have not yet been institutionalised.

National urban planning of the BMBau

The official view, as expressed in the national German report to the Habitat conference in Istanbul in 1996, saw five urban policy challenges with which Federal (and states) governments were confronted. These five challenges of urban development policies have been described (BMBau, 1996a: 45–48).

1 *Sustainable resource utilisation:* 'Efforts have to be made to increase the density of cities, to alleviate the functional division of labour within a city, and to aim for a higher degree of urban polycentrality.' Paradigms such as 'decentralised concentration', 'urban consolidation' or 'short distance city', to 'promote settlement density, and protect free space' are developed (Stark, 1997: 49–50).
2 *Socially compatible urban development:* The creation of socially compatible settlements and urban development is seen as critical in order to try to cushion the most negative social impacts of the free market economy, and to avoid the emergence of socially disadvantaged urban neighbourhoods and urban slums. According to BMBau, 'the remedy is seen in an urban policy which strengthens the self-organisation of disadvantaged neighbourhoods, the creation of affordable social infrastructure, and in better targeting housing policies to minority groups'.
3 *Affordable housing for everybody:* 'Improvement of fringe benefits for private housing investment, the provision of cheap land for private development, the promotion of owner-occupied housing at affordable costs, and the modernisation and partial privatisation of obsolete housing stock in the former East Germany etc.'
4 *Sustainable urban infrastructure:* 'Decentralised service structures are favoured rather than highly centralised utilities and facilities, enabling

private–public partnerships and improving the involvement of local residents and households.'

5   *Attractive cities:* The physical attractiveness of cities and the local investment climate are seen to be essential factors in attracting inward investment and for retaining the existing local business community. BMBau: 'Given the financial constraints of local authorities, and being caught in the obvious conflict between modernisation and sustainable development, ways and means are being sought to compromise between well justified urban development objectives.'

In recent years, the federal government in Germany has intensified its efforts to formulate future-oriented urban development principles and guidelines. This was caused by the need to incorporate the spatial and urban development of the new East German states into the territory of the reunified Germany, and the process was additionally strengthened by European and international initiatives. The outcome has been a few documents aimed at both policy areas; that is, at national spatial planning (*Raumordnung*) and at the local urban development planning (*Stadtentwicklungsplanung*). In these two highly interrelated policy fields, a number of documents have been launched which have pulled various policy strings together.

Environmental policy of the BMU

The Rio conference saw Germany speaking as the 'teacher's pet' (*Musterknabe*) in environmental conservation, with chancellor Helmut Kohl promising the reduction of CO_2 emission in Germany by 25 per cent. This reduction was to be implemented in full by the year 2005 through a programme involving and applying one hundred emission-cutting options and 'voluntary commitments' (see below). However, the government itself criticised this programme as regards its capacity to protect our climate, and it has become obvious that on the basis of past policies Germany will not be able to keep up with the expectations raised. Whether the new government, committed to energy savings and a nuclear phase-out, will be able to meet the target they explicitly reconfirmed remains to be seen (see SPD/Bündnis 90/Die Grünen, 1998). At least the higher energy taxation announced (compensated by a decrease in social security payments) and strict standards for construction may help to reduce emissions to some degree.

During the last decade the reunification of East and West Germany and the subsequent economic turmoil dominated national policies, strengthening a trend away from sustainability policies, which has been dominant in the national government since 1991. Besides this special German problem, however, the national business community hesitates to make what they regard as a 'solo run' because, its leaders argue, of the competitive pressures resulting from the globalisation of the economy which are of special relevance for the World's second ranking export nation. Although today the share of foreign trade in the GNP has just reached the levels of 1912 (a historical background often ignored, as in other countries), with Germany gaining many benefits from globalisation ever since, the argument of 'Globalisation Pressures' has been used politically to promote deregulation of environmental protection and social security 'in order to combat unemployment'.

In 1997, unemployment in Germany reached a historic peak of 4.5 million, with 25 per cent in some subregions of Germany, and unprecedentedly high figures in some traditional industrial cities. Unemployment, preserving Germany's economic attractiveness for foreign investors, and concern about the increase of the GNP were considered to be more important than caring for the environment. And being regarded as international developments, these arguments have provided perfect excuses for past government failures. A sustainable lifestyle that is compatible with the environmental and social requirements of tomorrow has been denounced as a luxury. Whereas the old government set clear priorities for the economic interests, the new tries to strike a balance between economic growth, environmental protection, and social justice – and is strongly opposed by conservative (i.e. neoliberal) politicians, media and business representatives.

Naturally enough, concerns and priorities vary among different interest groups. Private industries, the business community and professionals, aided by their supporting institutions and clientele, complain about costly comprehensive environmental and land use control mechanisms and (against empirical evidence) about slow bureaucracies in granting building permissions and business licences. They also deplore the deteriorating 'competitiveness' of airports, motorways and infrastructure facilities, the business-damaging effects of inner city traffic calming initiatives and the tight control of new developments on virgin land – briefly, all urban environmental achievements since the 1970s are considered 'unnecessary costs' and all new initiatives, regardless of their real effects, are being opposed.

The more environmentally conscious loudly argue for a total reformulation of economic growth policies, towards more efficient (and thus less) energy consumption and organic small-scale agriculture (the agricultural structure still dominating in the South), reducing unnecessary additional mobility instead of promoting it, reconsidering any further extensions of airports and motorways, and making the consumption of open land more difficult. Some of those promoting this line are now in the government. Others, more concerned with the growing social problems in cities, point to increasing poverty and homelessness, to crime and drug problems in certain urban quarters, to increasingly visible social segregation and to the spatial concentration of social and economic problems in urban backwater districts as a consequence of increasing unemployment. This group now forms the majority of government members: both have made important elements of sustainability their main concerns. Hopefully they will search more successfully for win-win-situations than get split over the possible trade-offs.

Although today more people are working in the environmental technology sector than in the automobile industry, a large proportion of conservative politicians and managers are still sceptical. They complain that instead of establishing general guidelines or ecological and social standards, there are command-and-control detailed regulations – a complaint which has some merit.

Preventive environmental conservation in Germany uses two principal instruments: the enacting of laws, and the co-operation between political and economic spheres through 'voluntary commitments'. Economic instruments due to the lack of political will and liability legislation due to the structure of the legal system play no significant role in German politics. The conservative government relied all too often on 'voluntary commitments' of business sectors as a substitute for legal or fiscal measures, often without proper monitoring and enforcement. However, since the German political culture (as compared e.g. to the Netherlands) has no tradition of contractual instead of administrative regulation, they hardly ever got anything beyond the implementation of the results of technological progress that had been on the business agenda anyway. The new government has announced that such commitments as suggested by international experience (Spangenberg and Verheyen, 1996) will be backed up with monitoring and legal enforcement in cases where the outcome of the agreements proves to be unsatisfactory, thus substituting 'command and control' policies for 'agree and control' approaches.

The economic motivation to commit oneself to less-polluting production and management is mainly stimulated by the hope of a better image, which would aid expansion on the (world) market, rather than by real cost saving, since with proper control transaction costs (monitoring, co-ordination, communication, enforcement) have to be carried by the business sector and tend to increase with the new tasks shouldered, with the risk to overcompensate gains from the increased efficiency of allocation. Consequently, it remains to be seen how much action follows 'the gospel' of sustainability, now that this motivation will be supported by increasing taxes on energy. The core of the thinking behind ecological tax reform comprises two dimensions: an overcoming of environmental market imperfections, on the one hand, and the use of the market to enhance competition for environmental performance, on the other.

Euphoric advocates of the market, who reduce the ecological issue to a struggle for future markets and technologies, easily forget the simple fact that economic expansionism outweighs any technological efficiency gains, and the demolition of cultural barriers through uncontrolled division of labour on a global scale is exceptionally destructive in terms of personal identity and social cohesion. Market sceptics, on the other hand, focus on the latter and underestimate the creative potential in the (sustainably regulated) unfolding of market forces. Society, economy, environment are interlinked, non-deterministic complex systems, each with its own dynamics, time-frame and logic. Imposing the logic of one on the other is undermining the balance and thus sustainable development (see Spangenberg, 1999).

On concrete targets for Germany, the new government has confirmed the general goal to reduce CO_2 emissions by 25 per cent by 2005, but has – as yet – not set any sector-specific quantitative targets, a procedure used in many national environmental plans but not yet applied in Germany. So far, the only official paper that named concrete targets for sustainable development of the national economy as a whole has been published by the Ministry of Environment (BMU) in 1998. It was, however, a 'solo run' by the ministry and not a shared government position. None the less – or perhaps, rather, therefore – it might well serve as a starting point for the new government. According to this paper the priority topics of environmental policy (summarised in a system of sustainability indicators called the 'Environmental Barometer') should be energy saving, nature protection, non-polluting mobility, reducing land use and the reduction of material flows by a factor of 2.5 (BMU, 1998).

Since Rio, two-thirds of all industrial states have formulated national programmes for sustainability – but not so in Germany. However, based on a number of civil society institutions, the new government has announced such a programme, including targets, corresponding measures and a clear time schedule (Leitschuh-Fecht and Maier, 1998: 31–32).

Civil society activities

Today, many different social groups and political parties are discussing sustainability: non-profit organisations for environment and development, business, churches, trade unions, political parties. By way of illustration, we present some of these activities, though the account is in no way exhaustive.

New networks (http://www.oneworldweb.de/forum)

Six months after the UN Conference on Environment and Development, thirty-five organisations (environment, development, youth, women and other NGOs, trade unions, etc.) joined forces to found the German NGO Forum on Environment and Development. Already during the run-up to the conference in Rio de Janeiro some of the environmental organisations had been co-operating (an overview on the German environmental NGO movement is given in Spangenberg, 1994), and afterwards they decided to broaden the basis and set up a permanent structure to pursue the process of Agenda 21.

The major purpose of the Forum on Environment and Development is to prepare joint NGO position papers and strategies to campaign and lobby for new political perspectives. Therefore, working groups were established in which every member organisation of the Forum can participate. The Forum intends to promote the following goals:

- to take seriously the outcome of Rio and to do whatever possible to promote policies to eradicate poverty world-wide and to protect the environment;

- to lobby both at national and international level to implement the decisions passed in Rio, particularly Agenda 21, the Climate Convention, the Biodiversity Convention, the Desertification Convention and the Forestry Agreement;
- to establish working groups which would, for example, develop position papers on the most pressing issues in the wake of Rio and in preparation for CSD meetings, LoPs, etc.;
- to increase pressure on government and legislative bodies by joint NGO actions;
- to act as a contact for international partners.

It is one of the main goals of the Forum to educate the German public on the link between environment and development. Moreover, it promotes change in wasteful habits of consumption and patterns of production in the industrialised countries which deplete natural resources and deprive millions of people, particularly in the South, of their right to live in dignity.

The Forum has, besides its working groups and the biennial plenary meetings, one crucial institution: a professional secretariat, financially supported by BMU and BMZ. It is the Forum's voice as well as its tool for co-ordination, disseminating the publications of the working groups, maintaining contacts with organisations in developing countries and monitoring the international Rio follow-up in the context of the UN system. The tasks of the Secretariat are defined by the Steering Committee which advocates the positions and demands of the NGO Forum on Environment and Development towards government and public.

A number of development NGOs are crucial actors when it comes to maintaining the political impact of the Rio process (e.g. WEED – World Economy, Environment and Development; Oro Verde; Bread for the World), but one organisation ('Germanwatch') is somehow special in that is has been newly founded.

New organisations: Germanwatch (http://www.germanwatch.org)

Germanwatch is a non-profit, non-partisan, non-governmental North–South initiative founded by environmental and development activists after the Rio summit 1992 explicitly to lobby and campaign for the Rio-follow-up.

Germanwatch works for the structural changes of global society necessary to give southern countries the chance for self-determined development. It argues that instead of claiming the wealth of the world for itself alone, reorientation in economics and ecology is necessary in the North so that people in the North and in the South may enjoy a high level of well-being and dignity. The work of Germanwatch includes public and media information, as well as intensive dialogue with politicians and co-operation with business people, including common initiatives for ecoefficiency with the winning sectors of industry. It is engaged in sustainability education and training, providing easy to understand information; it goes into schools and other institutions, arranges conferences, etc. Currently Germanwatch is under fierce criticism from NGOs and trade unions for its all too close alliance with certain business groups.

The issues are: German and EU development policy, climate change, food security and the socially and ecologically sound development of world trade. Besides the national level work, a number of regional groups work with autonomously set priorities. Germanwatch is funded by its members and sponsors, and the organisation is not supported by government.

New campaigns: sustainable Germany

In 1995, the Catholic Church's development organisation Misereor and the Friends of the Earth Germany (BUND, Germany's largest environmental NGO) commissioned a study by the Wuppertal Institute for Climate, Environment and Energy to answer the question of how Germany can become sustainable (BUND/Misereor, 1996). The institute had already been in charge of the framework study *Towards Sustainable Europe*, commissioned by Friends of the Earth Europe (Spangenberg, 1995), which was the basis for similar investigative and campaign efforts in 31 countries of Europe.

Misereor was founded in 1958 as a 'Campaign against Hunger and Disease in the World' by the German Bishop's Conference. Today, it is one of the largest development NGOs. It offers co-operation, projects, advisory services and financial aid to people in the South in need, irrespective of their race, religion or nationality. In Germany their main activities are the promotion of the 'One-World-Vision', awareness raising, and fund raising. Founded in the 1920s, with 270,000 members and 2,800 local groups, today BUND is the largest environmental NGO in Germany and since 1988 a member of the Friends of the Earth International network.

The study deals with the question of how an industrial country can become sustainable to save its natural resources for future generations. Therefore dates have been identified and concrete targets proposed for the sustainable use of resources. Furthermore the study presented '*leitbilder*' – visions and advice – for sustainable politics, economy and society (see Sachs *et al.*, 1998: 93).

The study was a bestseller, translated into three languages and, with 60,000 copies sold (plus more than 100,000 copies of the summary), became a focus of public debate (including both critical voices and positive echoes), more than 2,000 conferences and public events, hundreds of local and regional projects, TV shows, and parliamentary debates. Even the conservative government applauded but called the concepts 'unrealistic'. It remains to be seen, how former enthusiastic supporters, now in government, will pick up on the concepts.

Within the last years, a number of studies and research projects carrying the concept of sustainability a step further have been published by NGOs, trade unions and academia, but with limited response from the government. Several of them are mentioned in the references at the end of this chapter.

For example, at the Wuppertal Institute scientists are analysing 'Eco-efficient Services', 'Sustainable Consumption', 'Indicators for Institutional Sustainability' and 'Resource Management'.

In a common appeal, the Reformed Churches and the Catholic Church have called for an integrated approach to social, economic and environmental problems, with many detailed proposals (Evangelische Kirche Deutschland, 1997). The national Trade Union Confederation during its last general assembly made the call for a 'socio-environmental reform strategy', a key element of its political agenda (DGB, 1996).

The general framework of local authorities in Germany

In Germany, the municipalities (with the exemption of Berlin, Hamburg and Bremen, which are Federal States themselves) are part of the Federal States. As such they have to implement laws and measures, where regulation, being the responsibility of the Federal States, has not been availed of at the state level or where the specific tasks have been delegated to the municipalities. In such cases the municipalities have both the constitutional right of self-determination and financial authority. Municipal tasks are taken over on (1) a voluntary basis (e.g. museums, theatres, parks, town halls), (2) a compulsory basis without directives (e.g. schools), (3) a compulsory basis with directives (where specific instruments or devices must be applied, e.g. municipal elections, social security) and (4) to fulfil state or government tasks such as police (see also ICLEI, 1996). So there is, theoretically, a great scope for development beyond the sphere of energy, waste, sewage, transport, air quality, urban planning and local economy advancement, topics of great importance for Local Agenda 21.

Scope for policy development

Compared with the situation of, for example, British municipalities, German municipalities have much scope for political and financial action. Five main challenges of sustainable development are in focus, which, in connection with the Local Agenda 21, are very important for local authorities. However, the exploitation of these opportunities will always be an administrative and financial capacity problem for municipalities.

1   *Municipal environment protection, planning, and education.* Municipal environment policy is, in contrast to other European countries like Great Britain, but similar to the Netherlands or the Scandinavian countries, deeply rooted in German municipalities. This area of responsibility can be integrated in Local Agenda 21 processes and must be combined with economic and social aspects.
2   *Urban planning.* The constitutionally guaranteed right of municipal autonomy is also manifest in urban planning which is one of the main

tasks of local authorities. It is a task over which both federal and state government have only an indirect influence by developing framework legislation. The '*leitbild*' of sustainable development must be embodied in urban planning, and these tasks integrated in the Local Agenda 21 process.

3   *Municipal economic support.* To some degree the German municipalities see themselves more and more as an economic location competing with other municipalities in Europe. So municipal economic support becomes more and more important (and therefore the Multilateral Agreement on Investment was much opposed). Municipal economic support and the right to fix some kinds of taxes (e.g. trade taxes) are opportunities to exert much more influence on the municipal economic situation than is the case in other European countries. But far too often economic interests are played off against ecological interests. It is the task of Local Agenda 21 processes to prevent this.

4   *Municipal foreign co-operation.* Understood as exchange of views and concrete support, this is an opportunity to live up to the global claim of Local Agenda 21. By setting up partnerships with other municipalities world-wide the German municipalities have to hand a good instrument for foreign co-operation. Such co-operation is of course quite under the pressure of financial restrictions, because the profit from such co-operation is ethical and cultural, but not financial. A number of German municipalities work together in a forum for North–South co-operation (the so-called 'Mainz Forum'), while others are active members of the Towns and Development (T&D) initiative of the International Union of Local Authorities (IULA).

5   *Participation and public relations.* On the one hand municipalities have great freedom of decision in this area, if there is the political intention to spend money for professional moderation. On the other hand the people's lack of interest in politics and decision-making restricts this freedom (Valentin and Spangenberg, 1999).

The idea of a 'leitbild'

The concept of a *leitbild* has been used in German comprehensive urban planning since the 1970s. In its actual meaning *leitbild* is an operational, but not too detailed vision of how for example a city should be developed in the long-term (compare the idea of 'planning doctrine' developed in the Netherlands). The development of a *leitbild* is a subjective process of weighing different interests and has to be agreeable to all participants, because many different aspects of municipal living are involved. A *leitbild* is an abstract vision which has to be implemented by action plans. To evaluate the progress of the implementation of a *leitbild* indicators are often developed (BMBau, 1996a).

The National German Report to the Habitat conference in Istanbul in 1996 can be understood as proposing such a *leitbild*. Developed by the

Federal Ministry for Regional Planning, Building and Urban Development (BMBau 1996b), the report – as mentioned above – outlines five challenges of urban development policies, i.e. sustainable resource utilisation, socially compatible urban development, affordable housing for everybody, sustainable urban infrastructure and attractive cities.

The institutionalisation of Local Agenda 21 in German municipalities

It is not easy to get the general idea of Agenda 21 on the municipal agenda for implementation. There is no campaign or network on the federal level where all information is collected; instead there are different networks, which depend on the municipalities starting a Local Agenda 21 process getting in contact with them. Since 1996 the networks have tried at regular intervals to get to know how many local authorities have started a Local Agenda 21 process and what are the main topics and problems. However, most assessments so far are based on the municipalities in contact with certain networks. None of these inquiries can claim to be representative.

In 1997 the German Institute of Town Planning (Difu) carried out a poll by sending packages of questions to the members of the Deutscher Städtetag (DST) – the German Association of Cities – which is possibly also not representative, but provides a good overview of the main topics and implementation problems of Local Agenda 21 in Germany. Of the 153 municipalities which answered the questionnaire (out of a total of 17,000 in Germany), 75 per cent claimed to support a Local Agenda 21 process. In 38 per cent of the answering municipalities there was a declaration of the municipal council to support the Local Agenda 21 process. Also the experiences of other German networks, especially ICLEI, underline the need for a commitment by the municipalities to support a Local Agenda 21 process. The content and degree to which such declarations are morally obligatory vary, but none of them are legally mandatory. So no reliable conclusions can be drawn from these declarations about how seriously the political intention to support a Local Agenda 21 is, and in particular to what degree the residents are involved in participation. Therefore it is particularly difficult to find out much about the *quality* of the Local Agenda 21 processes since, for practical reasons, for the most part only the municipal administrations can be asked, and these have a strong self-interest in presenting the municipality positively. NGOs or individuals who participate in the process are not organised in a way that they could serve as a source of information (see ICLEI, 1998).

According to Difu, in 59 municipalities extra staff were placed to co-ordinate and moderate the process (both by means of hiring new staff or by internal regrouping), and in 30 municipalities additional funds were allocated – however, so far in Germany, as opposed to Sweden for example, no Federal Government grants were available. In the majority of municipalities processes are co-ordinated by civil servants working in the environmental sector.

Sometimes the process is organised by the urban planning department. Rarely, but in more and more cases, the municipal economic department is involved. But the participation of social welfare offices seems to be on the decline. In some cases a cross-departmental Agenda 21 office is established and placed directly under the mayor to cope with the integrative claim of Local Agenda 21. Thus, it has become obvious that the traditional structure of administration is not really suitable to co-ordinate a Local Agenda 21 process because of its strictly separated spheres of responsibility and its hierarchical character. Regarding the organisation of a Local Agenda 21 process (with the focus here on environmental issues, as in the majority of cases), see Figure 6.1.

The issues of Local Agenda 21 in German municipalities

Since 1996 a dynamic movement in support of Local Agenda 21 has emerged (Beuermann, 1997). The number of municipalities engaged in the Local Agenda 21 process has risen from 12 in 1995 to over 50 in 1996, over 100 in 1997, 400–500 in 1998 and 900 in the spring of 1999. However, compared with the total of about 17,000 municipal authorities in Germany, the movement is still very small. Although, due to many non-governmental activities, a certain change of public consciousness can be observed, the *'leitbild'* of ecological sustainability has not been integrated into political decision-making.

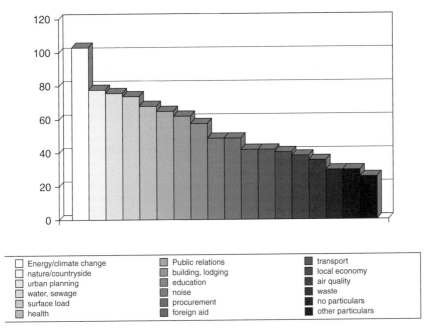

Figure 6.1 Main topics of the implementation of Local Agenda 21 in Germany
Source: BMBau 1998:7

The Difu survey of topics of policy-making showed that environmental topics play the most important role in Local Agenda 21 processes. Most of the German municipalities recognise Local Agenda 21 simply as a new environmental programme. One of the main reasons is probably that public opinion, the press and the federal government misunderstood the 1992 conference (UNCED) in Rio de Janeiro as a summit simply on environment and climate change.

Furthermore, the German Association of Cities (DST), one of the three municipal umbrella organisations in Germany, also promotes the Local Agenda 21 through its guidelines as an environmental programme. Finally, German municipalities concentrate their Local Agenda 21 activities on environmental topics because they have established a long tradition and gone through many experiences in municipal environment protection. To some degree, the same holds true for North–South co-operation with twin cities all over the world.

Consequently, it is mainly environmental grassroots movements and some development groups that initiate Local Agenda 21 processes. It is a widespread experience that in the latter case economic and social aspects are integrated more easily. On a few occasions churches have played an important role in initiating a Local Agenda 21 process. Labour unions or business associations, however, are almost never engaged in the process from the very beginning. In this context much window-dressing occurs: in a number of cities (with Hamm, Nord Rhein Westfalen, being a case in point) the administration created a Local Agenda 21 which does nothing but list many old environmental projects mixed in with some new ones, including mandatory tasks like waste collection, sorting and disposal. New methods for participation such as citizens' councils, future workshops, open roundtables were not considered necessary.

Next to environmental topics, urban planning, long considered an important municipal task, is a very significant topic of German Local Agenda 21 processes. Municipal economic support, education and municipal foreign aid, however, are topics of little salience. This is rather surprising, because the municipal economy especially is given the highest priority at local level in Germany. A Difu poll from 1995 showed clearly that the consolidation of finances, the development of economy and unemployment, as well as transport, are seen as the most important municipal responsibilities. However, except for transport, these topics are not understood as significant subjects for Local Agenda 21 processes. The unbalanced interpretation of Local Agenda 21 as an environmental programme in this connection also shows up and is one reason for the lack of municipal interest in implementing Local Agenda 21.

Problems of implementation of Local Agenda 21 in Germany

The main problem of Local Agenda 21 processes in Germany is the lack of financial support and, in connection with this, the poor level of staffing for the co-ordination and moderation of the Local Agenda 21 process. In contrast with the Netherlands or Sweden, there is only small central support for Local Agenda 21 by the federal or state governments in Germany. This financial support becomes more and more important as municipalities have to bear the high costs of the economic crisis and structural change in Germany.

The second important problem, the varying level of priority given to the programme, can be explained by the false interpretation, as mentioned above, of Local Agenda 21 as an environmental programme. The reason for the lack of information could be, along with the inadequate financial support, the lack of a national campaign to promote Agenda 21 at municipal level. Apart from a few research projects the federal government has not given any support or advice to municipalities on how to implement Local Agenda 21, which comes to most as a completely new challenge. A short time ago ICLEI published a guide to LA 21 for local authorities on behalf of the federal ministry of the environment, but new networks are no substitute for direct federal financial involvement.

To know more about the obstacles in the way of the implementation of Local Agenda 21 (Figure 6.2) it is a good idea to take a look at the expectations of Local Agenda 21 initiatives in the process itself. Within the scope of a federal environment department's (UBA) research project 'Initiatives for sustainable future: new methods of dialogue and communication in connection with the Local Agenda 21', Anke Rheingans analysed Local Agenda 21 initiatives in Berlin. The focus of attention was on the moderation and co-ordination of the process, followed by the longing for participation in decision-making and structural improvements of the Local Agenda 21 process – a topic, which is directly connected with the longing for co-ordination and moderation. The wish expressed for more public attention is expected to be realistic, whereas the longing for co-ordination and moderation seems then to be unrealistic. Municipal representatives interviewed by the Bundestag's *enquête* commission 'Protection of the human being and the environment' mentioned similar ideas. Also the claims for a new policy between the federal/state and the municipal level (equalisation of burdens) and an ecological reform of taxes played an important role. The subsidy policy should be

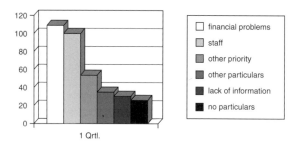

Figure 6.2 Main problems of Local Agenda 21 in Germany

orientated on criteria of sustainable development, and more exemplary projects should be supported. The municipalities also claimed better public relations by the federal government, a central network (probably organised by ICLEI) and a better scientific support.

Structural changes at local level after the Rio Declaration

Looking superficially at the local level the Rio Declaration does not seem to have produced much change, with only a small number of municipalities engaging in Agenda 21 processes. But this first impression is misleading. Structures have changed in two ways: the inter-municipal level with its networks and the local authorities themselves.

Changes at the inter-municipal level: the networks

Although there is no central government network to organise the exchange of views between municipalities, there is a growing number of networks and umbrella organisations in Germany and Europe (with actual effect upon German municipalities) which have been constructed in the aftermath of the Earth summit. There are three kinds of network.

First, there are those which have sought to generate a new municipal consciousness for Local Agenda 21. The Climate Alliance of the German Cities working with the indigenous people of South America (Klimabündnis) have done an important job to widen municipal understanding by connecting climate change programmes with other global aspects (including support for indigenous people in South America). Although this network was set up before the Earth summit in Rio de Janeiro in 1992, it corresponds closely with the idea of Local Agenda 21. Many members of the Climate Alliance have started a Local Agenda 21 process. Also the DST played an important role in promoting Local Agenda 21. In 1996 this umbrella organisation of the German cities published a guide to Local Agenda 21 and what municipalities can do to implement it. Unfortunately the guide concentrated on environmental problems and did not integrate social and economic aspects. But in the period following its publication a dynamic movement of municipalities starting a Local Agenda 21 process was set up.

Second, there are networks which link the Local Agenda 21 municipalities. The most important of these networks is probably the International Council for Local Environment Initiatives (ICLEI). Set up before the Earth summit, ICLEI participated in the preparation for the conference. Today ICLEI is the most important network consulting with municipalities and providing staff training and scientific support. Since 1996/97 the states (*Länder*) have built up their own networks which assist municipalities by organising workshops on Local Agenda 21 topics. The state of Nordrhein-Westfalen also supports its municipalities financially. The fund for municipal foreign aid in Nordrhein-Westfalen is a good example. The state government generally gives 0.50 DM per inhabitant to each municipality to support foreign aid and education projects, as well as for the co-ordination of the Local Agenda 21 process.

Third, there are networks for collecting information. For example the Aalborg Charter Campaign of the European Towns and Cities Towards Sustainability mostly collects information about the implementation of Local Agenda 21 in Europe. The Aalborg Charter has only a symbolic function for German municipalities: the member municipalities obligate themselves to initiate a Local Agenda 21 process with widespread participation of the urban population. Most of the German municipalities engaged in Local Agenda 21 processes are members of the Aalborg Charter. Organisations such as the Healthy Cities Network (WHO), the network of bicycle-friendly cities, or the North–South council of German cities have no immediate relevance for implementing Local Agenda 21 in Germany although they are concerned in a general way with important Local Agenda 21 topics.

Structural changes within the municipality

Some structural changes are apparent within municipalities. These changes cannot be generalised, because German Local Agenda 21 processes vary too much, but some kind of trend can be observed.

Public participation, as it was practised in urban planning since the 1970s in German municipalities, has failed. It is true that each resident could participate in the procedure, but only certain interested experts took the opportunity, and even then with fading impact. Under the customary procedure the local administration typically developed plans and then presented them to the public. But in practice suggestions for improvement made by the residents could not be implemented because the plan was usually decided in advance. So the traditional procedure of participation just seemed to be a way to win people's acceptance for municipal planning. Municipalities therefore started to experiment with new methods of participation (e.g. councils, open space conferences, future workshops) around clearly defined topics. By means of these new processes residents have the opportunity to participate in the development of certain areas of planning policy right from the beginning. Professional moderation of these processes is very important in successfully resolving conflicts of interests, though again financial constraints can be a problem.

Local indicators for sustainable development are increasingly considered necessary for successful implementation of Local Agenda 21. An international investigation by ICLEI in 1996–97 shows that most advanced Local Agenda 21 processes used some kind of evaluation method – mostly local indicators. German municipalities need experience with 'targets of environmental quality' (*Umweltqualitätsziele*), which have been developed in some municipalities since the end of the 1980s. Some municipalities have developed a scientific system of indicators using much data to describe with some precision the actual situation and the progress towards sustainable development.

The choice of indicators is always a subjective decision, which has to be supported by as many concerned persons as possible. So municipalities try to develop systems of indicators in co-operation with local associations or research institutes. For small municipalities without statistical departments it is often more difficult to implement indicators than for larger ones because of the inadequate database.

Summing up

Sustainability forms a central part of the rhetoric of all the programmes of the political parties forming the new government. Overall this concern has resulted in a plethora of documents and conferences on sustainable development, which have in turn kept community activists, researchers and number crunchers quite busy. Hence the movement for ESD has developed significant momentum locally, regionally and nationally. This momentum has created a favourable climate for sustainable urban development all over Germany. Wherever committed personalities took their chance they found they could achieve considerable progress towards sustainability. Clearly, much more has to be done, and just as clearly green activists still have reason to complain. Yet what has been achieved since Rio can undoubtedly be seen as progress.

At present, however, the situation is opaque. Ministries and regional support institutions are communicating the message and advising those wishing to adopt Agenda 21. Environmental consultants and research institutes are selling the concept along with their knowledge of how to introduce and monitor the process. Community groups organise 'round tables'. Training centres offer courses. Undoubtedly, all these efforts will contribute to raising awareness of the resource-conserving message. With the advent of the new government, there is reason to hope for change.

If the local authorities in Germany and the federal and state governments realise the integrative significance of Local Agenda 21 and its importance as a strategy to solve both environmental and economic problems, especially unemployment, Local Agenda 21 has a real chance of being implemented in the majority of German municipalities. On the other hand if the participants misunderstand the importance of Local Agenda 21 and do not provide financial support, the dynamic movement of 1997–98 may break down. Which direction the German Local Agenda 21 movement will choose cannot be forecast with any certainty.

Despite Germany's explicit claim, ecological sustainability has not in the past been a defining feature of federal policy, especially not the cross-sectoral approach called for in the Rio process as well as in the European Union's Amsterdam Treaty. The former German federal government did not even try to get all ministries to check their policies against the paradigm of sustainable development. Both environmental and development policy were treated as narrow departmental responsibilities with limited influence and excluded from 'hard' policy areas.

This might change: sustainable development and Agenda 21 (still understood with an environmental bias) are at the core of the new government's policy guidelines, at least in writing. However, the German report to the CSD's 1999 session, published in November 1998, is as unintegrated, uncritical and greenwashing as ever. Is this a legacy of the old or an indicator of the new government's approach? At national level the future seems quite open at the moment.

Note

1 The authors would like to acknowledge that this chapter is based partly on the paper given by Prof. Dr Klaus Kunzmann at the Conference: 'Environmental Justice: Global Ethics for the 21st Century' at the University of Melbourne October 1–3 1997.

References

Beuermann, C. (1997) 'Local Agenda 21 in Germany (I) – Five years after Rio and it's still uphill all the way?', *Wuppertal Papers No. 68*, Wuppertal: Wuppertal Institute for Climate, Environment and Energy.

Bohnet, M. (1997) In Forum Umwelt und Entwicklung: Fünf Jahre nach dem Erdgipfel. Umwelt- und Entwicklungspolitik auf dem Weg ins nächste Jahrtausend. Dokumentation des Symposiums, Bonn, pp. 44–46.

BFUB (Bundesverband für Umweltberatung) (1998) *Lokale Agenda 21 in Deutschland* (Local Agenda 21 in Germany), Bremen.

BMBau (Bundesministerium für Raumordnung, Bauwesen und Städtebau) (1996a) *Raumordnung in Deutschland* (Spatial Planning in Germany), Bonn.

—— (1996b) Habitat II, *Human Settlements Development and Policy*, Bonn.

BMU (Bundesministerium für Umwelt, Naturschutz und Reaktorsicherheit) (1998) *Nachhaltige Entwicklung in Deutschland. Entwurf eines umweltpolitischen Schwerpunktprogramms* (Sustainable Development in Germany. Draft Environmental Priority Programme), Bonn.

—— (1997) *Towards Sustainable Development in Germany*, Bonn.

BRBS (Bundesministerium für Raumordnung, Bauwesen und Städtebau) (1996) *Nationaler Aktionsplan für nachhaltige Siedlungsentwicklung* (National Action Plans for Sustainable Housing Development), Bonn.

BUND/Misereor (Hg.) (1996) *Zukunftsfähiges Deutschland, Ein Beitrag zu einer global nachhaltigen Entwicklung* (Sustainable Germany, a Contribution to Sustainable Global Development), Basel.

DGB (Deutscher Gewerkschaftsbund-Bundesverband) (1996) *Die Zukunft gestalten. Grundsatzprogramm des Deutschen Gewerkschaftsbundes* (Shaping the Future, Programme of Principles of the German Trades Union Congress), Düsseldorf.

Enquete Kommission des 13 Deutschen Bundestages (1998) 'Schutz der Menschen und der Umwelt: Konzept Nachhaltigkeit, Vom Leitbild zur Umsetzung', Nr. 4/98.

Evangelische Kirche Deutschland (Hg.) (1997) *Für eine Zukunft in Solidarität und Gerechtigkeit. Wort des Rates der Evangelischen Kirche in Deutschland und der Deutschen Bischofskonferenz zur wirtschaftlichen und sozialen Lage in Deutschland*, Bonn.

Eudrukaitis, E. (1997) In *Forum Umwelt und Entwicklung, Dokumentation des Symposiums*, Bonn, pp. 37–38.

Forum Umwelt und Entwicklung (1997): *Fünf Jahre nach dem Erdgipfel. Umwelt-und Entwicklungspolitik auf dem Weg ins nächste Jahrtausend. Dokumentation des Symposiums* (Forum on the Environment and Development: Five years after the Earth Summit. The politics of environment and development on the way to the next millennium: documentation of the symposium), Bonn.

Fues, T. (1997) 'Rio Plus 10. The German Contribution to a Global Strategy for Sustainable Development', *Policy Paper No. 6*, Bonn: Development and Peace Foundation.

Government of Germany (1998) *Germany: Country Profile. Information provided to the United Nations Commission on Sustainable Development, Seventh Session, New York April 1999*, Bonn.

ICLEI (International Council for Local Environmental Initiatives) (1996) Report to the Bundestag's Enquete Kommission 'Schutz des Menschen und der Umwelt': *Stellungnahmen der Sachverständigen im Rahmen der öffentlichen Anhörung zum Thema 'Kommunen und nachhaltige Entwicklung'*, Kommissionsdrucksache 13/3b (Expert opinion for the public hearing on the theme 'Municipalities and Sustainable Development' of the Federal Parliament's Enquiry Committee 'Protection of Humankind and the Environment'), Bonn.

—— (1998) *Lokale Agenda 21 Deutschland: Kommunale Strategien für eine zukunfts-fähige Entwicklung* (Local Agenda 21 in Germany: Municipal strategies for sustainable development), Heidelberg: Springer-Verlag.

Leitschuh-Fecht, H. and Maier, J. (1998) 'Nationale Umweltpläne', *Forum Umwelt und Entwicklung: Lokale Agenda 21 und Stadt- und Regionalentwicklung. Rundbrief Nr. 3* ('National Development Plans', Forum on the Environment and Development: Local Agenda 21 in Municipal and Regional Development. Circular No. 3), Bonn.

Sachs, W., Loske, R., Linz, M. *et al.* (1998) *Greening the North. A post-industrial blueprint for ecology and equity*, New York: Zed Books.

Spangenberg, J.H. (1994) 'The Role of NGOs in German environmental policy – history, organisation and activities of different environmental NGOs in the Federal Republic of Germany', in Puschra, W. and Chung, Ch., *Environmental Policy Towards the Year 2000*, KDI Publ., Seoul (Paper from the Joint KDI/FES Conference. Korea).

—— (1995) *Towards Sustainable Europe. A Study from the Wuppertal Institute for Friends of the Earth Europe*, Luton: Friends of the Earth Publications (summaries available in 20 languages).

—— (1999) 'Investing in sustainable development', in *International Journal for Sustainable Development*, Vol. 1 (in press).

Spangenberg, J.H. and Verheyen, R. (1996) *Von der Abfallwirtschaft zum Stoffstrom-management* (From Waste Treatment to Material Flows Management), Bonn.

Spangenberg, J.H. and Bonniot, O. (1998) *Indicators for Sustainable Development*, Wuppertal Paper 81, Wuppertal.

SPD/Bündnis 90/Die Grünen (1998) *Koalitionsvereinbarungen* (SPD/Bündnis 90/Die Grünen: Coalition Contract), Bonn: http://www.spd.de

Stark, S. (1997) *Lokale Agenda 21, Hemmnisse–Risiken–Chancen* (Local Agenda 21, Impediments–Risks–Opportunities), Wuppertal Paper No. 73, Wuppertal: Wuppertal Institute for Climate, Environment and Energy.

Valentin, A. (1998) *Indikatoren für zukunftsfähigen Lebensstil* (Indicators for Sustainable Consumption Patterns), in Forum Umwelt und Entwicklung, Rundbrief 4/1998.

Valentin, A. and Spangenberg, J. (1999) *Leitfaden zur Lokalen Agenda 21* (Guide to Local Agenda 21) VHS Materialien, Wuppertal/Frankfurt (in press).

7 Japanese urban policy

Challenges of the Rio Earth Summit

Fukashi Utsunomiya and Toshio Hase

Introduction

The history of Japanese environmental policies may be divided into three distinct periods (Utsunomiya, 1995: 112–13). In the first period, ranging from about 1955 to 1975, environmental policies centred on the regulation of pollution. During the rapid economic growth of this early period, both the public and private sectors invested in industrial and urban infrastructure with little or no consideration for environmental effects. Japan was subjected not only to increased pollution and natural destruction but also to serious health problems resulting from pollution, such as Minamata disease (mercury poisoning), Itai-Itai disease (cadmium poisoning) and asthma. Environmental policies were developed in response to such environmental degradation. From around 1973, as a result of strict regulations, environmental pollution of this sort seemed to be well controlled. In the second period, between 1976 and 1987, came the age of amenity in which the goal of public policy was the creation of comfortable surroundings. Once the serious pollution problems had been dealt with the nation seemed to lose its concern for the environment. In the third period, from around 1988, however, environmental problems at the global scale suddenly emerged as a great issue threatening even the existence of humankind. Political leaders around the world paid renewed attention to the environment. In Japan this situation created another rise of concern among the people, industries and government.

The impact of the United Nations Conference on Environment and Development held in 1992 in Rio de Janeiro is felt both at national and local government levels as well as among the people and industries in Japan. In this chapter we first review how recent economic change affects both the environment and environmental policy. Secondly, we consider the emerging environmentalism in Japan, taking into account the change of citizens' consciousness and renewed action in environmental conservation. We look at how the people, industries and the government are co-operating to create a new ecological society. Thirdly, we discuss developments and trends in environmental policies at central and local levels of government after the Rio Conference.

The present economic situation and the environment

Over the last fifty years Japan has developed from being a newly industrialized production economy to a post-industrial 'service' economy sharing all the characteristics of the top rank of developed nations. However, in common with the trend in most post-industrial societies, with the maturing of the Japanese economy in the 1980s the growth rate fell and the economy stagnated. After recovery from the recession of 1986, the following five years saw a boom, called in Japan 'the Heisei boom' after the newly enthroned emperor. In this period the average annual GDP growth rate was 5.1 per cent. Many enterprises and individuals invested in stocks and land and the banks expanded lending, throwing caution to the winds. Especially in the Tokyo area the price of land quadrupled (OECD, 1994: 21). The boom has also been called the 'bubble economy'. In 1991 the bubble burst with a sudden fall of the inflated prices of land and stocks. Many investors went bankrupt; banks and investors were loaded with junk bonds. Since then the Japanese economy has experienced a very low growth rate, compared with earlier times, and an increase in the rate of unemployment (see Figures 7.1 and 7.2).

The Rio Conference on Environment and Development came at a time when Japan urgently needed an economic recovery. The official interest rate is the lowest on record and the Government is now tackling the difficult tasks of administrative reform, deregulation and decentralization. Moreover, the Japanese population is rapidly ageing. Today 15 per cent of the population is aged over 65. In 2020 people over 65 will account for about one-quarter of the population (Yano, 1996: 72).

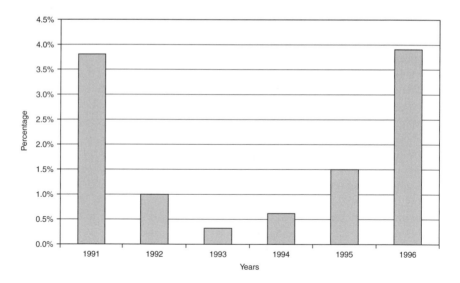

Figure 7.1 GDP growth in Japan, 1991–6

Source: Economic Planning Agency.

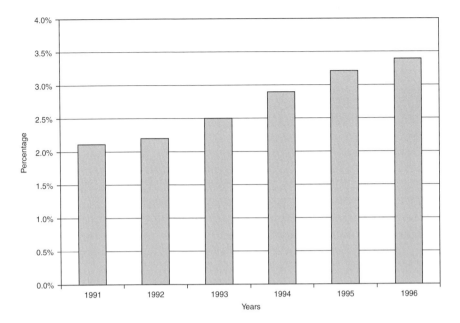

Figure 7.2 Unemployment rate in Japan, 1991–6
Source: Statistics Bureau of Management Co-ordination Agency.

Today Japan is the world's largest importer of food (Yano, 1996: 391). Japan imports more than half of its food requirements from all over the world, including fresh food such as shrimps and fish, meat, grains, vegetables and fruit. Agricultural land in Japan is rapidly being lost to industry and housing and the associated needs of the urban population. In 1994 land available for grain growing was 52 per cent less than in 1955 (Brown, 1995: 56). Taiwan and South Korea show the same tendency. From the 1960s onward rapid urbanization occurred. Cropland was not the only rural land to be consumed for urban purposes, but also forests, wetlands, river banks and natural beaches. The cities absorbed huge numbers of people from the countryside. The cities became great consumers of energy and natural resources. From a distance of several hundred kilometres nuclear power plants send electricity to the Tokyo Metropolitan area, while huge dams in the back country supply Tokyo's water.

Some conscientious people in the remote countryside have begun to oppose the big projects constructed to serve the metropolitan centres. In August 1996 in Maki in the Niigata prefecture the siting of a new nuclear power plant was rejected by citizens' votes. In the referendum to decide whether to accept the nuclear power plant 80 per cent of voters expressed the negative opinion and the project has now been blocked. But in July 1997 at another site – at Kashiwazaki city in Niigata – the fifty-first nuclear reactor (of 1,350,000 kw) in Japan began operation. The city, where employment opportunities are limited, needed the government subsidy and gift money from the electricity company. The trade-off is that metropolitan Tokyo gets the electricity and the poorer city of Kashiwazaki gets money to sustain its local economy.

Both the increased dependence on travel by car and the method of disposal of waste pose environmental problems. In 1996 Japan's automobile industries produced over ten million cars and sold seven million in Japan. The Toyota car company made its highest profits in ten years. More highways are being built to accommodate the increased numbers of cars. Industrial production adds to the amount of garbage produced and this waste is disposed of mainly by incineration. Because of the incineration of waste, pollution of the air and soil by dioxin is notorious throughout Japan. There are currently disputes over some 300 sites of garbage disposal.

The growth of environmentalism

In the Edo period Japan was a closed country for 250 years under a feudal agricultural economy. After this period, in the late nineteenth century, Japan began to open its doors to the Western world and the nation has industrialized rapidly ever since.

Between the Meiji restoration of 1867 and the end of the Second World War a number of serious cases of pollution occurred, but the protests against this pollution were suppressed by the authoritarian government. In 1893, for example, the notorious Ashio copper mine began operation using a new method of smelting which emitted greatly increased quantities of sulphur dioxide. This new pollution destroyed the village of Matsushita upriver from the mine. Pollution from the mine also destroyed the nearby forest, which caused greatly increased runoff of water – flooding the mine. Taninaka village was submerged by the pond which the government had to construct to prevent flooding. The copper-rich wastes damaged crops around the mine. The people of another village in the locality migrated to Hokkaido to avoid the unbearable damage. Since the copper mine was of strategic importance to the Japanese military-industrial complex, listening to the protestors' demands to close the mine was simply unthinkable.

After the Second World War, Japan concentrated its efforts on economic recovery and expansion, and succeeded in achieving high economic growth. Between 1950 and 1974 the annual GNP growth rate was 10 per cent (Sasaki, 1996: 125). As industrial growth continued through the 1960s, serious pollution of the air, water and food supply resulted. Many popular movements were organized to protest against the pollution. These protests focused on various issues: mercury poisoning in Minamata and Niigata; severe episodes of air pollution in Yokkaichi; Kawasaki and Osaka; cadmium pollution in Toyama, and the noise created by airports and highways in many places. People suffered and died from the pollution. In these times anti-pollution movements spread all over the country and their struggle led to the initiation of pollution control measures by central and local governments which were effective in causing pollution to diminish. For this reason the anti-pollution movements in Japan should be considered to have played a very important role in the fight against pollution.

The Japanese environmental movement is composed of many different *ad hoc* volunteer groups formed to fight for local environmental values. Some groups demanded that views and cultural monuments be preserved. Others sought the preservation of nature. Opposition to development was an important theme of many: dams, highways, airports, the reclamation of the sea, golf courses, nuclear power plants all encountered opposition from environmental groups. Another branch of the environmental movement is consumer-orientated. These groups asked for safety of food, water, and consumer products, and proposed alternative lifestyles. The government, however, appeared to be authoritarian and paternalistic and local community leaders were rather conservative. To stand up and do something for the environment was difficult under such circumstances.

There are several thousand local environmental movements in Japan, but there are no ecological parties – either in the Diet or in the local assemblies. The environmental movement is fragmented and politically weak at the national level. The biggest organizations have large memberships – the Japan Society for the Conservation of Nature, for example, has 15,000 members. But most groups remain local in character and apolitical. Compared with America, which has such powerful organizations as the Sierra Club, the tradition of volunteer organization is somewhat lacking. Moreover most environmental organizations are not 'persons in law'. The Japanese legal system does not permit groups of volunteers to acquire such standing at all easily. The property of the environmental organizations is always in the name of particular individuals, and, lacking legal status, donations to the movement are not tax exempt. Professional activists are few, simply because the financial limitations prevent the movement from employing full-time professionals.

In 1997 the government and political parties presented bills to the Diet to give legal status to such movements, but, as of December 1997, the bills are still pending. New types of problems now face the nation: dioxin poisoning, climate change, ozone layer depletion, acid rain, electromagnetic pollution, and the problem of disposing of nuclear waste and industrial and household garbage. The environmental movement cannot ignore such issues, and new groups are being organized to fight these new types of problem. An example is the Kyoto Environmental Citizens' Foundation, founded in 1991. It aims to enhance environmental education, promote more ecologically sustainable lifestyles, and involve local governments and industries in ecological methods of management. In 1997 the Foundation had about a thousand members and provided an organizational focus for NGOs at the 1997 Kyoto conference on climate change (Third Conference of the Parties, COP 3).

The Rio Conference aroused the interest of many Japanese environmental groups and revealed new types of problems. Being small in membership and local in origin, most Japanese groups could not send delegations to Rio de Janeiro. But they understood that international contacts are inevitable and necessary among NGOs. So some Japanese NGOs went to the Rio Conference in 1992 anyway. The Japanese government, however, distanced itself from the Japanese NGOs, and no NGO delegation was included in the Japanese official delegation. This was simply a reflection of the Japanese domestic situation where the weak NGOs and powerful government were antagonistic to each.

In June 1997 the site for the World Exposition of 2005 was finally decided. Two hundred and fifty hectares of forest in the Kaisho region of the municipality of Seto are to be denuded to make the site of the World Exposition. The Aichi prefectural government and local industries welcomed the decision. The Aichi Prefecture competed with Calgary in Canada for the 2005 EXPO. Environmentalists have opposed the candidature of 'Kaishonomori Forest' and have lobbied to prevent it, but in vain.

The average Japanese has become materially rich in the past 20 years and it now seems difficult to recover the Japanese tradition of care for Nature. Present Japanese lifestyles require the consumption of huge amounts of resources as post-war Japan pursues the American way of life. However, the Rio Conference certainly gave a strong message that our planetary environment is in crisis, and the message was heeded by the public. According to a survey conducted by the *Asahishimbun* (a national newspaper) on the 6th and 7th of July 1997, 84 per cent of Japanese people think that carbon dioxide should be reduced despite the cost. Only 9 per cent of those interviewed disagreed. As to environmental taxation, 58 per cent of respondents to the survey felt the need for such taxation to be introduced, while only 29 per cent disagreed (*Asahishimbun*, 21 June 1997).

We should also pay attention to two other kinds of activism which have emerged in Japan. First, new methods of citizen participation are being practised at the local level. Concerned citizens have played a leading role, while local administrations have played a supporting role, in environmental conservation and amelioration. Cases of this type include a citizens' movement to conserve a row of historical streets in Kawagoe City, Saitama Prefecture, where shops and houses with storerooms stand. Another example is the National Trust movement in Kanagawa Prefecture whose purpose is to conserve the unique natural environment through public participation (Utsunomiya, 1996: 1, 9–39). Secondly, there is hope now that a new co-operative network among citizens, governments, business and NGOs is getting organized, and we expect that the establishment of such social partnership among various organizations and individuals might contribute to the solutions of complex environmental problems, including global environmental problems.

Environmental legislation and national policy

National environmental legislation and Agenda 21

In 1971 the Environmental Agency of Japan was established to co-ordinate environmental administration in a comprehensive way. Since Japan lacked a comprehensive law on the environment, which included measures to cope with global environmental problems, in December 1991 the Director of the Environment Agency asked both the Deliberative Council for Nature Protection and the Council for the Environment to consider a new environmental policy with a global focus. These two Deliberative Councils provide advice to the Environment Agency's director who is a Minister of State. Environmental problems of global scope were defined by the government as acid rain, climate change, marine pollution, rain-forest destruction, desertification, ozone layer depletion, transfrontier toxic substance movement, depletion of endangered wildlife and pollution in developing countries.

In November 1993 the Environmental Basic Law was brought into force. This law provides the framework for comprehensive environmental policy in the global era. It is addressed to central and local levels of government, industry, and citizens. It replaced the pollution control basic law of 1967 and the nature preservation law of 1972. The fundamental principles on which the law is based are defined as: (a) appreciating the fruitful benefits conferred on humanity by the environment, and conserving and transmitting the benefits of the environment both for the present and future generations, (b) fostering a society which encourages sustainable development and imposes less of a burden upon the environment, and (c) promoting the protection of the global environment through international co-operation.

The law calls for partnership between governments, industries, and citizens in order to bring about better results through co-operation. It includes particular measures to realize its fundamental principles. One of the measures is the making of an environmental basic plan – described in detail below. Another was the need for environmental impact assessment legislation, which Japan was lacking in at the time. In this connection, in June 1997 the Environmental Impact Assessment Law was finally passed by the Diet which covers all government projects and designated projects such as those of electric utilities, railway construction, and construction of forest roads. The Director of the Environmental Agency has the right to check all such projects and express his opinion. The Law also mentions environmental education, the utilization of economic instruments, and the encouragement of NGO activities. But it did not consider the impact of projects for which the Japanese government's official development aid is given.

One might say that the environmental basic law is inspired by the Rio Summit, because it contains the ideas presented and adopted in the Rio Conference in their entirety. Facing the deteriorating earth environment and realizing the effect upon future generations, the protection of the environment becomes an urgent and common challenge for mankind. Up to 1992 Japan developed piecemeal pollution control laws and nature protection laws, but the situation needed more comprehensive action and well-organized environment policies. The Rio conference provided the opportunity to review existing laws and restructure them (Kitamura, 1996: 217).

In response to the call, contained in Chapter 38 of the Rio Agenda 21, for national policies the government started to create a National Agenda 21 for Japan. The Environmental Agency and the Foreign Ministry met with all the government ministries concerned to draft the policy, and they also received various comments from NGOs and industries. In spite of opposition from NGOs, the policy discussed the further development of nuclear power to alleviate the greenhouse effect. The national policy emphasized the making of a sustainable society within Japan and contributed to international society by increasing official development aid to developing countries. In December 1993 the government presented the Japanese National Agenda 21 to the United Nations.

The government decided to take the lead itself in implementing the National Agenda 21. A government action programme, based upon the National Agenda 21, involves the expenditure of 2.2 per cent of GDP each year in purchasing and contracts (Environmental Agency, 1997). Under the programme the government purchases environmentally sound products in order to influence businesses and consumers and demonstrate itself to be a model 'green consumer'. Demonstrating the government's commitment to the interests of international society, Japan sponsored the third Signatories' Conference on climate change in Kyoto in December 1997.

The environmental basic plan

In December 1994 the cabinet approved the environmental basic plan under article 15 of the Basic Law. This plan, promoting comprehensive and systematic environmental conservation, was regarded as quite a new measure. Before making the plan, the government had reviewed the precedent of other governments' measures. Among these the Netherlands' National Environmental Policy Plan (NEPP) was considered to be a good model for Japan.

The environmental basic plan expresses four principles: recycling, coexistence, participation and international co-operation. Recycling means increased re-use of materials such as metal, paper and glass. Coexistence means that humans and the environment are one, and that we respect our ecological relationship with other creatures. Participation in environmental decision-making is essential to achieve the goal of environmental protection. Citizens, governments and industries must co-operate to protect the environment. International co-operation is also needed to solve the problem of the environment.

To achieve these four long-term objectives, the plan states the basic measures necessary to achieve the goals. In the decision-making process for the plan, and after publishing an interim report, the Deliberative Council listened to many people from different walks of life and held public hearings at nine places throughout Japan (Environmental Agency, 1997: 422). It is not usual to have such public participation in planning at national level and it was an innovation for the Council to seek such consultation with the public. But this was necessary to create more understanding of the plan on the part of citizens. In five years the plan will be reviewed.

Other measures

There was much governmental activity following the Rio Conference: legislation and new policies. These movements can be interpreted as positive evidence of the effect of Rio. For example, Japan signed the Biodiversity Convention in Rio, and formally approved the Convention in May 1993. In April 1994 a law concerning the protection of species was put into effect. Under the law the government designates particular plants and animals in danger of extinction and takes measures to protect them. A strategy on bio-diversity was adopted to promote the convention and the law. While metal and glass were already being recycled, a law was passed to promote recycling of papers and PET plastics. The Non-profit Organization Law was presented to the Diet in 1997 to give non-profit organizations legal standing (that is the status of a 'person in law').

However, decisions concerning earlier projects with a potentially adverse environmental impact have not necessarily been reviewed. In April 1997 the Isahaya Bay reclamation project, whose aim is to create new farming land, went ahead. The government closed the bay completely in order to reclaim the land for agriculture. A large area of significant wet lands (3,400 hectares) has been lost forever. Despite opposition and criticism the government clearly had no intention to alter that decision. The government recognizes that dioxin pollution, as a result of lack of regulation, remains a serious problem. Incineration is the most common and practical method of garbage disposal because Japan lacks the space to dispose of the waste in landfill sites. But dioxin is emitted by most municipal garbage incineration plants. Probably Japan is the most dioxin-polluted country in the world. The emission standard for municipal incineration plants in Japan is 80 mg per m^3, whereas the European Community directive stipulates a maximum of 0.1 mg per m^3 (see: The Council Directive on Incineration of Toxic Wastes, 94/67; The European Community Directive, 1994, Article 7[2]). The government has only very recently begun to consider further regulation of dioxin emission.

In the White Paper on the Environment issued in June 1997 the government has admitted its failure to ensure the reduction of CO_2 emissions to the 1990 level by the year 2000, as the action plan on climate change stipulates. At the Denver summit of G7 held on 20–22 June 1997, Japan supported the United States in not setting numerical goals for the reduction of CO_2. And in the UN Special Assembly on the Environment and

Development held from 23–27 June 1997, Japan again worked hard to avoid setting specific numeric goals. Domestically, there is no consensus within the government on the matter. The Ministry for International Trade and Industry claims there is great difficulty in meeting the goals of the CO_2 action plan, and opposes the Environmental Agency's estimations. Although Japan sponsored the third Conference of Parties to the UN Convention on Climate Change in Kyoto (December 1997), Japan itself had yet to set up pertinent policies for climate change.

Finally, the existence of the Environmental Agency itself has been challenged. The Agency has had the role of a co-ordinating rather than a regulatory and project management agency. Under a plan to reform government administration overall, the government is considering sweeping changes to the administrative structure, and some thought was given to merging the Environmental Agency with other Ministries. Environmentalists expressed alarm that comprehensive independent administration of the environment would be lost if the reform went ahead. Fortunately, however, the government decided to create a separate Ministry of the Environment.

Local environmental legislation and policy

Japan has two levels of local government: prefectures and municipalities (city, town and village). There are 47 prefectures, each containing a number of municipalities. There are also 12 designated cities. A 'designated city' is an urban municipality with more than 1 million people, and has a status almost equal to that of a prefecture.

The present system of local government is to some degree centrally controlled. Between 30 and 70 per cent of the budget of a local government body comes from the central government. Although both the head of the prefecture and the local assembly members are directly elected by the people, central government allocates some executive officers to prefectural governments. In most cases the central government promulgates laws, and local government executes them with the direction and guidance of the central government. But there always exists some scope for local discretion which enables local government to play a major role. This was the case with environmental legislation.

Local Agenda 21

'Agenda 21', formulated at the Rio Summit, appealed to local governments around the world to make Local Agenda 21 plans for the purposes of ecologically sustainable development. Japan's Environmental Basic Law of 1993 also mentioned the role of local government. In this statute local government is requested to follow the lead of central government in pursuing environmental policy, especially through comprehensive planning and management of the environment according to the new law. Towards the end of the 1980s some local governments already understood some of the looming problems of the global environment. Soon after the Rio Summit the Environmental Agency recommended that local governments make Local Agenda 21 plans. In 1994 the Environmental Agency issued a report on Local Agenda 21 and also a manual on the implementation of Local Agenda 21 strategies.

A year later (1995) the government passed the law on decentralization under which local governments have increased independence and also added responsibilities. Pending the detailed redistribution of powers between central and local tiers, local government has been allocated increased responsibility for the environment.

Environmental management plans

Non-statutory local planning for the environment began in the 1970s. The Osaka prefectural government in 1973 made an 'environmental management plan' (nicknamed the 'big plan'). The Osaka 'big plan' was the first local environmental management plan to be made, and by the 1980s many prefectural governments and designated cities, and even some smaller local governments, had made such plans. The purpose of a management plan is to provide a total assessment of environmental resources in a locality, and to manage these resources appropriately. Management plans were made by the administrative arm of a local authority without explicit legal statute, and they were not legally binding. Thus they were felt not to be very effective.

Local environmental basic ordinances

More recently some local governments have, in fact, advanced faster than the central government in the development of environmental policy. For example, towards the end of the 1980s a number of local governments which were facing very grave pollution problems decided that they had to act immediately pending the emergence of central government measures and new national legislation. Some governments have produced innovative environmental policies and pollution control agreements. In order to cope with increasingly complicated environmental problems, these governments began to institute their own 'environmental basic ordinances' ahead of the central government.

The 'environmental ordinance', a law made by local government, is regarded as a sort of umbrella or framework law to cover all environmental policy. In 1990 the Kumamoto Prefecture brought into effect an environmental basic ordinance (which did not, however, include an environmental basic plan; see below). In 1991 the designated city of Kawasaki created an environmental basic ordinance. These ordinances were made prior to the central government's legislation introduced in 1993 – though, following the national legislation, most other prefectural and designated cities followed suit. The national legislation does not explicitly oblige local governments to have an environmental basic ordinance, nor did the central government provide a 'model' ordinance. Nevertheless, by 1997 37 prefectural governments and ten designated cities had brought into law their own environmental basic ordinances (Environmental Agency, 1997: 218).

The new local ordinances enshrine the people's *right* to the environment, a right not mentioned in the national legislation. For example, Kawasaki City's Environmental Basic Ordinance stipulates that the environmental policy of the city shall ensure and realize the right of the citizen to a safe, healthy and comfortable environment (Article 2). However, the wording remains abstract and does not immediately give any particular right to the citizens. Without practical realization, the right to the environment may well become like a decorative façade.

The ordinance also stipulates amelioration of the global environment, and, by this means local government agencies have been able to gain a legal footing for coping with global environmental problems. It is also surely praiseworthy that the ordinance requires partnership between the citizen and the local government in working on the issue. The local environmental basic ordinance has comprehensive coverage of the policy field and a fundamental character, and it stipulates the making of a local environmental basic plan. The basic ordinance obliges the head of the local government agency to make an environmental basic plan.

Local environmental basic plans

The environmental *management* plan differs from the later environmental *basic* plan in two major respects. First, the environmental basic plan includes long-term goals and specific measures to realize these goals, whilst the environmental management plans aimed simply to manage environmental resources. Second, as noted above, the environmental basic plan is, in most cases, legally binding whereas the management plan is not.

Among prefectures and designated cities, Kawasaki City was the first to make a local environmental basic plan (1994). By March 1997, following the promulgation of local environmental basic ordinances, most prefectural governments and designated cities had made environmental basic plans. Much the same patterns were repeated as in the case of the environmental basic ordinance. In the Japanese context the environmental basic plan is regarded as an 'action plan' like Agenda 21. The basic plan is not difficult to introduce since it is not necessary to have the plan approved in the local assembly. The plan is simply approved by the mayor who listens to people's opinions and comments before making the decision.

It would seem that solutions to environmental problems should be fitted to the local situation and thus local government is the appropriate tier to cope with local environmental problems through regulation. Thus decentralization is essential. However, global problems require close co-operation among many tiers of government: local, central and at international level.

While local government is the appropriate level for detailed regulation, it may also be that today new approaches to governance are required at the international level which provides the context for both national and local action.

Case studies of local environmental law and policy

We present three examples of cities whose environmental policy is very advanced and innovative in coping with urban environmental problems. They are the cities of Kawasaki, Sendai and Sagamihara (see Figure 7.3).

Figure 7.3 Location of case studies of cities in Japan with plans for ecologically sustainable development

Kawasaki

Kawasaki City lies between Tokyo and Yokohama, the two largest concentrations of population in Japan. The heavy industries on the sea coast and the dense network of highways pose serious air pollution problems. In this city the rate of asthma sufferers in the population is 9.9 persons per 1,000 inhabitants – 13,518 asthma patients in September 1994 (Serizawa, 1994: 152). Asthma patients in Kawasaki appealed to the courts in 1982, and the Yokohama District Court ordered industries located in Kawasaki to pay compensation for causing air-pollution-related diseases in January 1994. The second and third litigations presented by the patients still continue. The government and the highway authority are also defendants for causing air pollution in these cases because they operate highways and let polluting cars pass through the area. So far the government and the highway authority have not been judged responsible for causing asthma.

The city of Kawasaki has been very active in tackling environmental problems, mainly seeking to control pollution and preserve greenery, and has served up a menu of new ideas on environmental administration to the rest of Japan. However, in recent years, owing to a shift in Japan's industrial structure and the concentration of urban services and infrastructure, subsequent environmental problems emerged which have created new challenges. These include pollution by motor vehicles, disposal of urban wastes, high-technology pollutants, and the difficulty of providing urban dwellers with fulfilment and peace of mind. Additionally, global environmental problems such as global warming and ozone layer depletion have become crucial challenges for local as well as national government.

To address these new challenges, the city of Kawasaki, in September 1990, set up the Kawasaki City Committee composed of environmental specialists with a brief to consider an integrated environmental administration system. In August 1991 the Committee proposed that: 'it is necessary to integrate the entire corpus of administrative policy measures relating to the environment, and for the entire city government to establish a system that can deal with environmental matters in a consistent manner'. On the basis of this proposal, in December of that year the city passed and promulgated the Kawasaki City Environmental Basic Ordinance which took effect from 1 July 1992.

The Environmental Basic Ordinance has the objective of preventing pollution, making a better environment, considering the problem of the global environment, and achieving ecologically sustainable development. The Ordinance stipulates the establishment of a five-member committee on environmental policy which has the right to advise the mayor on environmental issues. It also emphasized the importance of public participation on its implementation. An environmental basic plan shall be made which insists that environmental policy take precedence over other policy areas and shall be given maximum priority. To ensure its effectiveness, a co-ordinating committee was established within the city government comprised of the heads of the major departments. The Ordinance regards protecting the earth's environment to be one of its five basic principles. Most noteworthy is the inclusion of 'taking the global environment into consideration'. Specific measures are to be taken to prevent climate change, stop the depletion of the ozone layer, and exchange preventive technology through international co-operation.

The degree of public participation in the process of making the Environmental Basic Plan was extraordinary in Japan. The initial plan was made public for eight weeks. Explanatory meetings were held at three places at which 190 people attended. The city received 52 letters concerning 388 points (Kawasaki City, 1993: 123). The committee reviewed the comments made by the citizens, and, in the report to the mayor, discussed how they were taken into account. In March 1993 the city formulated its Environmental Basic Plan in accordance with the Ordinance.

The city monitors the emission of CO_2 and methane gas, and takes steps to reduce the emissions from industry and cars. The city committed itself to reducing CO_2 emissions to the level of 1990 by the year 2000. It also takes measures to ensure efficient energy use and develop new energy sources. Recycling of wastes and the planting of trees are also being pursued. The city of Kawasaki has already eliminated chlorofluorocarbon gas from its facilities (Kawasaki City, 1996: 5). In 1994 the city could report that it had collected 316 kg of chlorofluorocarbon gas from 6,776 abandoned refrigerators (Kawasaki City, 1996: 5). The mayor reported on the progress of the Environmental Basic Plan in January 1996, and made public the report for six weeks during which citizens could read the report and voice their opinions. The Committee reviewed the comments made and on 1 May 1996 submitted the opinions to the mayor. The committee proposed that the city government needs to reinforce measures on air, water, greenery, resources, garbage disposal and the earth's environment (Kawasaki City, 1996: 5).

The Ordinance adopted an environmental review system which is designed to investigate environmental impacts and reconcile them with other considerations at an early stage of planning; that is to say, when formulating primary policy measures, projects, guiding principles and the like. The system prescribes the procedures for properly arriving at decisions. Environmental Review Guidelines were also developed, which serve to guide the implementation of the environmental review system. This system became effective in October 1994. Recently the mayor has requested the Committee to review all the environmental laws of Kawasaki in terms of the Environmental Basic Ordinance. The city would like to streamline environmental policy and considers that it is time to review the whole legal system pertaining to the environment.

Sendai

Sendai City lies on the north-east of the mainland (Honshu), and is the capital of that region. In 1989 it became a designated city. Its population is about 980,000. There are concerns about the worsening situation of pollution from cars and the increase of garbage, as well as the global issues of climate change and ozone layer depletion (see Figures 7.4 and 7.5). Even while we have been discussing the problem of increasing consumption of materials in the last ten years, the city of Sendai has seen a 30 per cent increase in numbers of cars, and a 25 per cent increase in garbage being dumped. One might well point here to the increasing environmental consumption of a big city. Under these situations the city of Sendai legislated in March 1996 the Environmental Basic Ordinance.

The Environmental Basic Ordinance outlines four major ideas: (a) making a pleasant environment where a healthy, safe and comfortable life can be possible for present and future generations; (b) realization of the human being as a part of nature and respect for nature shared by all living things; (c) making of an ecologically sustainable city through the promotion of the proper management and cyclical use of resources, acknowledging the environmental capacity to be limited; and (d) encouraging the participation and action of all citizens to protect the global environment. It should be noted that the Ordinance is very conscious of the need for ecological balance and of the problem of the global environment. The Ordinance may be said to take an ecocentric perspective.

In March 1997 the mayor announced the Environmental Basic Plan as stipulated in the Ordinance (Sendai City, 1996: 14). It is worth noting that in the city's Environmental Basic Plan specific numerical targets were set,

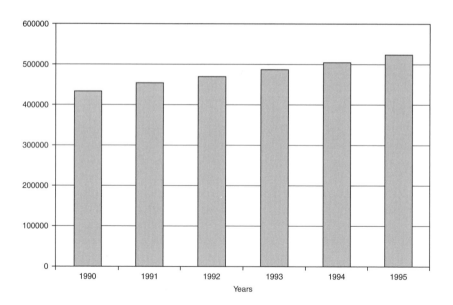

Figure 7.4 Number of cars in Sendai, Japan, 1990–5

Source: Sendai City Environmental Basic Plan (1997: 9).

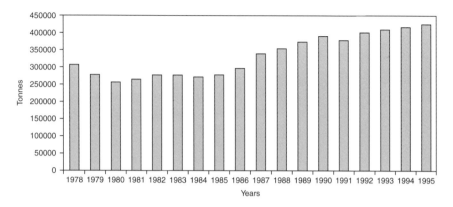

Figure 7.5 Garbage disposal in Sendai, Japan, 1978–95
Source: Sendai City Environmental Basic Plan (1997: 18).

with CO_2 emissions set to be stabilized to the level of 1990 by the year 2010. Other numerical targets relate to the prevention of toxic materials dissolving in the water, limits of water consumption per person, the recycling ratio for garbage, the park area to be allocated per person, opportunities for contact for people with other living creatures, achievement of the environmental standards established by the government, reduction of nitrous oxide emitted by cars, and the percentage of hybrid cars owned by the public service. All these numerical goals should be met by the year 2010 (Sendai City, 1996: 14). Another characteristic of the Plan is the introduction of the concept of an 'urban growth control system'. This plan aims to control the spread of urban growth within the limits of the environmental capacity of Sendai. These methods represent rather new approaches for the Environmental Basic Plan.

Another speciality of the Sendai City Environmental Basic Plan is the CO_2 reduction plan. In June 1992 the city decided to formulate policy concerning the protection of the global environment. It has developed several measures, those concerning climate change being among the most important. In September 1995 the city decided upon the 'Sendai City Global Warming Action Plan'. This plan aims at the reduction of emissions and absorption of CO_2. The council for the promotion of anti-climate change, an official organization constituted by the mayor, is implementing measures in which citizens and businesses can participate and act. Environmental education is also being enhanced to encourage people to take care in protecting the environment.

Since the city of Sendai has a large population and many industries, and consumes a huge amount of energy, the city felt that it should be its responsibility to initiate the climate change policy. The city must be the first to tackle the problem of climate change. Thus the city of Sendai seems to be quite progressive in the making of new environmental policy. The climate change issue is very well articulated in the Sendai City Environmental Basic Plan.

Sagamihara

Sagamihara City, located 40 km from Tokyo, has developed as the dormitory town of Tokyo, Yokohama and Kawasaki. The population of the city has increased rapidly and the rate of urbanization of Sagamihara has been remarkable. In the past 40 years its population has increased sixfold, and the green space in the area has been massively lost. The pollution of the air, the land surface and underground water has become serious.

The city adopted a 'declaration on the environment' in November 1992, which demonstrates a real awareness of the relationship between humanity and nature and reflects contemporary environmental ethical priorities. One might say that, as in Sendai, the declaration reflects an ecocentric attitude. It expressed the obligation of the present generation to contribute to an environment which can be proudly handed to the future generation. All citizens should promote the quality of the environment in which human and non-human nature can coexist.

Based on the spirit of the 'declaration', the city decided upon the Sagamihara Environmental Plan in March 1993. The Environmental Plan aimed to improve comprehensive environmental policy to cope with the increasingly complex environmental problems, including global environmental issues. According to the Plan, global environmental problems are ozone layer depletion, climate change, acid rain, decrease of the tropical forests, desertification, extinction of wild animals and plants, marine pollution, the transfrontier movement of toxic waste and the pollution of developing countries. The Environmental Plan may be regarded as a landmark in environmental policy-making.

The Environmental Plan points out three objectives for the environment. The first is to increase public understanding of the environmental problem and, thus, to lead the three main actors (the city administration, citizens and businesses) to conscientious action. For this purpose the 'centre of learning' (a training institution for environmental education) was established to promote environmental studies and to assist in the move towards a better understanding of environmental problems. Secondly, the Plan posits that institutional reorganization is needed in order to manage the broad systematic impact of environmental problems which do not confine themselves neatly within existing administrative boundaries. The Plan proposes that such reorganization should be worked out by the Sagamihara administration through co-operation between the central and prefectural governments, plus neighbouring local governments. Thirdly, monitoring of acid rain and toxic wastes in the city is essential. The Plan insists that the city must create a monitoring system to provide regular, specific information on the changing situation concerning the effect of pollutants.

Local government has taken considerable interest in the global environmental problem. The dictum 'Think globally and act locally' is being put into practice. Since it is the everyday lifestyle of citizens that creates toxic gas and chlorofluorocarbon emissions, the city administration considers that small local government has an important role to play to solve the global problem. Being closer to the citizens' daily life, local government can help change people's lifestyles to make them less environmentally damaging. Local government can also lead industries towards more environment-friendly behaviour.

In March 1995 the city created an 'action programme' for the administration to implement the environmental plan and, the following year, action programmes for citizens and businesses. These programmes propose detailed and specific actions which should be taken for the sake of the global environment. The action programme is the city's Local Agenda 21 and is thus a direct response to the appeal of Agenda 21 and the Rio Convention. The programmes cover six fields: water management, green space protection, conservation of energy and resources, garbage disposal, protection of the earth environment, and public participation. Policies for the protection of the earth environment, are threefold:

1   *Reduce damaging uses.* Enterprises are to reduce emissions of chlorofluorocarbons, turn to non-chlorofluorocarbon products and recycle the gas; they are encouraged to reduce, or if possible cease using, tropical timbers.

2   *Promote international exchange.* Citizens can learn more about environmental problems in other countries through exchanges with non-Japanese people and participation in NGO activities. Enterprises are to engage in technological co-operation, accepting trainees from abroad and sending specialists overseas. They can also study how environmental impact assessment is practised in operations abroad. The city administration promotes international co-operation, exchanging people with sister cities, sending delegates to international conferences, and participating in international networks.

3   *Develop environmental education.* The city administration promotes environmental education in schools and through adult education courses, and encourages research on environmental problems. Citizens are encouraged to discuss the global problem at home with children, attend seminars on the environment, organize meetings, and monitor acid rain, air quality and water pollution in their neighbourhood. Through education, enterprises are encouraged to develop new technology to reduce impacts on the environment and increase the efficiency of energy use. Education also encourages recycling technology and the creation and maintenance of green space.

The Plan also includes a system for regular review which appears to be effective. The mayor monitors the achievement of the action programme and makes regular reports. The council checks the report and makes recommendations to the mayor. The mayor then modifies the action programme.

This action programme resembles the Netherlands National Environmental Policy Plan, though without the latter's numerical goals. The city brought into legislation the Environmental Basic Ordinance in November 1996. By designating the Plan as an 'environmental basic plan' the Ordinance gives the Plan a legal foundation.

Sagamihara City has developed a very detailed and thorough set of environmental policies. Its action programmes consider every possible action that the city administration, citizens and business enterprises can take. Furthermore the city systematically reviews each year and modifies the programme in flexible ways. These approaches show that a city government has much to offer for the solution of the global environmental problem.

Conclusions

The Japanese economy has become very large, mature – and stagnant. Its influence on the world economy is far from negligible. Japanese official aid to developing countries has now reached the highest level in real terms since 1991. The responsibility of Japan for the global environment is equally large. The Japanese people, having become materially rich, are well aware of the environmental issues the world now faces, and they would support rather strict control and protection of the exploitation of nature. Since there are many problems, Japanese environmentalism, albeit relatively small and local, is active everywhere.

The huge consumption of resources and energy by cities is noticeable everywhere. The deteriorating quality of the air, the water pollution, the garbage disposal problems are serious. Greenery is disappearing from urban areas. As awareness of global environmental problems before and after the Rio Summit strengthens, consuming cities have now come to relate their environmental problems to the global ones. In Japan most big cities and prefectural governments have legislated ordinances to cope with the new situation from a more comprehensive, planning-orientated, long-term perspective.

The legislation of environmental basic ordinances, and the making of environmental basic plans are now common styles of Japanese local administration. Some Japanese designated cities and prefectures are ahead of the central government in environmental control. They are sensitive to the declarations and recommendations of the Rio Summit. Some local governments have acknowledged that their consuming power is not negligible and have started to behave as conscientious consumers. They want the same thing to be practised by industries as well.

The environmental behaviours of consuming cities are of vital strategic importance for the nation's capacity to control the ecology of its environment. Any industrialized country might be essentially an urban society where resources and energies are consumed in a wasteful way. But the impact of the Rio Summit has not been small in Japan. Local governments have taken new initiatives for the solution of global environmental problems, and Local Agenda 21 is a means to realize such contributions by local governments.

The Rio Summit appealed directly to local governments to participate in the process of solving global problems; Agenda 21 explicitly so requested. Kawasaki, Sendai and Sagamihara local governments seem to understand the negative impact the cities make on the environment, and they have stepped forward to accept the challenge by making environmental basic ordinances and environmental basic plans. The information that the Rio Summit conveyed has reached the consuming cities of Japan. Most environmental basic plans made by local government include measures to solve the problems of the environment and include an expression of environmental ethics and fundamental ideas on the environment. This general direction can be interpreted as encouraging. It is now up to consuming cities to prevent further deterioration of the global environment and to give hope for the twenty-first century.

References

Asahishimbun (Asahi Newspaper) (1997) 'Chikyuukankyo Yoronchousa' (Public opinion survey on global environment) 21 June, Tokyo edition, pp. 8–19.

Brown, L. (1995) *Who Will Feed China?*, New York: W.W. Norton.

Environmental Agency (1995), *Kankyokihonkeikaku* (Environmental Basic Plan), Ministry of Finance Publishing Office, Tokyo.

Environmental Agency (1997) *Kankyohakusho* (White Paper on the Environment), Ministry of Finance Publishing Office, Tokyo.

Kawasaki City (1993) *Kawasakishi kankyokihonkeikaku* (Kawasaki City's Environmental Basic Plan), Kawasaki City Government, Kawasaki City.

Kawasaki City (1996) Kawasakishi kankyokeikaku houkokusho (Kawasaki City's Environmental Basic Plan Review), Kawasaki City Government, Kawasaki City.

Kitamura, Y. (1996) *Jijitai Kankyogyouseih* (Local Environmental Administrative Law), Ryoshofukyuukai Publishing House, Tokyo.

OECD (1994) *OECD Environmental Reviews Japan*, Paris: OECD.

Sagamihara City (1996) *Sagamihara kankyoplan koudoukeikaku* (Sagamihara Action Programme for Citizens and Industries), Sagamihara City Government, Sagamihara City.

Sasaki, T. (1996) *Seiji Keizai* (Politics and Economics), Tokyoshoseki Publishing House, Tokyo.

Sendai City (1996) *Sendaino kankyou* (Sendai's Environment), Sendai City Government, Sendai City.

Sendai City (1997) Morinomiyakono Kankyo Plan (Environmental Plan of the Capital City of Sendai), Sendai, Sendai City Government.

Serizawa, K. (1994) *Kensho Kawasaki Kogai* (Pollution of Kawasaki City), Kawasaki: Kanagawashibumsha.

Utsunomiya, F. (1995) *Kankyorinen to Kanrino Kenkyu* (Environmental Philosophy and Management), Tokaidaigaku Press, Tokyo.

Ustunomiya, F. (1996) *Kankyosouzo to Juminsanka* (Environmental Participation and Citizen Participation), Sanreishobo Publishing Co., Tokyo.

Yano, T. Foundation (1996) *Nihon Keizaizue* (Japanese Economic Statistics), Kokuseisha Publishing Co., Tokyo.

8 China's urban environmental sustainability in a global context

Yongyuan Yin and Mark Wang

Introduction

With its economic reforms and the adoption of open door policies in the last two decades China has surprised the world: a striking economic growth rate of around 10 percent, dramatically improving living standards, and bold market reform policies. The World Bank lists China as its number one success story for post-socialist transition, in contrast to the difficulties experienced by Russia (Muldavin, 1997). China is a vast new market, a place to invest East Asia's surplus capital, and a huge resource base. China is also the world's largest developing country with over 22 percent of the global population. It is increasing in size by about 16 million people every year. The huge population has exerted great pressure on the global environment and on local development. With a large proportion of its population still living in rural areas, China is accelerating its urbanisation process through economic reform. The continued rapid urban growth, which appears to be an inevitable consequence of China's economic and industrial development, is greatly expanding China's consumption of energy and its contribution of global pollutants.

China's periods of reform have added new environmental problems to those of the period of 'collective socialism', representing a major obstacle to both medium- and long-term efforts to introduce sustainable development policies. These problems on the horizon are arising not just from the improving living standards, and hence increased consumption of the billion-plus population, but from the way this consumption is stimulated and managed by a capitalist world market. This is particularly evident in coastal cities. Foreign business know-how and monetary investment combined with inexpensive Chinese labor and land have begun to turn many cities into giant building sites. Skyscraper office blocks and modern shopping centers have literally appeared overnight on land that just a short time ago was farmed. Restaurants are today filled with business executives shouting into their mobile phones between mouthfuls of exotic food, a scene not dreamed of under the old regime. Traffic jams are the norm and luxury hotels teem with prostitutes. There are theme parks, discos, karaoke bars on every second corner, golf courses and even racing tracks (Robinson and Goodman, 1996). It is difficult for the Chinese government to resist pressure from the global market. But, if the rapidly growing Chinese economy does not develop in a sustainable manner, then the entire world will feel the economic and environmental consequences. In fact, China's development now under way poses enormous environmental questions for the world, and especially for the Chinese people.

By focusing on urban environmental issues, this chapter will first discuss the necessity for ecologically sustainable development in China. This will be achieved by an examination of a range of environmental, economic, and social challenges, and of the numerous connections among population growth, urbanisation, resource depletion, and environmental pollution in China. The chapter will then present some major initiatives towards ecological sustainability. These initiatives reflect the implications of the Rio summit for Chinese environmental and economic decision-making.

Population growth

In 1982 China became the first country to reach a population of 1 billion. And in 1990, the Chinese census recorded a population of 1.134 billion. China, which had 1.224 billion people at the end of 1996, is expected to have 1.5 billion population by the year 2033. With its restrictive one-child family program and the increasing awareness of the population problem in urban areas, China's population growth rate has dropped significantly since the late 1970s (see Figure 8.1). Chinese demographers have reported a sharp drop in the birth-rate from 37 per thousand in 1962 to about 17 per thousand in 1996. Such a drop in birth-rate has considerable global effects. In simple terms of reducing population growth and ignoring the social side-effects, China's population policy can be considered a success.

However, due to the demographic momentum of China's large population of women of child-bearing age, population growth will continue for several decades. In 1996, with a natural population growth rate of 10.42 per thousand, China added roughly 12.68 million people to the world (Qu and Li, 1994; SSB, 1997a). Although China's population policy has succeeded in checking population growth since the 1970s, much work is still needed to meet future goals, one of which is to limit its population to 1.3 billion by the end of this century. The long-term goal in China is to bring its population within the range of 1.4 to 1.5 billion in the next century.

After nearly two decades of economic reform, population control in China is facing new challenges, and family planning programs have gradually lost some of their effectiveness. Privatisation and deregulation in China have substantially diminished central political control and loosened the state's control over people's lives. Since the 1980s, rural-to-urban migration has been increasing dramatically. Tens of millions of Chinese peasants have migrated to cities in search of higher incomes, in particular to the coastal cities where there are economic opportunities. It is possible for these newly

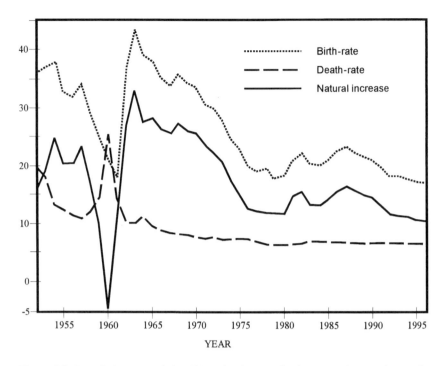

Figure 8.1 Population growth in China: birth-rate, death-rate and natural growth-rate

Source: *China Population Statistics Yearbook* (1997); *China Statistics Yearbook* (1997).

created floating populations to have more than one child per couple because of the difficulties the government has in implementing the family planning measures and population control. Based on mini-census data collected on October 1, 1995 by the State Statistical Bureau (1997b), China's 'floating population' can be estimated at 75–90 million people.

Urbanisation in China

With a large proportion of its population still living in rural areas, China is accelerating its urbanisation process through economic reform. The Chinese planners suggest that, 'urbanization is a necessary consequence of the economic development of society, whatever the country, whatever the societal system, admitting absolutely no exception' (Kirkby, 1985: 221). Cities are political centers with concentrations of cultural, social, and educational institutions which are usually not available in rural areas. Currently, the free market system has been introduced to remedy some shortcomings of the central planning system. Many regional development projects in China are designed to put the market system to work and to provide Chinese people with adequate income, better social services, and higher living standards. Economic reform, with an ever-expanding private sector and more flexible government policies, is the precondition for rapid urbanisation.

There is little doubt that China's urban growth before 1978 was carefully monitored and strictly controlled. Since the 1980s, the establishment of special economic zones, 14 coastal 'open cities' and the open-door policy have attracted foreign investment inflow which has become a new driving force of economic growth and urbanisation. Rapid economic growth in China has created enormous numbers of new jobs with higher wages in the manufacturing and service sectors. The new opportunities flowing from higher urban wages and the establishment of private enterprises attract people from rural areas to cities.

As a result, both the urban population and the number of newly established urban centers have increased dramatically.[1] The urbanisation level (the proportion of people living in cities and towns) increased steadily from less than 18 percent in 1978 to about 30 percent in 1997 (see Figure 8.2). Its cities and towns are now accommodating an average of about 14 million new residents annually. This dramatic increase has been attributed to a combination of rural–urban migration, newly established urban centers, expansion of the existing urban areas, and natural increase of the urban population. Chan (1994) suggests that about three-quarters of China's urban growth

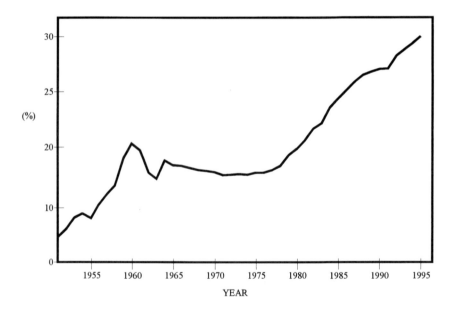

Figure 8.2 Urbanisation level in China, 1949–96

Source: *China Population Statistics Yearbook* (1997); *China Statistics Yearbook* (1997).

Note: Urbanisation level = urban population as a percentage of the total population.

was attributable to migration and urban reclassification, leaving about one-quarter attributable to natural increase.

Many new urban centers are emerging in China. Between 1985 and 1997 the number of Chinese cities increased from 324 to 666. It is expected that, by the end of this century, there will be 800 cities in China. The World Bank estimates that China will add over 18 million in urban population per year between 1990 and 2020. This accounts for 39 percent of the world's annual urban population growth (Kingsley *et al.*, 1994). Chan (1994) esti-mates that China may double its urban population from the present 350 million to about 700 to 800 million in 2010 (Chan 1994). In both cases, China's urban population will grow faster than the total natural population increase.

The population pressure on urban areas is much higher if the floating population is counted. According to the Chinese government definitions, most of the floating population is not officially categorised as 'urban' because the household registration status of these people is still rural even though they live in cities. The floating population is a post-1978 phenomenon which is pushed and pulled by different forces. In rural China, the farming respon-sibility system (or contract responsibility system) initiated by Deng Xiaoping after 1978 has resulted in surges in agricultural productivity. Under this new system, farmers' incentives have been changed. Farmers are able to contract with the collective for the use of land over a long period, and they can sell their surpluses on the free market for profit. So, with the increased labor productivity which this incentive stimulates, the agricultural sector generates more surplus labor in rural areas. Meanwhile, changes in govern-ment policies in China, such as a more liberal interpretation of the household-registration system (permitting increased population mobility), have further facilitated the rural-to-urban migration (Chinese Academy of Social Science, 1985).

Urbanisation and rising consumption

The primary goal of Deng Xiaoping's economic reform was to raise per capita income, the level of consumption, and the living standards of the people in a relatively short time period. The Chinese government wants to achieve the so-called 'four modernisations' through fast economic development. Continued rapid urban growth is greatly expanding China's consumption of natural resources and energy, and its contribution of global pollutants.

The World Commission on Environment and Development suggests that cities account for a much higher share of resource use, energy consumption, and environmental pollution than rural areas. Big cities usually have a long reach and draw their natural resources and energy supply from distant rural areas, with considerable impact on ecosystem sustainability (WCED, 1987). Dramatic population growth and urban expansion are increasing the pressure on China's limited resource base and environment for food, housing, highways, and wastes deposit sites. There is increasing concern about how to manage scarce resources to meet ever-growing demands. A more fundamental question is whether China's natural resource base and environmental assimilation capacity will be adequate to satisfy the ever-rising material expectations associated with a growing population living in the cities and towns. Ehrlich's (1996) simple equation will be helpful in answering this question:

$$I = P \times A \times T$$

where I is the total societal impacts, P is the population size, A is the living standard (or per capita consumption), and T is the technology used.

While unchecked population growth is now viewed by the Chinese government as a fundamental problem for societal (and ecological) sustainability, the increasing *consumption* level is seen as beneficial. In recent years in China, city planners, decision makers, and the public increasingly aspire to the western urban lifestyle: highways, low density single dwellings, cars, colour TV, air conditioning, and profligate water use. In fact, raising per capita consumption levels or living standards is considered the primary goal of economic reform. Per capita household income in China has been increasing steadily in the period of economic reform. Per capita income of urban residents was 5,620 Yuan (RMB) in 1996, up about 13 percent from the year 1995, with a real growth of 3.8 percent after excluding inflation. Rapid economic growth and urbanisation is creating a middle-income class and even a consumer class in Chinese cities. With high standards of living, people in urban areas demand more housing space which often translates into sprawling urban development and costly services. In addition, city residents desire more open space for recreation and public parks.

Table 8.1 shows the differences in per capita income and consumption between urban and rural households between 1990, 1993, and 1996 in the Yangtze River Delta of China. The Delta, one of China's most developed areas, consists of Shanghai, the southern part of Jiangsu province, and a small part of Zhejiang province. It is now the location of 11 large and medium-size cities, and 61 counties or county-level cities. In Table 8.1, the strong income–consumption relationship is clearly illustrated. The income effect on the consumption level is most pronounced. It is obvious that urban households with much higher income consume more. Gaag (1984) suggests that not only does accelerated economic growth in China increase the total demand for consumer goods, the improved living standards also substantially change the composition of consumer expenditures. Using data provided by the Statistical Yearbooks of China, a joint study by staff of the Chinese Academy of Social Sciences and the World Bank calculated the income elasticities for various consumer goods. The study results indicated that a rise in per capita income would significantly change consumption patterns in China. In particular, the shares of non-staple food consumed, such as meat and wine, and of housing, would increase (Gaag, 1984).

Table 8.1 Household per capita income and consumption change in the selected regions (Yuan)

Region		1990	1993	1996
China average	Urban income	1,387*	2,336*	4,838
	Rural income	686*	921*	2,806
	Urban consumption	1,686	3,027	5,620
	Rural consumption	571	855	1,756
Shanghai municipality	Urban income	2,198	4,297	7,196†
	Rural income	1,989	3,148	4,860†
	Urban consumption	1,936	3,530	7,699
	Rural consumption	1,591	2,559	4,339
Jiangsu province	Urban income	1,852	3,337	5,454†
	Rural income	1,182	1,478	3,290†
	Urban consumption	1,470	2,615	4,456
	Rural consumption	787	1,058	1,936
Zhejiang province	Urban income	1,769	3,371	5,718†
	Rural income	1,099	1,746	2,960†
	Urban consumption	1,122	1,653	5,926
	Rural consumption	975	1,305	2,138

Sources: *Statistical Yearbooks of Shanghai, Jiangsu and Zhejiang* (1996).
Notes: Data in current price; *net income; †data in 1995; Yuan: Chinese currency or Renminbi (RMB).

Chinese people, especially urban residents, are moving up the consumption ladder at a remarkable rate. In most Chinese cities, even though people still eat a large amount of starchy staple food such as rice, and travel by buses or bicycles, they are now consuming more meat, eggs, milk, butter and ice cream. A large proportion of urban households now have washing machines, refrigerators, air conditioning, and other durable goods. More and more people can afford to buy private automobiles and travel in aeroplanes. Table 8.2 indicates that an increasing number of people in China are now joining the ranks of western-style consumers. For example, in Shanghai, spending on electronic equipment increased significantly from 1990 to 1996.

Urbanisation and resource depletion in China

As indicated in Table 8.1, an average urban household consumes much more than a rural household. Wackernagel and Rees (1996) use the concept of the 'ecological footprint' to estimate the resource consumption and waste assimilation needs of a defined human population (e.g. of a city) in terms of the corresponding area of productive land (and water) required to meet it indefinitely. The results of their calculations show that, for modern cities, the ecosystem area required for supplying resources and energy, and for depositing wastes, is many orders of magnitude larger than the area defined by their political boundaries. Particularly, most industrialised countries ran a significant ecological deficit.

Table 8.2 Year-end possession of durable goods in Shanghai, per 100 households

Items	1990	1990	1993	1993	1996	1996
	Urban	*Rural*	*Urban*	*Rural*	*Urban*	*Rural*
Colour TV	77	25	94	36	113	53
Refrigerator	88	29	92	42	101	65
Washing machine	72	45	76	54	82	67
Air conditioning	n/a	n/a	5	n/a	50	n/a
Hot water shower	n/a	n/a	15	n/a	42	n/a

Source: *Statistical Yearbooks of Shanghai* (1996).
Note: n/a   not available.

Clearly, the influence of urbanisation and industrial development is felt in many rural regions of China. Cities rely on resources, energy, and other ecological goods and services drawn from distant places through natural flows and commercial trade. For example, the fruits in a Shanghai supermarket are likely to have come from farms located in Guangdong, Fujian, or even Xinjiang (see Figure 8.3). Peasants in China provide primary goods and materials produced within the 'ecological footprint' for the markets in cities and towns. Since the per capita consumption of energy and resources and the amount of waste generated in cities are much higher than those in the

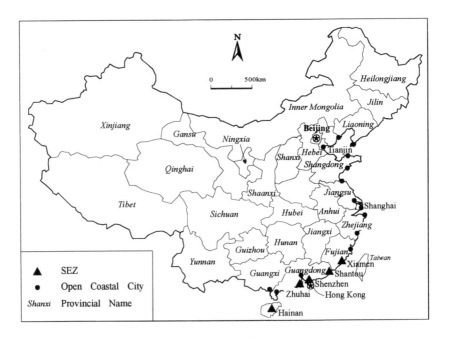

Figure 8.3 The municipalities, provinces, open coastal cities and special economic zones in China

rural regions, rapid urban growth will cause more environmental degradation and ecosystem damage. In the following sections, the environmental impacts of urbanisation and economic growth on land and water resources in China are examined in more detail.

Land deterioration

In China, agriculture is the foundation of the national economy. Providing enough food for the billion-plus population is always the first priority of the nation. Demand for agricultural products takes up a large proportion of national consumption. Chinese people, especially urban residents, are moving up the food ladder quickly along with increasing income. With higher living standards, people in cities consume more meat, packaged foods, and beverages. The conversion rates from grain to meat are: 2:1 for poultry, 4:1 for pork, and 7:1 for beef. All these factors mean higher demand for grain.

Historically, increasing food production in China is achieved mainly by two approaches: (1) increasing arable land by means of reclamation; and (2) improving yields by adopting advanced technology. Increasing cultivated land has contributed to about a half (49.4 percent) of China's food production improvement since 1949. Since today there is very little potential land to bring into crop production which is not already in use, the only choice left for China is to raise yields (Zhao, 1994).

The pressure of higher food requirement in China is accompanied by increasing stress on the land resource base, which has significant adverse effects on the ecosystem. Many economic development activities do not pay sufficient attention to land resource depletion and environmental deterioration. Land degradation, such as erosion, salination, desertification, and land compaction have caused a decline in crop yields and an increase in production and environmental costs.

Urban and industrial development requires land for construction of housing, factories, warehouses, commercial centers, roads and highways, and other infrastructure. Chinese cities and industrial regions are now dependent for their growth upon a vast and increasingly large hinterland of ecologically productive areas. Many of these developments occur in productive farmlands, forestry lands, and areas which are perceived to be of natural, historical, cultural, scenic, or scientific importance. For example, in large Chinese urban centers, enormous amounts of farmland have been converted to urban uses over the period of economic reform. In the Yangtze Delta region, the built-up areas of three important cities, Suzhou, Wuxi, and Changzhou, were expanded 224 percent, 175 percent, and 151 percent respectively from 1957 to the early 1990s (Yao and Chen, 1996). China's so-called 'Economic Development Zones' (EDZs) have been established all over the country. These occupy a large amount of good farmland. Two well-known examples are the Pudong and Suzhou EDZs which initially used 26.5 and 6.5 square kilometers of high-quality farmland respectively. They will be expanded to

100 square kilometers and 45 square kilometers respectively by the year 2010. These zones are located in fertile rice paddies. Cropland loss in the Pearl River Delta, another important agricultural production region in China, is probably even greater than in the Yangtze Delta. By using geographic information systems (GIS) and remote sensing technologies, Yeh and Li (1996) confirmed that in Dongguan, in Guangdong province, 62,602 ha (or 35 percent) of croplands were converted to urban uses between 1988 and 1993. Township and village enterprises present another conflict between economic growth and ecosystem degradation. In Guangdong province, rural brick factories are using up land and damaging the top soils.

It is estimated that China's area of arable land is declining at 0.33 million ha per year. Per capita arable land declined from 0.19 ha in 1949 to only 0.08 in 1991. In one-third of the nation's provinces and cities, per capita arable land is under 0.06 ha (Feng, 1997). Therefore, slowing down the conversion rate of agricultural land to urban uses is critical for agricultural sustainability. The Chinese government has applied stricter policies of farmland protection since the early 1990s. In 1994, the State Council issued a 'Capital Farmland Protection Regulation' to control land conversion from agricultural to other uses. The regulation requires all the local governments to establish a strict land-use approval system. To further emphasise the government's concern on the issue of farmland conversion, the Central Communist Party Committee and the State Council published a special document in May 1997 entitled 'Further Improving Land Management and Protecting Arable Land'. The central problem is how to implement these new regulations successfully.

Depletion of forests

Two big mistakes Mao made between 1949 and 1976, the 'Great Leap Forward' (1959–1962) and the 'Cultural Revolution' (1966–1976), accelerated China's forest depletion. During these two periods, China suffered severe deforestation as a result of using forests for steel production and expanding agricultural land to achieve food self-sufficiency. The post-Mao reform periods also witnessed forest depletion. In fact the pace of deforestation has picked up again since the 1978 economic reform because of the skyrocketing demand for forest products. Industrial development has become an important driving force affecting deforestation. Rapid economic growth and urban development require enormous quantities of forest products to provide construction materials, housing, furniture, books, newsprint, and fuel-wood. Since 1978, many forests in the coastal regions of China have been depleted for new industrial development, road construction, urbanisation, and agricultural expansion.

The conflict between the need for ecological conservation and the demand for forest products will continue into the foreseeable future. According to the official forestry inventory, China's forest cover is about 131 million ha, which was about 13.6 percent of the national territory in 1993. Per capita forest cover is 0.11 ha – only 11 percent of the world average. Forest stock is 8.4 cubic meters per person, compared with the world average of over 80 cubic meters. However, these figures may have been be overstated for political reasons. Remote sensing data collected in the 1980s showed that forest cover in China was 74 million ha, which was roughly 8 percent of the country (Edmonds, 1994).

In 1986, the Chinese government issued a new version of the Forestry Law which included regulations invoking criminal penalties for the first time. In addition, the Forestry Law sets a target of reforestation–afforestation. With the implementation of the national reforestation plan to establish five wind-break forests, China hopes to increase its area of forests for windbreak, fuel, and cash significantly by the year 2000. The proposed national forest cover will be 164 million ha by the end of this century. If this target is achieved, forest cover in China will reach 17 percent of the land area, which will improve ecosystem sustainability.

Water shortage in China

Water resource shortage is another problem facing Chinese decision makers. In China, agriculture, industrial use, and domestic use account for about 87.5 percent, 10.6 percent, and 1.9 percent of water consumption respectively. Due to the uneven spatial and temporal distribution of precipitation and runoff in China affected by the summer monsoon, the country has an uneven distribution of water resources. About 81 percent of the nation's water resources is found in areas south of the Yangtze River, which contains 36 percent of the arable land of China. In contrast, the North, an area accounting for 64 percent of the nation's arable land, only has 19 percent of China's water resources (Zhao, 1994).

Water shortage has become an alarming situation in China. Per capita water use in 1988 was about 455 cubic meters, or about one-fifth of that of the USA, and three-fifths of the world average (Zhang *et al.*, 1992). Population increase has placed a heavy burden on water resources. China is now consuming about 470 billion cubic meters of water annually, and is expected to increase consumption to 710 billion cubic meters by the end of this century. It seems that the overall efficiency of water use is not very high. The reality is, however, that the use-efficiency is unevenly distributed in different regions. In northern China, 43 to 68 percent of the surface water and 40 to 84 percent of the groundwater are consumed annually. The severity of the water shortage problem can be gauged from the annual utilisation rates of shallow groundwater resources: 93 percent in the Haihe plain near the city of Tianjin and 54 percent in the Yellow River Basin.

Severe droughts in various regions of the country cause enormous damage. China has been identified by the UN Food and Agriculture Organization (FAO) as one of the 26 most water-deficient countries in the world (Song, 1997). There is not enough water for irrigation. The area receiving irrigation and the periods of irrigation are shrinking every year. Agriculture experiences a water shortage of 40 billion cubic meters annually. As a result, grain production is reduced by about 10–15 billion kg annually. In rural areas of northern and north-western China, approximately 70 million people and 30 million domestic animals do not have reliable supplies of water. Chinese industry also suffers an annual water shortfall of 6 billion cubic meters. This costs the nation about 200 billion Yuan (RMB) per year in industrial output value. Many northern and coast regions have to exploit their groundwater resources excessively to ease the water shortage. Consequently, the water table in these regions is declining dramatically and this decline forms many so-called 'funnel areas'. Moreover, some cities are even experiencing cracking ground or subsidence, as in Shanghai and Tianjin (Liu and Lu, 1996).

China's urbanisation process and huge floating population further increases water demand dramatically because of much higher rates of per capita water consumption in cities than in rural areas. About 300 Chinese cities have water supply shortage problems; 110 of these are severe. The shortfall in daily water supply can be 10 million cubic meters or more (Gao, 1997). The impacts of water shortage on industries and residents in northern cities are considerable.

Current technologies which have frequently been applied in China to deal with water shortage are mainly focused on the supply side of the resource use system, ignoring water demand management. For example, one major approach adopted in China is inter-basin transfer. In the 1970s, China carried out water diversion projects to transfer water from Huanghe (Yellow River) to Tianjin via Shengli Canal in Henan province, Weihe and the Grand Canal over a total distance of 850 kilometers. Under increasing pressure of water shortage in northern China, China has been working on a huge inter-basin water transfer project: the South–North water transfer to divert water from the Yangtze River to the Beijing–Tianjin region. However, some scholars argue that this inter-basin water diversion project will create many environmental problems such as waterlogging and salinity in areas along the South–North water transfer canal.

Facing the serious water shortage in China, the Standing Committee of the National People's Congress recently issued a call for a tougher enforcement of the Water Law which was enacted in 1988. The Water Law attempts to improve water conservancy facilities and water management. The Chinese government found that some provinces failed to enforce the law. China's Agenda 21 response indicates that improvement of water resource management should be achieved through more effective and efficient water supply *and demand* control.

Energy pressures

In general, while China has enormous supplies of conventional energy resources, the per capita energy resource availability is relatively low (except for coal). However, when the per capita figure is multiplied by a factor of 1.2 billion, China's total energy consumption was ranked third in the world (Lu, 1993). Economic growth and energy production in China is also correlated (Lu, 1993). Since 1949, China's energy production has been increasing rapidly, along with its economic development. For example, over the period 1953–1989, the national income and energy production in China grew 895.7 percent and 1958.5 percent respectively (Lu, 1993). Another source indicates that between 1953 and 1994 energy consumption increased about twenty-two times in China (*The Economist*, 1994). In the last twenty years, the pace of industrialisation and economic growth have been faster than the Chinese planners expected. The goal to quadruple its 1980 GNP by the year 2000 was achieved by 1996. This translates into a huge increase in energy consumption. The traditional labor-intensive economic structure is now being transferred to an energy-intensive one. China is heavily reliant on increasing the supply and effective use of energy to retain its high economic growth rate and to realise its modernisation. Its energy consumption will continue to rise because both industrialisation and urbanisation are energy intensive.

Urban development has accelerated building and road construction. Public transportation systems are established and new vehicles are used to carry people to and from work, and to distribute goods and services. All these increase the demand for energy. Cities also require energy to transport concrete and steel, to supply electricity for various appliances, and to power firms for production.

One concern arising from the improving living standard in China is the increase in car ownership. In the early 1980s there was only a handful of privately owned cars or trucks running on Chinese roads. Stimulated by a government policy issued in 1994 to promote the development of the domestic automobile industry, China's civil vehicle fleet increased from 8.2 million in 1993 to more than 28.7 million in 1996 (SSB, 1997a). Private vehicles increased from 1.5 million to 2.9 million, nearly doubling over the three-year period. Meanwhile, the total Chinese vehicle fleet is estimated to reach 70 million motorcycles, 30 million trucks, and 100 million cars by 2015, and the scope for further growth remains huge (*The Economist*, 1994). The streets of Chinese cities are now suffering from severe traffic congestion and air pollution, attributed to the increasing numbers of motor scooters, cars, and trucks on China's roads which were previously populated mostly by bicycles and public buses.

Increasing numbers of vehicles will account for a significantly larger contribution of carbon dioxide emissions. China, once an oil exporter, became a net oil importer at the end of 1993. A fleet of this size requires enormous amounts of motor fuel and will add to the already heavily polluted atmosphere. The White Paper setting out the Chinese government's response to the Rio Declaration and Agenda 21 (for details see below), challenges the automobile-dependent transport system, arguing that the country cannot afford to import large quantities of oil, and does not have enough land to provide highways, roads, and even parking lots (Brown, 1996). Chinese cities, it is argued, should be designed to meet the needs of bicycles and public transport.

China is a major coal-burning country. Its total coal consumption is ranked third in the world, even though its per capita consumption is only a quarter of the world average and one-tenth of that of the developed world. Coal counts for 75.6 percent of the total national energy consumption. To diversify energy supply, China is increasing its oil and natural gas production. China has abundant oil resources and probably still contains some large undiscovered fields. For example, in Xinjiang Uygur Autonomous Region, experts estimate that there is a reserve of 70 billion barrels of oil in the Tarim Basin, and yet another 70 billion barrels in Karamay located in northern Xinjiang. With this prospect, China's oil reserves may be second only to those of Saudi Arabia. The per capita oil reserve in China, however, is quite low. Moreover, the two potential oil fields in Xinjiang are located in the Gobi desert – a very harsh physical environment. It is suggested that some difficulties, such as the fractured geology, deep discoveries, and a long distance from consuming regions, will increase the costs of oil production and transportation considerably (*The Economist*, 1994).

Rising environmental concern may also restrict both coal and oil use in the future. Hydro-power and nuclear energy have the potential to substitute for coal in the future. China is also working on developing new sources of energy, including solar and other forms of renewable energy. The hydro-power potential in China is tremendous because of its mountainous terrain. To date, only 6 percent of the hydro-power resources has been developed. The Chinese government has made various efforts to promote hydro-power development. One example is the well-known Three Gorges Dam project. The project has generated much controversy and debate all over the world. While many people who favor the project argue that it will help in flood control, inter-basin water transfer, navigation improvement, and hydro-power production (Zhang, 1991), many others, including international environmental organisations, describe the project as ill-conceived. They argue that the project will cause dislocation of millions of people, threaten the Yangtze's unique species of sturgeon and dolphin, sacrifice cultural landmarks, increase reservoir siltation, deprive downstream agriculture of water, and threaten salination of the Yangtze estuary (Fearnside, 1988). In 1992, the National People's Congress of China, assisted by the huge flood of 1991 in the lower parts of the Yangtze river region, approved the project which is now under construction.

While in most western countries, nuclear programs are declining, China is now entering its nuclear era. The Chinese government has launched an ambitious nuclear power program to generate energy and heat. The target is to build nine nuclear power reactors (Lu, 1993). The safety concerns on nuclear energy development are common all over the world. How to ensure the safe operation of reactors and the safe disposal of radioactive wastes is the major challenge. Lu (1993) suggests that Chinese scientists and the nuclear industry have developed the advanced technology and accumulated the experience to deal with the two safety problems. The domestically designed 300 MWe pressurised water reactor (PWR) nuclear power station was built in Qingshan, 70 miles south of Shanghai. The first unit of the commercial 2×900 MWe PWR nuclear power station was completed in 1991 in Daya Bay, not very far from Hong Kong.

Environmental pollution, economic growth, and urbanisation

China is increasingly identified as a newly industrialised country (NIC) (Muldavin, 1997). The emergence of many new cities and a middle class (new rich), along with a free market system and the open door policy, has led to explosive growth in demands of natural resources, energy, and almost everything. In the next several decades, China will rapidly increase its industrial production capacity, particularly in sectors such as energy and transportation, and in steel, chemical, and construction material industries. These sectors and industries are big pollution producers.

Burning 1.4 billion tonnes of coal annually by 2000, China will emit more pollutants into the atmosphere. The environmental costs of rising fossil fuel consumption in terms of carbon dioxide emissions, acid deposition, and air pollution are already immense in China. The World Bank (1997a: 23) estimates that 'The effects of excessive pollution – in the form of premature deaths, sickness, and damage to productive resources – are estimated to cost China about $54 billion a year.' It is obvious that China is facing a challenge in dealing with industrial pollution in the next century.

Acid rain

Acid rain has become one of the major environmental problems in China. It is now recognised that sulphur dioxide generated from coal burning will increase atmospheric acidification and transporting. The south-west basin of China is the major region suffering severe acid rain damage. A value of pH less than 4 has been observed in precipitation. The area affected by acid rain in China is growing rapidly to about one-third of the national territory. The frequency of acidic precipitation has reached up to 70–80 percent (Gao, 1997). Acid rain in regions of southern and south-west China may have reduced crop and forestry productivity by 3 percent (World Bank, 1997a: 23).

Greenhouse gas emission

China's carbon dioxide emission increased from 80 million tonnes (1.3 percent of the global total) in 1950 to about 2.4 billion tonnes (2.4 Gt), or 11 percent of the global share in 1990. Thus, China is now the second largest emitter of carbon dioxide in the world (after the USA with 20 percent). China, with more than 1.2 billion people, is expected to pass the United States in carbon dioxide emissions within about twenty-five years. Methane emission in China from rice paddies, animal metabolism, coal mining, natural gas extraction and transportation, biomass burning and landfills is estimated at 45 million tonnes 1990, or 15 percent of the world total (Smil, 1994).

National plans to double energy production in 2000 from the 1980 level suggest that China's global share of carbon dioxide emissions will continue to rise for some time to come. Since China is unlikely to slow down its energy production and consumption, and since coal will be China's major energy source in the future, the prospects for long-term greenhouse gas emission reduction in China are limited. China and India, together with some other developing countries, argue that since poverty is the main cause of environmental degradation in developing countries, it is not reasonable to expect poor countries to slow down economic development which would result from cutting their energy consumption.

Of course this is not to understate the government's efforts in dealing with global warming and acid rain. The Chinese government, however, is more interested in adopting measures which can improve energy efficiency and industrial structure. For example, Shi Dazhen, Minister of the Chinese Power Industry, made an announcement in Beijing on December 29, 1997 that China planned to shut down hundreds of small thermal power stations which were major contributors to greenhouse gas emissions, and big producers of acid rain (*China Daily News*, 1997). A total of 10,860 megawatts of installed capacity would be eliminated over a three-year period starting in 1998. China now has the opportunity to get rid of inefficient and dirty generating facilities. Power supplied by these small stations will be replaced by clean hydro-power facilities and by nuclear power.

Urban air pollution

China's air pollution is mainly caused by consumption of fossil fuel, especially by the burning of coal. High air pollution is mostly found in cities. The major air pollutants are suspended particulates and sulphur dioxide. Chinese industry released about 13.9 million tonnes of sulphur dioxide and 13.2 million tonnes of particulates into the atmosphere in 1996. Air concentrations of sulphur dioxide observed daily are nearly double the World Health Organisation (WHO) standards. And the average suspended particulates in cities were measured at 309 micrograms per cubic meter, which is a long way above the WHO standards of between 60–90 micrograms per cubic meter. Less than one percent of Chinese cities met the first grade objective of the national air quality standards. As a result, cases of both respiratory diseases and cancer increased in 1996 (NEPA, 1997).

Water pollution

Industrial discharges are the major contributors to water pollution. Toxic chemical waste water and heavy metals are the main concern. Statistics show that in 1994, the volume of total waste water released was about 41 billion tonnes in China (Environmental Yearbook Editorial Committee, 1995). Among them, 25 percent of the 21.6 billion tonnes of industrial waste water (excluding township and village industries) are directly discharged into rivers, lakes, and oceans without treatment.

The 1996 Environmental Report (NEPA, 1997) shows that up to 70 percent of the water in major waterways and lakes equalled or fell below the lowest Chinese water quality standards. About 133 out of 138 urban river sections were polluted to some extent in 1996. Water quality for the main channel of the Yellow River is generally poor, with only 8 percent of the water meeting China's top water-quality standards. The lower reaches of the Yellow River were completely dry for 136 days in 1996 and for more than five months in 1997. The report also indicated that water quality for several large rivers, such as the Huaihe, Liaohe, and Hai, was still in a serious condition. The three most polluted lakes in China are Dianchi, Taihu, and Hongze which are choked by phosphorus from detergents, soaps, fertilisers, nitrogen, and other organic matter.

Solid wastes

In China, volumes of industrial solid wastes and urban garbage are increasing rapidly. Solid waste treatment and re-use rate is relatively low. In 1990, the re-use rate for industrial residuals was only 29 percent. There were about 6.64 billion tonnes of accumulated industrial solid wastes and urban garbage piled up on 560,000 ha of land near big cities and river banks. Urban garbage release is increasing at a rate of 8–10 percent per year. It is estimated that by the end of this century urban garbage release will reach 140 million tonnes annually (Chen, 1994). Toxic solid wastes are also increasing with the rapid development of the Chinese chemical industry.

Noise

Urban noise pollution generated by industrial operation and construction is quite serious. In 1990, traffic, domestic, and industrial noise contributed 32.7 percent, 40.6 percent, and 26.3 percent of the total urban noise respectively. Noise levels in most urban areas surpassed the national noise emission standards (Chen, 1994). Thirty percent of the workers are operating in noise polluted environments. Forty per cent of urban residents are suffering noise pollution. Noise levels in most urban zones exceed the standards and continue to rise.

Moves toward sustainable urban development in China

It has become clear that economic development, population growth, and urbanisation in China have caused a series of environmental and social problems. These problems are some of the main reasons for uncertainty concerning the sustainability of the nation to support future societal needs. What has been implied throughout the preceding discussion is that major policy gaps exist which inhibit a satisfactory resolution of many problems related to urban development.

The Chinese leaders acknowledge that there are environmental problems. As the Premier Li Peng said in his speech at the Fourth National Environmental Protection Conference held in Beijing in 1996: 'We clearly are aware that the situation of the environment in our country is still quite severe, since China is now still at the stage of rapid industrialization and urbanization' (NEPA, 1996: 7–23). China was one of the first countries to ratify Agenda 21. In 1994, the Chinese government published 'China's Agenda 21 Report' as a response to the Rio Declaration and Agenda 21 (Chinese Government, 1994). Since then, many officials and experts from various countries and international agencies (including Maurice Strong, General Secretary of both the 1972 Stockholm Conference and the 1992 Rio Summit) have been invited to Beijing to discuss the Report. The main purpose of the Beijing meeting was to present the Report to the international community and to seek international loans or other financial support to initiate co-operative projects listed in the Report.

The Report was formulated with reference to the framework and structure of the UNCED Agenda 21, and was adjusted to China's reality. The main target of China's Agenda 21 is to set development strategies, policies, and measures in dealing with issues of natural resources, environment, social aspects, law, finance, and institutional structure. In China's Agenda 21, four main themes were identified: (1) overall strategies for sustainable development; (2) societal sustainability; (3) economic sustainability; and (4) resources and environmental protection.

The Report was approved by the State Council of China on March 25, 1994 as the White Paper on China's population, environment, and development towards the twenty-first century. The term 'sustainable development' is now commonly used by Chinese officials and can be found in many newly published government documents. Bradbury and Kirkby (1996) suggest that the relatively rapid publication of China's Agenda 21 demonstrates China's wish to engage more fully in international affairs and to co-operate with other countries on global environmental issues. Chinese governments at state, provincial, and local levels are beginning to incorporate the concept of sustainability in decision making processes. For example, at the municipal level Guangzhou City has recently completed a Local Agenda 21 (Guangzhou's Agenda 21, Leading Group Office, 1997). The format of the document follows the national Agenda 21 and contains the same four parts as the national Report.

China's Agenda 21 should not be viewed in isolation from other development addressing the environmental problems in the country. China is also enhancing its laws to improve environmental quality. A set of new laws and tough policies regarding environmental protection has been initiated since the Agenda 21 Report. For example, in 1994, the Chinese government published the Action Plan for the Environment (Chen, 1994). The Environmental Protection Law (Trial) was initiated in 1979, which established a general legal framework in China to regulate environmental problems. But the first real government effort was the approval of the ninth five-year-plan for Environmental Protection for 2010 by the State Council in 1996. The plan also included two supplementary documents: (1) Total Emission Quantity Control Plan for Major Pollutants, and (2) Trans-century Green-Engineering Plan. In the same year, the Chinese government also amended the Water Pollution Prevention and Control Law by adding provisions for controlling river basin pollution and imposing tougher regulations for protecting sources of drinking water (World Bank, 1997b).

Since then, the Chinese government has struggled to monitor and smoothly implement these regulations and laws. China will use 1.5 percent of its gross domestic product (GDP) for environmental protection during the last years of this century (*Beijing Review*, 1997). According to information provided by NEPA, an average of 65 billion Yuan (RMB) will be spent annually on green projects. Tough actions include shutting down about 7,000 small factories, most of which are township and village enterprises (by order of the State Council effective on April 17, 1997). These small plants, including paper mills, tanning factories, printing and dyeing mills and coking plants, were using outdated equipment and technologies in production. It was reported that by the end of February 1997 polluters on the lists of the most provinces across the country, a total 62,561 operations, had been shut down. NEPA and the Ministry of Supervision of China have been carrying out nationwide inspections of plants with serious pollution problems.

There remain enormous problems in implementing strategies laid out in the Report and it is too early to judge whether a real transition to sustainability is now under way. However, it is clear that China has stepped up efforts to deal with its environmental problems. One improvement in China is the remarkable change in the perceptions of policy makers. At the Fifteenth Party Congress, Chinese President Jiang Zemin (1997) indicated that, as the world's most populous nation with a relatively scarce natural resources supply, China must carry out a sustainable development strategy in its modernisation. The key items concerning sustainable development in Jiang's opening address included: (1) upholding the fundamental policies of family planning and environmental protection; (2) reconciling population, resources, and environment; (3) combining resource exploitation and saving; (4) improving resource use efficiency; (5) enforcing resource and environmental regulations and laws; and (6) promoting measures for pollution treatment, reforestation, soil erosion and desertification control, and ecological improvement. These items represent a serious effort on the part of the Chinese government in its struggle to achieve ecological sustainability. Under the tremendous pressure of population growth, increasing consumption, and rapid industrialisation, China needs to engage more in international co-operation and develop yet tougher policies to ensure its societal sustainability and to leave its children a blue sky and clear water. Hopefully, the ratification of Agenda 21 represents the beginning of an environmental era in China.

Note

1   In China, 'urban' includes two categories: city and town. 'City' refers to a settlement with the approval of the central government (normally with a population of more than 100,000). 'Town' is designated by the governments of provinces, autonomous regions, or municipalities directly under the central government (like Beijing, Shanghai, Tianjin, and Chongqing). Prior to 1963, a town was defined as a settlement with more than 2,000 permanent residents, of which 50 percent or more were non-agricultural population. A revision of the definition was made in 1964. By the new definition, a town was a settlement with more than 3,000 permanent residents, of which 70 percent or more were non-agricultural population, or a settlement with permanent residents between 2,500 and 3,000, of which 85 percent or more were non-agricultural population. Further adjustment was made in 1984: a town is defined as (1) a settlement being the location of a county-government agency, or (2) a township with a total population less than 20,000, where the non-agricultural population in the location of a township government exceeds 2,000, or (3) a township with total population of more than 20,000, where the proportion of the non-agricultural population to the total population in the location of a township government is greater than 10 percent; or (4) a remote area, mountainous area, small-sized mining area, small harbor, tourism area, or border area with non-agricultural population less than 2,000. For detail see Chan (1994), and State Statistical Bureau, *China statistical yearbook 1997*, p. 86.

References

Beijing Review (1997) 'Big budget inked for environment', February 17–March 3, 1997, p. 5.

Bradbury, I. and Kirkby, R. (1996) 'China's Agenda 21: a critique', *Applied Geography*, 16(2): 97–107.

Brown, L. (1996) *Tough choices: facing the challenge of food scarcity*, New York: W.W. Norton and Company.

Chan, K.W. (1994) 'Urbanization and rural–urban migration in China since 1982', *Modern China*, 20(3): 243–281.

Chen, Jinghua (1994) *Environmental action plan of China: 1991–2000*, produced by National Environmental Protection Agency and State Planning Commission, People's Republic of China, Beijing: China Environmental Science Press.

The China Daily News (1997) 'China to close polluting thermal power plants', December 29, Beijing.

Chinese Academy of Social Science (ed.) (1985) *Population yearbook of China*, Beijing: Zhongguo Zhehui Kexue Chubanshe (in Chinese).

Chinese Government (1994) *China's agenda 21: white paper on China's population*, *environment, and development in the 21st century*, Beijing: China's Environmental Science Press.

The Economist (1994) 'Energy: the new prize', June 18, pp. 5–10, 13–20.

Edmonds, R.L. (1994) *Patterns of China's lost harmony: a survey of the country's environmental degradation and protection*, London: Routledge.

Ehrlich, A.H. (1996) 'Toward a sustainable global population', in D.C. Pirages (ed.) *Building sustainable societies: a blueprint for a post-industrial world*, Armonk, New York: M.E. Sharpe, Inc.

Environmental Yearbook Editorial Committee (1995) *Environmental yearbook of China*, Beijing: Chinese Environmental Yearbook Press.

Fearnside, P.M. (1988) 'China's Three Gorges Dam: "fatal" project or step toward modernization?', *World Development*, 16(5): 615–630.

Feng, J. (1997) 'China protects cultivated land', *Beijing Review*, March 31–April 6, pp. 15–18.

Gaag, J. van der (1984) *Private household consumption in China*, World Bank Staff Working Papers, No. 701, Washington, DC: The World Bank.

Gao, G. (1997) 'Post Deng's China – the thoughts of paramount political consultation: an interview with the members of the National Political Consultative Conference', *Huasheng Yuebao* (in Chinese) 6.

Guangzhou's Agenda 21 Leading Group Office (1997) 'Implementation of sustainable development strategies and policies in Guangzhou', Paper presented at the Workshop of Sustainable Development in the Pearl River Delta and the Georgia Basin, 8–10 December 1997, Guangzhou City, China.

Jiang, Zemin (1997) 'Hold high the great banner of Deng Xiaoping theory for an all-round advancement of the cause of building socialism with Chinese characteristics to the 21st century', A Report at the 15th Party Congress, Beijing, China.

Jiangsu Statistics Bureau (1996) *Jiangsu statistical yearbook 1996*, Beijing: China Statistical Publishing House.

Kingsley, G.T., Ferguson, B.W., Bower, B.T. and Dice, S.R. (1994) *Managing urban environmental quality in Asia*, Washington, DC: The World Bank.

Kirkby, R.J.R. (1985) *Urbanization in China: town and country in a developing economy 1949–2000 AD*, London: Croom Helm.

Lin, G.C.S. (1997) *Red capitalism in south China*. Vancouver: UBC Press.

Liu, Chunzhen (1996) 'On the sensitivity and vulnerability of water resources to climate change in China', in Y. Yin, M. Sanderson and G. Tian (eds.) *Climate change impact assessment and adaptation option evaluation: Chinese and Canadian perspectives*, Beijing.

Liu, W.D. and Lu, D.D. (1996) 'Industrial and urban water use and policy', in C.M. Liu and X.W. He (eds.) *China's water resources strategy in the 21st century*, Beijing: Science Press (in Chinese).

Lu, Yingzhong (1993) *Fueling one billion: an insider's story of Chinese energy policy development*, Washington, DC: Washington Institute Press.

Muldavin, Joshua S.S. (1997) 'Environmental degradation in Heilongjiang: policy reform and agrarian dynamics in China's new hybrid economy', *Annals of the Association of American Geographers,* 87(4): 579–613.

NEPA (National Environmental Protection Agency) (1996) *Selected documents from the fourth national environmental conference*, Beijing.

NEPA (National Environmental Protection Agency) (1997) *1996 Annual Report of China's Environmental Quality*, Beijing.

Qu, Geping and Li, Jinchang (1994) *Population and the environment in China*, Boulder: Lynne Rienner Publishers, Inc.

Robinson, R. and Goodman, D. (eds) (1996) *The new rich in Asia: mobile phones, MacDonald's and middle class revolution*, New York: Routledge.

Shanghai Municipal Statistics Bureau (1996) *Statistical yearbook of Shanghai 1996*, Beijing: China Statistical Publishing House (in Chinese).

Smil, Vaclav (1994) 'China's greenhouse gas emissions', *Global Environmental Change*, 4(4): 325–332.

Song, Jian (1997) 'No impasse for China's development', *Beijing Review*, May 12–18, pp. 18–22.

SSB (State Statistical Bureau of The People's Republic of China) (1997a) 'Statistical communique on socio-economic development in 1996', *Beijing Review*, March 10–16, pp. 23–33.

SSB (State Statistical Bureau of The People's Republic of China) (1997b) *Tabulations of China 1% Population Sample Survey, National Volume*, Office for National Population Survey, Beijing: Statistical Press of China (in Chinese).

SSB (State Statistical Bureau of The People's Republic of China), *China statistical yearbook* (various years), Beijing: China Statistical Publishing House (in Chinese).

Wackernagel, M. and Rees, W.E. (1996) *Our ecological footprint: reducing human impact on the earth*, Gabriola Island, B.C.: New Society Publishers.

WCED (1987) *Our common future, report of the world commission on environment and development*, Oxford and New York: Oxford University Press.

World Bank (1997a) *China 2020: a supplementary report*, Washington, DC: The World Bank.

World Bank (1997b) *Clear water, blue skies: China's environment in the new century*, Washington, DC: Oxford University Press.

Yao, Shi-mou and Chen, Shuang (1996) 'The trend of urban spatial evolution in the Yangtze river delta', Paper presented at the International Conference on Rural–Urban Transition and Development in China, December 10, Guangzhou, China (in Chinese).

Yeh, A. G. O. and Li, Xia (1996) 'An integrated remote sensing and GIS approach in the monitoring and evaluation of rapid urban growth for sustainable development in the Pearl River Delta, China', *International Planning Studies*, 2(2): 195–222.

Zhang, G. D. (1991) 'Earlier construction of the Yangtze River Three Gorges project is appropriate', *People's Daily* (Overseas Edition), December 21 (in Chinese).

Zhang, Z., Chen, B., Chen, Z. and Xu, X. (1992) 'Challenges to and opportunities for development of China's water resources in the 21st century', *Water International*, 17: 21–27.

Zhao, Songqiao (1994) *Geography of China: environment, resources, population, and development*, New York: John Wiley & Sons, Inc.

Zhejiang Statistics Bureau (1996) *Zhejiang statistical yearbook 1996*, Beijing: China Statistical Publishing House (in Chinese).

9   Agenda 21 and urban India

Asesh Kumar Maitra
and Arvind Krishan

Introduction

India contains 16 percent of the world's population with 2 percent of the total land mass. There is, therefore, tremendous competition for the use of land. Urban areas and human habitation occupy 6 percent of this land and, due to the growing population, this percentage will increase. Therefore a trade-off between urban and non-urban land uses is inevitable. This severe competition for land is also causing environmental degradation and deterioration in the quality of life.

Although India is still largely rural (74 percent of the population), the urban population (214 million in 1991) is growing rapidly and most of the urban centres have become bloated in terms of population without commensurate improvement in infrastructure and other facilities. This situation creates a deteriorating urban environment. A massive programme to improve India's capacity to manage the urban environment is under way. Equally significant are the economic reforms initiated in India since 1991, designed to provide added momentum for economic growth. Briefly the economic reforms are aimed at (a) liberalization of the investment regime, (b) liberalization of trade to encourage foreign investment, (c) skilful and significant liberalization of the financial sector, and (d) simplification and strengthening of the tax regime.

In this chapter we first briefly consider the basis of environmental conservation in India's cultural traditions and Constitution. We then outline some of the conservation programmes which have been introduced since Independence, touching on their urban dimension. We discuss the immense dilemma which massive and rapid urbanization poses for the nation and its environmental responsibilities, and we review problematic aspects of the urban environment. Finally we consider aspects of the government's programme in response to the Rio Earth Summit and some potential pathways to ecologically sustainable urban development, which remain as yet far from fully developed.

Environmental conservation in tradition and in the constitution

Conservation of environmental resources is part of India's traditional ethos. The ancients recognized Panch (five) Bhutas (elements); namely, Kshiti (earth, soil, etc.), Aap (water), Tej (the sun's energy and energy sources in general), Marut (air) and Bom (the sky and climate) as the constituents of the physical environment. These elements commanded respect. Social mores helped to protect the environment and its biological diversity. For example, the Himalayas were considered to be the abode of the gods, and this belief prevented denudation of the forests, thus saving the fertile plains of northern India from siltation and flooding. Similarly, restriction on the consumption of riverine fish in certain months ensured that fishing was restricted during the spawning season. Although many such restrictions appear as taboos, the wisdom behind them is being rediscovered. These restrictions, which promote the conservation of nature, are accepted by the people as part of living.

Cognizant of these traditions, the Constitution of India (1949) provides for conservation of the environment. Among the 'fundamental duties of citizens' are the following: (1) 'to value and preserve the rich heritage and composite culture' (clause 51A (f)), and (2) 'to protect and improve the natural environment, including forests, lakes, rivers and wild life, and to have compassion for living creatures' (clause 51A (g)). In India's federal structure responsibilities in different spheres of activity are defined for the Union and for the States. The 'Concurrent List' enables both levels of government to legislate in the same area. However, with the consent of a majority of the States, the will of the central government may supersede the provisions enacted by a State. In 1976 an additional clause was inserted in the Constitution under the Directive Principles of State Policy: 'The State shall endeavor to protect and improve the environment and to safeguard the forests and wild life of the country' [clause 48A]. This amendment (the 52nd Amendment to the Constitution) was brought under the Concurrent List (Seventh Schedule) and thus enabled the Parliament of India and the Indian Government to legislate on these matters. This provision encourages a consistent policy to conserve the environment through uniform legislation all over the country.

Immediately after the 1972 Stockholm conference, a National Committee for Environmental Planning and Co-ordination (NCEPC) was set up, chaired by Indira Gandhi, then Prime Minister. The committee was to be advised and aided by the Department of Science and Technology. In 1974 and 1981, two important laws were passed: the Water (Prevention and Control of Pollution) Act, and the Air (Prevention and Control of Pollution) Act. These laws not only provided standards of permitted discharge, they provided powers to investigate and penalize offenders under the Indian Penal Code (IPC). An autonomous board to implement the legislation was established, called the Central Pollution Control Board (CPCB), with independent executive powers. Subsequently the Ministry of Environment, under a Minister of Cabinet rank, was established in 1986, and the Environment (Protection) Act, 1986, provided the Ministry with wide ranging powers.

India has been faced with the dilemma, which was recognized at the Rio Earth Summit, between ensuring an adequate pace of development and the need for conservation of the country's environmental resources. Today there is evidence to suggest that despite consistent efforts through legislation over the last fifty years, the rate of environmental degradation has escalated. It is also understood that the country's limited natural resource base is unable to sustain an ever-growing population. Population increase, employment and health needs, and the emancipation of women are part of the same web of development. The vast number of unemployed and underemployed, hinders the adoption of those more sophisticated capital-intensive technologies which might in the longer term benefit the environment. But at the same time the country has to guard against the dumping of environmentally unsafe technologies from the developed countries.

Environmental conservation programmes

Programmes for environmental conservation have been continuously introduced since Independence through initiatives taken by the Ministry of Environment, by other sectoral Ministries and Departments of Government, or by non-governmental organizations and entrepreneurs. India has 75 national parks and 421 wildlife sanctuaries which cover an area of 140,675 sq. km. Many urban centres are located close to national parks, wild life sanctuaries and wetlands (protected under the Ramsar Convention). Environmental impact assessment (EIA) is now obligatory for some types of urban projects listed in the Government Gazette, which is periodically updated with increasing awareness. Measurement of the 'carrying capacity' of urban areas is also undertaken as a tool for the planning of urban areas.[1]

The Government of India, has taken significant steps to reduce environmental pollution in a few of the highly industrialized states – Gujrat, Maharasthra, Tamil Nadu and Uttar Pradesh – through World Bank assisted projects. These measures included strengthening the Pollution Control Board, supporting the installation of pollution control equipment in industrial units, establishing Common Effluent Treatment Plants for small industries, and introducing cleaner technologies. Some of the major cities in India have established pollution monitoring stations. A new law dealing with storage, manufacture, disposal and transportation of hazardous substances has been enacted. Under the law industries identified as hazardous are not permitted within 25 kilometres of the boundary of an urban centre.

The Indian courts have been active in supporting environmental improvement. Existing industries of a hazardous nature are being asked to shift out of urban centres (Supreme Court judgment, 1996). For example 168 industrial plants in Delhi had been ordered by the Supreme Court to close down in 1996. Another 1,028 plants were asked to close down or shift by the end of 1997. Although this action causes considerable social tension, the government-backed by the judicial system, is quite firm in its resolve. The discharge of effluent into rivers and water courses must conform to specified standards, although the rate of prosecution for non-compliance is, so far, low. The municipality of Delhi has been ordered by the Supreme Court of India to stop the discharge of effluents into the Yamuna river by 1998.

Education and public awareness is seen as one of the main vehicles to ensure environmental conservation. India introduced environmental education in the school curriculum in the early 1960s, through the National Council for Educational Research and Training (NCERT), in the lower and middle school levels. At university level, there has been a dramatic rise in environment related courses with over sixty universities and institutions offering degrees with environmental specialization. Yet there is no attempt to infuse environmental concerns into the conventional subjects, and as a result the level of environmental awareness among engineers, doctors, physicists, and chemists remains poor. Non-formal education on the other hand has received considerable support. The government started non-formal education in the mid-1960s. The National Environment Awareness Campaign started by the Ministry of Environment and Forests has attracted participation by more than 2,500 voluntary organizations across the country (Warrier and Kothari, 1992).

Energy production is one of the most critical sectors of the Indian economy. It is estimated that energy demand will be more than double the production capacity in the foreseeable future. Since conventional technologies of energy production are environment unfriendly, emphasis has shifted to production of energy through non-conventional sources such as wind, sun, etc. Several initiatives are being taken to reduce consumption of energy by designing energy-responsive buildings and habitats (see below).

The urban dilemma

The volume and pace of urbanization in India contributes to the degradation of urban environments. In the last three decades, the urban population has more than doubled. The number of cities with more than 100,000 population has nearly trebled. The process of urbanization consumes ever greater amounts of environmental resources: land (for construction and construction material), water (for consumption and for irrigation), and air (as a sink for emitted gas). Since land is finite, there is intense competition for its use. As monetary land values increase, the poor are invariably priced out. Low-lying lands, riverbeds, unstable slopes, and garbage heaps at the edge of the city have become the habitats of the urban poor. All major Indian cities accommodate between 30 and 40 percent of their urban poor in slums and scarcely habitable settlements. Yet India is on the threshold of further massive urban growth.

Urbanization in India is generated from small settlements in a diverse topography and climate. Traditionally settlement planning has been rich in its response to nature and the environment. The ancient settlements of Varanasi, Patliputra, Takhsashila, Indraprastha and many others testify to careful settlement planning with nature. The cities of Jaisalmer and Jaipur in Rajasthan, India, are unique examples of cities which have developed in a very hostile climate but have survived over the centuries as vibrant urban centres.

In the first 30 years of this century the 'urban' proportion of the total population was less than 12 percent. Since then, however, urbanization has increased steadily from about 17 percent in 1951 to over 25 percent in 1991. The pace of urban growth has also been high during this period, and the number of urban centres has increased from 2,795 in 1951 to 3,609 in 1991. Of course we must be clear about how 'urban' is defined. An area is considered 'urban' in national census calculations if it fulfils the following criteria: (1) it has a population of more than 5,000, (2) a population density of not less than 400 persons per sq. km., (3) at least 75 percent of its male working population is engaged in sectors other than agriculture, (4) it has some form of local government.

While urbanization in the above terms is increasing, so also is the number of metropolitan cities (cities with more than one million population). In 1901 there was just one such city, in 1991 there were 23 (Table 9.1). There has been, moreover, a marked shift in the population *balance* between rural and urban areas (Table 9.3). Projections of the urban population for the next three decades vary, but most suggest that the urban population will reach 40 percent of the total population by 2011. At that point there will be an urban population of approximately 500 million. Containing this population explosion presents the most significant problem in formulating India's Agenda 21 policies.

Table 9.1 The number of metropolitan cities in India 1901–1991

Year	Metro-cities	Population ('000)	Increase	Proportion of metro-cities as % of India's population	
				Urban pop. (%)	Total pop. (%)
1901	1	1,510	0	5.8	0.6
1911	2	2,764	1,254	10.7	1.1
1921	2	3,130	366	11.1	1.3
1931	2	3,407	277	10.2	1.2
1941	2	5,308	1,901	12.0	1.7
1951	5	11,747	6,439	18.8	3.3
1961	7	18,102	6,355	22.9	4.1
1971	9	27,831	9,729	25.5	5.1
1981	12	42,122	14,291	26.4	6.2
1991	23	70,661	28,539	32.5	8.4

Source: Census of India (1991).

Table 9.2   Size classes of cities

Classification	Population
Mega-cities	5,000,000 and above
Metropolis	1,000,000 and above
Class I	100,000 to 999,000
Class II	50,000 to 99,000
Class III	20,000 to 49,999
Class IV	10,000 to 19,999
Class V	5,000 to 9,999
Class VI	Less than 5,000, with a pronounced urban character

Source: Census of India.

Cities tend to form centres for all major economic activities and they attract migratory population from their hinterlands. Cosmopolitan cities like Delhi have a very high rate of immigration. According to one estimate, Delhi attracts almost 150,000 people annually from the rural areas. Population growth due both to immigration and natural growth of the existing population results in a tremendous pressure on the entire urban infrastructure, consuming the hinterland and generating environmental degradation within metropolitan and smaller cities.

Studies of different classes of cities (Table 9.2) reveal that larger cities grew at a faster rate while cities of class IV to VI are registering negative rates of growth (Census of India, 1951, 1961, 1971, 1981, 1991). The population balance between metropolitan cities and smaller cities is beginning to break down with detriment to the level of services provided by the urban centres of metropolitan hinterlands. This phenomenon was noticed in the 1970s and the Indian Government promulgated a programme to encourage the growth of cities below 100,000 population by providing several fiscal incentives through the 'Integrated Development of Small and Medium Towns' programme (IDSMT). But these fiscal incentives were too meagre to make any meaningful impact on the deteriorating conditions, and any improvements in this respect were not due to IDSMT. Cities of 80,000 (plus) population have registered higher rates of growth of population in general. Also there was no consistent policy in selecting urban centres to be supported. The ad-hoc nature of policy resulted in investment without significant effect.

The mega-cities of India (namely, Mumbai, Calcutta, Chennai, Hyderabad and Bangalore) have been brought under a separate Mega Cities project to improve their infrastructure, provide habitable housing, control pollution and improve the living environment in general. For Delhi a 'National Capital Region' (NCR), has been delineated by Act of the Parliament of India, and a National Capital Region Planning Board has been constituted, with the Prime Minister as its Chairman and the Chief Ministers of the Indian states as its members. The main objective of the NCR Plan is to distribute potential immigrants to priority towns identified within a 100 km radius of the

Table 9.3 Rural and urban population

Year	Population (millions)		Percentage of total population	
	Rural	Urban	Rural	Urban
1901	213	26	89.1	10.9
1911	226	26	89.7	10.3
1921	223	28	88.8	11.2
1931	245	33	88.1	11.9
1941	274	44	86.2	13.8
1951	298	62	82.8	17.2
1961	360	79	82.0	18.0
1971	439	109	80.1	19.9
1981	525	160	76.7	23.3
1991	629	218	74.3	25.7

Source: Census of India (1991).
Note: An area is considered 'urban' if it fulfils the following criteria: (1) it has a population of more than 5,000, (2) a population density of not less than 400 persons per sq. km., (3) at least 75 percent of its male working population is engaged in sectors other than agriculture, (4) it has some form of local government.

capital. Six other cities located at a strategic distance from Delhi, have been identified as growth centres. These are expected to take the pressure of population growth off Delhi. Regional development plans for all the metropolitan cities are under preparation. Unfortunately, however, the town planning objectives are not matched by the investment priorities of the government, and the initiatives promulgated have yet to show much positive impact.

Inadequate implementation of the physical aspects of town plans is compounded by the poor rate of investment from the 'liberalized' economic institutions. During the last three decades various public and private sector organizations, such as the Housing and Urban Development Corporation (HUDCO), Housing Development and Finance Corporation (HDFC), National Housing Bank (NHB), Life Insurance Corporation of India (LIC), etc. have provided financial support to development organizations in both public and private sectors. Although the total amount of finance available has been far less than required, the records show some impressive gains. In order to create greater liquidity in the debt market, several international agencies have been allowed to operate in the development of urban areas and infrastructure. Infrastructure Leasing and Financing Services (ILFS) is one of them, which aims to support financially viable urban projects. However, international investment in India during the last five years has not exceeded 5 percent of the total. Thus, in this sense at least, the hype of 'globalization' as saviour of India is more a myth than a reality.

Indeed some have argued that globalization, where it has penetrated, has brought great ecological costs (Shiva *et al.*, 1997). For example, the export of cattle and meat as a result of trade liberalization has led to increased imports of fertilizers, fossil fuel, tractors and trucks. Liberalization has also opened up a trade in toxic and hazardous wastes, even though the export of such wastes is banned under the Basel Convention and by the Supreme Court of India. 'Toxic waste such as cyanide, mercury and arsenic is being shipped as "recyclable waste" – a deliberate attempt to mislead and disguise the true nature of the wastes' (ibid.: 50). In 1996 Australia exported 1,450 tonnes of scrap batteries, copper and zinc ash to Mumbai (see Daly, 1996: A9).

Globalization has had a detrimental effect on the urban situation in a more general sense. First, the smaller urban areas tend to be ignored by international investors in preference to larger ones. Secondly there has been an explosion in the number of vehicles on the roads as a result of opening up the vehicle manufacturing sector to international firms. These firms have dumped outdated vehicle manufacturing technologies in Indian cities. Old technology produces polluting engines. Together these factors have led to very high levels of air pollution and traffic chaos. The average speed of traffic across a typical metropolis is about 10–15 km/hr (e.g. Delhi 15 km/hr, Mumbai 10–12 km/hr, Calcutta 10 km/hr).

The industrial location strategy, which has aimed to redistribute major industries, has saved the majority of urban centres from environmental pollution through development of industrial towns in deprived regions, and industrial estates in carefully chosen out-of-city locations. The tragedy in Bhopal (1984) where Union Carbide had stored lethal poison is an exception. Studies of air pollution show that industries contribute a minor share of urban air pollution (see Central Pollution Control Board: *Air Pollution Study*). On the other hand the schemes to promote small-scale industries have resulted in uncontrolled industrialization in major metropolises and the resultant mixed land use has caused serious hazard.

The continuously deteriorating urban environment has been a cause of serious concern. The need for housing and other infrastructure arising from increasing population, coupled with the inability of municipal governments to deliver these goods, creates a major problem. Despite the constitutional amendment empowering local government to provide for such needs, funding is inadequate. Mere empowerment without financial provision is like asking a painter to make a picture without paints. The Indian Government has progressively decreased its share of investment in the urban sector (Table 9.4). The sector still does not constitute a political force to reverse the trend, and the few capital cities with political power siphon off most of the available investment.

The urban environment

Measurement of the status of the urban environment is difficult and the methods disputed. Although several parameters have been introduced for measurement of urban environmental quality, consistent data on all the factors is not available and the few data available are insufficient for objective prognosis. Therefore this review is restricted to basic urban infrastructure.

Table 9.4 Five-year plans: share of housing in total investment

Plan[a]	Public	Private	Total	Public	Private	Total	Share[b] (%)	Share[c] of Pub. Inv. (%)
	Total investment (Rs.100 million)			*Total investment in housing (Rs.100 million)*				
I	158	180	338	25	90	115	34.0	15.8
II	365	310	675	30	100	130	19.3	8.2
III	610	430	1,040	42	112	154	14.8	6.9
IV	1,365	693	2,058	62	317	379	18.4	4.5
V	3,140	1,616	4,756	104	363	467	9.8	3.3
VI	9,750	7,471	17,221	1,491	150	1,641	9.5	15.3
VII	16,814	16,000	32,814	245	2,900	3,145	9.6	1.5

Source: Five-year plans: Planning Commission in India, Government of India.
Notes:
a   Five year plans numbers I to IV.
b   Total investment in housing as a proportion of total investment.
c   Public investment in housing as a proportion of public investment.

Water resources

Water is a vital environmental endowment. On the basis of average rainfall, gross estimates of water resources indicate that there is sufficient water available in the country. However, due to diverse topography and the existence of several climatic regions, this resource is unevenly distributed. The careful management of water resources is thus critical for most of India. A government survey of 26 major urban centres in India revealed that the average supply of municipal water varies between 6 to 8 hours per day (Sivaramakrishnan, 1993). The study by the National Institute of Urban Affairs shows that only 44 percent of the urban poor have access to treated water supply, others use shallow tube wells (16 percent), open wells (21 percent) and the rest (19 percent) can only depend on totally unsafe water.

Even where a safe water supply is available, the disposal of waste water is of serious concern. Not even one city in the entire country has a fully hygienic waste water disposal system for the entire population. Most of the waste water is discharged untreated directly into rivers and other watercourses. A very large number of people are affected every year by water-borne diseases, with periodic epidemics of cholera, typhoid, bacillary dysentery, etc. The causes can be traced to the lack of sanitary disposal of waste, the non-availability of protected water supply and polluted sources of water.

The management of the urban environment is basically vested in the local level of government, but the overall effectiveness of environmental performance, and thus monitoring, is the responsibility of the national and state governments. However, because of the overlapping functions of different levels of authority there is delay in both decision and implementation. Providing potable water supply and safe disposal of waste is part of the municipal function. Protection of rivers and watercourses from pollution is a State function, since rivers flow through several territorial jurisdictions. Monitoring of river water quality and prevention of pollution thus became one of the major functions of the Central Pollution Control Board. A series of monitoring stations was established to assess water quality continuously over different stretches of the rivers. The data collected revealed that all the 14 major rivers of India are polluted to different degrees according to internationally accepted (e.g World Health Organization) criteria.

The level of pollution was greater near the urban centres (CPCB, 1994) although the *per capita* pollutant load of smaller settlements is less than that of larger cities (CPCB, 1994b). Studies of water pollution reveal that untreated discharge of effluent from industries, as well as from domestic sources, is the major cause of riverine water pollution. A survey undertaken by the CPCB shows that, on average, 90 percent of the water supply of Class II cities is polluted (*ibid.*). Only 1.6 percent of urban waste water is treated before discharge.

India has a coastline extending over 6,000 km. Coastal and estuarine water quality data show that estuaries adjacent to major industrial centres are highly polluted, e.g. Thana creek and Mahim creek in Mumbai, with high concentrations of lead and cadmium (820 microgrammes per litre and 336 microgrammes per litre respectively). The Kerala coastal water shows a high concentration of petroleum hydrocarbons. Protection of the coastal environment is one of the major priorities of the national government, for which separate sets of legislation and guidelines have been issued (CPCB, 1996). Cities located along the coast are now given high priority for control of the discharge of effluents.

Air quality

The Air Act 1981 prescribed the desirable ambient air quality and also the permitted maximum levels of discharge of pollutants into the atmosphere. Monitoring stations have been set up by CPCB and NEERI (National Environmental Engineering Research Institute) in some major cities and environmentally critical areas. It has been observed that the air quality of some of the major metropolises is below the prescribed norms. In non-metropolitan areas, CPCB has identified 22 problem zones which are prone to industrial pollution. Among these certain areas attract special attention: Korba in Madhya Pradesh, Talcher in Orissa, Visakapatnam in Andhra Pradesh, Manali in Tamil Nadu, Udyogmandala in Kerala, Chembur in Maharasthra, Dhanbad in Bihar and Durgapur in West Bengal.

The major source of urban air pollution is dependent on the activity and affluence of the particular city. In Delhi, for example, most air pollution is contributed by automobiles, while in Calcutta domestic coal burning ovens are the worst source of pollution. However, except in a few cities, the level of air pollution is much lower than in industrialized nations (pollution data in Udaipur/Jodhpur CPCB report). To some degree this is a reflection of

Table 9.5 The percentage of urban households and population covered by water supply

India/States	Households covered by Safe drinking water (%)		Population covered by water supply (%)	
	1981	1991	1987	1992
India	75.06*	81.38†	79.20	84.65
Andhra Pradesh	63.27	73.82	62.40	78.00
Arunachal Pradesh	87.93	88.20	100.00	100.00
Assam	n/a	64.07	37.50	40.00
Bihar	65.36	73.39	63.60	70.00
Goa	52.31	61.71	81.50	92.50
Gujrat	86.78	87.23	93.40	98.00
Haryana	90.72	93.18	100.00	100.00
Himachal Pradesh	89.56	91.93	92.90	100.00
Jammu and Kashmir	86.67	n/a	95.00	98.00
Karnataka	74.40	81.38	98.70	96.00
Kerala	39.72	38.68	65.60	75.00
Madhya Pradesh	66.65	79.45	80.50	88.00
Maharashtra	85.56	90.50	99.70	98.00
Manipur	38.71	52.10	75.50	71.00
Meghalaya	74.40	75.42	49.50	100.00
Mizoram	8.79	19.88	18.60	79.00
Nagaland	57.18	45.47	19.90	63.00
Orissa	51.33	62.83	37.10	50.00
Punjab	91.13	94.24	71.20	71.00
Rajasthan	78.65	86.51	54.40	100.00
Sikkim	71.93	92.95	67.10	74.00
Tamil Nadu	69.44	74.17	88.20	49.00
Tripura	67.92	71.12	53.20	53.00
Union territories				
Andaman and Nicobar	91.95	90.91	100.00	86.00
Chandigarh	99.39	97.68	100.00	100.00
Dada and Nagar Haveli	54.35	90.97	73.70	100.00
Daman and Diu	67.04	86.76		100.00
Delhi	94.91	96.24	97.00	99.00
Lakshwadeep	3.65	18.79		
Pondicherry	84.18	86.05	100.00	95.00

Source: Annual Report, Ministry of Finance, Government of India.
Notes: *Excluding Assam. †Excluding Assam, Jammu and Kashmir.

low energy consumption. At the time of independence per capita consumption of energy in India was among the lowest in the world. Dependence on non-commercial energy sources, such as fuelwood, food residue, animal dung, etc. was considerable but is changing in favour of commercial energy

In order to progress, commercial generation of energy was seen as a major infrastructural support for development. In the pattern of the TVA in America (Tennessee Valley Authority), plans were made to harness the rivers of India to generate hydroelectric power, and at the same time to create a national grid for power distribution supported by thermal power stations. These thermal power stations were located close to coal mines in order to minimise transportation costs. Also, major thermal power stations were located near the metropolises to cater for the demand for electricity. Due to the high ash content of indigenous coal, as well as high sulphur content, these thermal power plants are major pollutants of air, water and land and the disposal of fly ash has become a major concern.

The main sources of air pollution in urban centres, therefore, are the thermal power stations and automobiles. Pollution by industries is comparatively lower, except in a few sensitive areas. On an international scale local air pollution in urban India, with some exceptions, is well within acceptable limits.

Land use and management

Although a substantial proportion of the land in India is hills and desert, nearly 54 percent of the land is used for agriculture. This is a very high percentage compared to the world average and has been achieved by converting wasteland, and other unused land, to agriculture. There is no further scope to bring more land under agriculture. The land under habitation on the other hand shows a sudden increase in the last two decades (e.g. from 4.09 percent in 1961 to 6.2 percent in 1981; the 1991 figure is available but requires computation from raw data). This increase has entirely happened by the conversion of agricultural land, and other ecologically sensitive land, to habitation.

While the growth rate of the population of India is slowly declining (2.14 percent between 1981–91, down from 2.22 percent in the previous decade), the urban population is growing at twice the rate of national population growth, severely straining the supporting environmental resources. Almost all urban centres are located on and are surrounded by good quality agricultural land, wet lands or other ecologically fragile ecosystems. The pressure for annexing more land for habitation is thus high. According to a 1986 study of land use by the Government of India, the per capita availability of land would decline from 0.89 hectares per person to 0.32 hectares per person at the current rate of population growth, while the land available per head of cattle would decline from 0.51 hectares per head to 0.24 hectares per head by 2021. The declining ratio of natural resources (especially for food production) *vis-à-vis* population is of serious concern.

Management of urban land is attempted through the application of town and country planning legislation borrowed from the United Kingdom. Although the system has been in operation for 40 years, it has failed miserably to solve India's urban environmental problems. It has failed to earn the trust of the people and is generally misused for the benefit of the few who are able to manipulate the system. On the other hand, the amendments to the Constitution of India have placed the responsibility for management of the urban environment on local government, which is not equipped with sufficient staff or financial resources to affect improvement.

The failure of conventional 'town and country planning' has resulted in several other facets and dimensions of the urban problem being taken into consideration, i.e. environmental resources of both the catchment region as well as within the urban area, categorization of land use based on environmental concerns, urban landscape, sustainable infrastructure development, employment, eradication of poverty, affordable housing and energy sources. However, bureaucratic inertia creates a serious bottleneck which is yet to be overcome. Despite Rio and later Habitat II, the planning, development and management of India's human habitat is still not adequately environment-conscious.

Solid waste disposal

The growth of the urban population has resulted in the production of growing quantities of solid waste. Until the packaging industry made its presence felt, urban garbage mainly consisted of domestic bio-degradable waste. Except in large metropolises (i.e. Delhi, Calcutta, Mumbai, Madras) there was no organized garbage collection, as the compostable garbage was used by households who had a considerable amount of land attached to their houses.

Increasing urban densities resulted in the need for organized garbage removal, which municipal authorities were not competent to undertake. The highly putrescible garbage created serious health hazards. Several cities have now handed garbage collection and disposal to private contractors. Cities such as Surat (90 percent), Rajkot (100 percent), Hyderabad (80 percent), and Chennai (80 percent), have succeeded in managing their solid waste disposal problems. Surat, which was one of the dirtiest cities of the country due to its poor management of garbage, has become the cleanest city, with co-operation among people and an efficient collection and disposal system using conventional methods over a period of two years only. Experiments with appropriate technologies (such as pelletization, bio-gas generation, and composting) are being conducted. Although it is premature to evaluate these measures, indicators suggest that in this particular area of urban services privatization may have been successful.

Energy production

Generation and distribution of electricity has been, in the main, kept in the public sector, although cities such as Calcutta (Calcutta Electricity Supply Corporation, CESC) and Mumbai (Bombay Electricity Supply and Transportation, BEST) had private companies which generated, supplied and earned revenue as commercial enterprises. One of the fallouts of liberalization has been the opening up of electricity generation to the open market, and particularly to international investors (there are several options for the form of privatization of electricity production and distribution).

Privatization of electricity production has been subject to intense criticism, particularly from the environment lobby. There is widespread suspicion that the differential between production cost and the revenue to be earned disproportionately favours the multinational companies. Comparison of electricity charges with those of developed countries tends to support this suspicion (e.g. in California electricity costs 9 cents per kw, while in India it costs 3 Rupees per kw (appoximately 10 cents per kw) although the labour is far cheaper). Political and popular dissension is delaying implementation of these projects. Thus the gap between production and growing demand is widening. Most cities do not have a 24-hour power supply and even major metropolises and mega-cities are subject to routine power cuts, despite the fact that consumption per capita is about one-tenth that of developed countries and at least 20 percent of the urban population are not connected to the power grid.

Administrative and legislative reforms after the Earth Summit 1992

In order to facilitate the growth of urban infrastructure and better management of the urban environment, two major initiatives have been taken by the Government of India. The first is to decentralize governance to achieve the objectives of Local Agenda 21, and, second, the introduction of a National Environmental Action Programme.

Decentralization of governance

Reform of local government to encourage the participation of people in decision-making is one of the most significant steps towards ecological sustainability and environmental conservation. The principle of 'subsidiarity' is fundamental to environmental governance and acknowledges that 'decision-making should be kept as close to the grassroots level of governance as circumstances permit' (Hempel, 1996: 189).

This reform was brought about by amendment of the Constitution to recognize local government as the legitimate decision-maker and implementer of policy at local level (Amendments 73 and 74). These amendments seek to strengthen local democracy and thus ensure decentralization of governance. It is worthy of note that this initiative was formalized on June 1, 1993, when the Constitution Amendment Act 1992 came into force – a good two years ahead of the Lisbon Protocol (May 1, 1995) which advocated Local Agenda 21 and the role that local government should play. The main objectives of the amendments are the following:

1   The municipalities to be duly elected governments, with due representation of the weaker sections of the population, and a third of the seats reserved for women.
2   The municipalities to enjoy fiscal autonomy.
3   Managing the urban environment to be the responsibility of the municipalities.

4   The focus of planning to be developmental, rather than regulatory.

Similarly at the village level, the governance and management of the village would rest on duly elected representatives who would form bodies called Panchayats.

By these objectives the third tier in the hierarchy of the federal structure is truly recognized. The amendment recognizes only three types of municipalities; namely, (1) metropolitan areas, (2) smaller urban areas, and (3) transitional areas, between urban and rural. The latter form a twilight zone of development and a potential haven for environmentally degrading activities. These authorities will have Municipal Corporations, Municipal Councils and Nagar Panchayats, respectively. Recognizing the dynamic nature of urban growth, the amendment allows the incorporation of neighbouring zones, which may be rural, into the ambit of urban management to ensure effective development. In order to allow for decentralization within large cities (300,000 population or more) there is mandatory provision for Ward Committees.

The XIIth Schedule of the Constitution lists the following among the functions of local government: urban planning; regulation of land use and construction of buildings; planning for economic and social development; public health, sanitation, conservancy and solid waste management; urban forestry, protection of the environment and promotion of ecological aspects; safeguarding the interests of the weaker sections of society, including the handicapped; slum improvement; alleviation of urban poverty; provision of urban amenities and facilities such as parks, gardens and playgrounds; promotion of cultural, educational and aesthetic aspects. The weak financial position of local government is to be ameliorated by the setting up of State Finance Commissions which would recommend, (1) distribution between the State and municipalities of the net proceeds of taxes, duties, tolls and fees leviable by the State and which may be divided between them; (2) municipal collection of certain taxes and duties, and (3) grants-in-aid from the consolidated fund of the State.

During the period 1993–1997 all States of the Union of India have amended their Municipal Acts to ensure the implementation of the constitutional provisions. Elections to most municipalities have taken place. It has, however, been discovered that the municipal workforce is, on the whole, not capable of undertaking the assigned jobs because of lack of training. Municipal 'capacity building' is now a major task confronting the country. A massive exercise is now underway, with this aim partly funded by external donor agencies.

The Environmental Action Programme

The Environmental Action Programme, following India's agreement on Agenda 21, is a statement of resolve to integrate environmental considerations into development programmes and projects to achieve ecologically sustainable development. The objectives of the action programme are as follows:

1   To assess the environment scene in India against the backdrop of changing economic policies and programmes.
2   To review current policies and programmes which address the various environmental problems of the country.
3   To identify the future direction and thrust of these policies and programmes to establish priorities and outline a strategy for implementation of these priorities.
4   To identify programmes and projects for a sustained flow of investment resources for improved provisioning of environmental services for better quality of life and for integrating environmental concerns into development projects.
5   To identify projects for organizational strengthening for better environmental management.

The most important aspect of these action programmes is the realization that control over population growth, eradication of poverty, acceleration of the pace of economic development and equitable distribution of developmental benefits are all integral parts of environmental policy. Environmental conservation and improvement is now viewed as a multi-faceted programme, involving government at all levels, NGOs, the lay public, and teachers and students. Improving the urban environment cannot be achieved by investment in infrastructure alone. It requires a much wider appreciation of the environmental paradigm. Control of industrial location, management of water resources, protecting forests, wildlife and biodiversity need close intersectoral collaboration.

The main task of implementing and monitoring the environment policies falls to the Minister for Environment and Forests and the Central Pollution Control Board. The Ministry's promotional activities are carried out through: (1) the Ganga Project Directorate which supervises the implementation of the National River Action Plan and promotes schemes to prevent the discharge of untreated effluent into the rivers and water courses, (2) the National Afforestation and Eco-Development Board which provides support for project preparation and interaction with the government on national social forestry programmes. Several other ministries of the government share the responsibility in their respective areas: the Ministry of Rural Development, the Ministries of Power and Non-conventional Energy Sources, the Ministry of Water Resources, the Ministry of Agriculture, the Ministry of Urban Affairs and Employment, the Council of Scientific and Industrial Research (CSIR) and other industrial laboratories. The States of the Union have complementary Departments of Environment which carry out the tasks of environment conservation, protection, and prevention of pollution. They are responsible for ensuring environmental protection at source, as well as ensuring an acceptable quality of life to the people.

However, in areas, such as exploitation of mines and extraction of minerals, forests, nationally important reserves, major projects, the States need to co-ordinate with the central government. Industries of 34 categories have been listed which need environmental clearance, and the Ministry of Industries cannot permit establishment of these industries without environmental clearance. Although considerable decontrol over industrial development has become part of policy, the environmental considerations rule the location. Similarly, major civil engineering projects, such as roads, bridges, and harbours need an environmental clearance.

The coastal area in particular is subject to very severe environmental scrutiny, and several Coastal Regulation Zones have been prescribed which determine the kind of development that may be permitted in these areas. The administrative machinery for effecting these controls is exercised by the Ministry and Departments of Environment, which have also learnt that protection of the environment is best done through popular participation. Thus voluntary non-governmental organizations are supported by the government, and in many cases the task of monitoring is entrusted to them.

The future: planning with nature

The potential of appropriate technology and design

Alternative economic methods of making urban development and management both financially attractive and ecologically sustainable are being examined, but so far these ideas have proved to be barren. In the authors' view placing too much faith in economic policies to achieve environmental improvement courts disaster. Instead the emphasis should be shifted to alternative technologies and citizen action. Examples of appropriate technologies with potential for development are renewable and non-conventional energy sources (Table 9.6), alternative sanitation techniques and energy responsive buildings.

No Indian city is fully covered by a sewerage network and treatment system. Most cities and towns depend on dry latrines and septic tanks. The number of households estimated to be without toilets is 14.28 million in 1991 (Census, 1991); that is, about 20 percent of the urban population. The 'two pit pour flush latrine' developed in India does not require a drainage system and has been tested by the World Health Organization and concluded to be safe. The system, in modified form, has now been adopted by the UNDP. So far only 0.73 million latrines have been installed, hardly 5 percent of urban households. Other biotechnologies, such as bio-digesters and root zone treatment, are being attempted to avoid the very expensive technology of conventional sewage treatment, which the country can ill afford.

Energy efficient buildings illustrate what can be done with technology coupled with ecological design. Some examples, three building complexes

Table 9.6   Renewable energy potential and achievements

Technologies	Approximate potential	Achievements (up to 31 March) 1997
Power generation		
1 Wind power	20,000 MW	900 MW
2 Small hydro power (up to 3 MW)	10,000 MW	141 MW
3 Biomass power	17,000 MW	83 MW
4 Solar photovoltaic power	0.20 MW per sq. km.	28 MW
5 Urban and municipal wastes	1,700 MW	
(a) Installed		3.75 MW
(b) Being taken up		90.00 MW
Thermal applications		
6 Solar thermal systems	35 MW per sq. km.	
(a) Solar water heating	400,000 sq. m.	
(b) Solar cookers	430,000 units	
7 Biogas Plants	12 million	2.5 million
8 Improved biomass stoves (Chulha)	120 million	23.7 million

Source: MNES, Government of India

designed by one of the authors (Krishan) in the 1980s are based on the ethos of sustainable architecture. An office building at Shimla (Himachal Pradesh), completed recently, is a zero energy consuming building during its normal hours of operation. It uses solar energy for heating, ventilation and hot water supply. Daylight distribution has been carefully designed to eliminate the use of artificial lighting during daytime.

Planning with nature

For planning of any major development, to improve our living standard and quality of life, each and every aspect of nature needs to be harmoniously integrated to achieve sustainable development within ecological limits. Today this approach to planning is recognized as a viable option. Some projects have been developed which further illustrate this trend, although these examples are few and far between. Located in environmentally fragile areas, these projects were entrusted to groups of architects and planners in the early 1970s who evolved a process of design which respected nature and the imperatives that nature imposed. Instead of flattening a hilly rugged site, the buildings were designed to fit into them. The layout was evolved after careful analysis of topography, geology, hydrology, climate and natural vegetation. Environmental considerations dominated the design options. Notable examples of the 'design with nature' approach are Jawaharlal Nehru University on the rocky spur of Aravalli Hills, the University of Hyderabad, and the North Eastern Hill University. A pioneering study was undertaken by one of the authors of this chapter, Maitra, to determine the redevelopment potential of an intensely developed area of Shimla respecting the natural ecosystems.

Rivers in India are part of socio-religious life from birth to death. In rituals there is no cognizance of the pollution level of the river, therefore restoring the chemical quality of river water does not awaken people's environmental consciousness. Unfortunately, due to the very flat terrain of northern India and extraction of river water upstream, most of the rivers are shadows of their mighty reputations, except during the monsoon. Although most cities are located on a river they seldom have a river-front, except at Mathura, Vrindabhan, Allahabad, Varanasi, and Calcutta. Therefore access to the river, by restoring, improving and creating river-front architecture, has been given priority. At Varanasi, Mathura, Vrindavan work involved conservation and restoration. At Allahabad it required new construction, which involved not only understanding nature but also discovering methods of preventing soil erosion using natural resources.

The scope for planning with nature was further expanded when the Delhi Development Authority (DDA) employed a group of consultants to prepare the plan for the area (Maitra was the urban planner) designated to accommodate further expansion of Delhi (Papankala new development area of Delhi Metropolitan Region). Analysis of the physiography revealed that there were very few pockets of groundwater aquifer, and the slope is such that during the monsoon period the site cannot be naturally drained. Careful examination also revealed that the villages in the area always drained the water into ponds, which acted as a balancing reservoir during the monsoon period. The problem of drainage was solved by creating a series of holding ponds interconnected into a system of waterways and ultimately leading to a large lake which would be an amenity as well as a holding basin. These holding ponds in each block could also be used for irrigating the open spaces with which they were integrated. These open spaces in turn formed the pedestrian system linkages within a block and in between blocks. Thus an environmentally sustainable settlement structure emerged.

The scope for planning with nature was further extended when analysis of data for the National Capital Region (NCR) of Delhi (30,000 sq. km) revealed that the status of the environment in the NCR leaves much to be desired. The situation in Delhi and adjacent Delhi Metropolitan Area (DMA) towns is critical. Land, water and air pollution has reached an alarming level. It has also been observed that the available water resources, both surface and sub-surface, are being rapidly depleted. It was, therefore, of utmost urgency that a comprehensive study be undertaken to evaluate the environmental status and draw up a comprehensive plan for developing a suitably pollution free and eco-friendly NCR within the extended time frame of 2011. The comprehensive study enabled measurement of the resource base, identification of critical areas, and mitigation measures, for planning the future development of the region.

The disposal of waste water poses a serious problem at urban level, particularly at metropolitan level. The case of the city of Calcutta is interesting, because it hit upon an ecologically acceptable recycling process accidentally, and which on scientific investigation appears replicable. A portion of the domestic sewage of the city of Calcutta (approx. 600 megalitres) runs though a system of channels into the vast wetlands in the east of Calcutta. This sewage is used for pisciculture and the growth of other aquatic flora and fauna, and the quality of water treated through this process discharged to the river reaches an acceptable quality. This technique of bio-degradation by natural flora and fauna has been successfully used in another area of Calcutta and many municipalities are using this technique for sewage treatment. In this indigenously developed technology of growing fish who find rich nutrients, the water is partially purified by the process of oxidation, radiation and biological degradation. The climate, and abundant solar radiation (250–600 langlays per day)[2] and shallow depth of water (less than a metre) have facilitated evolution of this unique ecosystem. This wetland provides food, fibre and fish, and also supports the food chain, improves the water quality and provides for climatic stability.

The efficacy of this biological technique has been successfully tested at Mudially in the heart of Calcutta City on an area of only 50 hectares and which receives toxic industrial waste in addition to domestic waste. Twenty-five megalitres of industrial waste is discharged into the area daily. The community living on this wetland has learnt to skilfully trap the rich organic nutrients of sewage to enrich fishpond water; but has also devised ways to filter and recycle the 25 megalitres of effluent and sewage through indigenously designed sludge tanks and filtering culverts with the help of purifying plants and wild weeds. This vegetation not only absorbs the heavy metal portion of the effluent but also allows the sludge to settle in the deep culverts. The toxicity level of the water through this indigenous process is reported to have come down from 250 to 15 BODs (units of biochemical oxygen demand).

Conclusion

The question of the quality of the urban environment has become critical. On one hand, rapid growth of the urban population has created a tremendous demand for basic service infrastructure. On the other hand, this growth is largely contributed by migrants from rural areas (pushed out by unemployment) who have no empathy with the urban way of life. There are thus two urban Indias in every major city: a population that is urbane, and another population whose heart and mind stays back in the rural villages from where they come. There is thus a curious mixture of the aesthetic and the vulgar. This is reflected in the built form of any city. Good examples of architecture are crowded by unsightly buildings; sporadic oases of good environment are interspersed by the appalling.

Gainful employment in urban areas is available for approximately a third of the population, and of these between 40 and 60 percent belong to the informal sector. Insecurity bred by this employment scenario is reflected in the growth of non-permanent structures of slums and shanties. Revenue for city management cannot be realized from these areas. Cities are, therefore, eternally short of funds. One of the major recommendations of Agenda 21 was decentralization of urban governance. Decentralization has been achieved by constitutional amendment as explained earlier. Several fiscal limitations have been contemplated. Rajkot and Ahmedabad have been able to increase their revenue base, although only time will tell if this will become possible for the whole of urban India.

Agenda 21 also emphasized the protection of environmental resources (particularly water, forests, bio-diversity, wetlands and mangroves), the mitigation of desertification and prevention of air pollution. The major impact of urbanisation has been on forests, water, wetlands and bio-diversity. A vast number of urban and rural households depend on fuelwood. Estimates now show that at the present rate of consumption the entire forest area of India will be destroyed by the year 2020 as additional urban areas demand timber for construction, furniture, etc. The future of the forests is therefore bleak. The most significant move, apart from massive programmes of afforestation, has been the ban on forest felling, the reduction of timber for construction and furniture and replacement of fuelwood by substitutes such as bio-gas, social forestry, solar cookers, etc. The impact of these measures cannot yet be assessed.

Water supply in urban areas is in serious crisis, despite the fact that the majority of the population does not receive satisfactory supply, and a significant proportion has no access at all to domestic water supply. Most Indian cities depend on groundwater for supply. The groundwater table has been going down rapidly, and in some cities alarmingly, such as Ahmedabad, Jodhpur, etc. In Delhi, available groundwater is brackish due to over-exploitation. In suburban areas around Calcutta, groundwater contains unacceptable levels of arsenic. Providing potable water to the entire population is not only on Agenda 21 but also in earlier UN resolutions. Various schemes of water harvesting, recharging of groundwater, recycling, pricing, etc. are being experimented with, without appreciable success so far. Some initiatives have been taken, some with international participation, for improving network management.

The Indian local agenda for the twenty-first century is thus multifaceted – some aspects deal with nationally supported programmes, others deal with meso and micro level initiatives. No one organization, ministry, or institute is entrusted with the entire task. The eternal price for ensuring sustainable development is of course vigilance. This vigilance is provided by large number of NGOs and Courts of Law, which of late have started taking serious note of environmental concerns. The success of the Environmental Action Programme, strict enforcement of legislative measures, and a vigilant population bodes well for the future of environment in India and may set an example for the developing world.

Notes

1   Measuring environmental endowments and their capacity to absorb exploitation within sustainable capacity.
2   'Langlay' is an Indian measure convertible to watts.

References

Central Pollution Control Board/Ministry of Environment and Forests (1994a) *Report on Ganga Action Plan*, New Delhi: Government of India.

Central Pollution Control Board/Ministry of Environment and Forests (1994b) *Report on Selected Small and Medium Towns*, New Delhi: Government of India.

Central Pollution Control Board/Ministry of Environment and Forests (1996) *Monitoring of India's Coastal Waters*, New Delhi: Government of India.

Daly, M. (1996) 'Australia's Toxic Trade', *The Age*, Melbourne, 13 September 1996, p. A9.

Government of India (updated from time to time) *The Constitution of India*, New Delhi: Government of India.

Hempel, L.C. (1996) *Environmental Governance: The Global Challenge*, Washington DC: Island Press.

Maitra, A.K. (1986–87) 'Report on Environmental Improvement of Saraswati Ghat at Allahbad', Unpublished, School of Architecture and Planning, New Delhi.

Maitra, A.K. (1994–95) *Report on Study of Environment and Ecology: NCR*, School of Architecture and Planning, New Delhi.

National Capital Regional Planning Board (1989) *The National Capital Region Plan*, New Delhi: Government of India.

Shiva, V., Jafri, A.H. and Bedi, G. (1997) *Ecological Cost of Economic Globalisation: The Indian Experience*, Delhi: Research Foundation for Science, Technology and Ecology.

Sivaramakrishnan, K.C. ed. (1993) *Managing Urban Environment in India*, New Delhi: Times Research Foundation.

Students (1995) *Report on Eco-Development of Doon Valley*, Unpublished class project in the Dept. of Environmental Planning, School of Planning and Architecture, New Delhi.

Supreme Court of India (1996) *Judgment of the Supreme Court on Public Interest Litigation*, New Delhi: Government of India.

Warrier M. and Kothari, S. (1992) *Study on EPA Litigation*, New Delhi: Indian Institute of Public Administration.

10 After Rio

Environmental policies and urban planning in Sweden

Rolf Lidskog and Ingemar Elander

Introduction

The official view in Sweden is that the UN Conference on Environment and Development (UNCED) in Rio in 1992 has had a strong impact both on Sweden's environment policy and its development assistance. No doubt the principles of the Rio Declaration and the integrated and broad-based recommendations in Agenda 21 have influenced environmental debates and policies. This applies, for example, to an Environmental Code, energy and traffic policy, the use of economic instruments, and the substitution of chemical products with less harmful ones. Also, Sweden's development assistance now has as an explicit goal to contribute to sustainable development in the receiving countries.

At the local and municipal level, Agenda 21 has provided a strong incentive for broad and active public participation in efforts to achieve sustainable development. Thus, although by the time of the Rio Conference the seeds for a greening of policies in Sweden had already been sown, the Conference was a trigger for more long-term work aimed at sustainable development. However, there is no one-way road towards sustainability. The project of modernity produced structures that do not always favour such development. Different ideologies and sectoral interests in state and civil society make the implementation of the Rio documents a continuously conflictual process, and when Sweden experienced economic recession in the 1990s its policies did not always give the highest priority to sustainability.

Nevertheless, there is a growing consciousness even among growth-orientated elites in Swedish society that a good environment is a prerequisite for economic growth. Therefore it remains urgently necessary and relevant to continue to pursue effective measures for sustainable development. The message from Rio concerning the importance of this work, both nationally and internationally, has resulted in a strong response in Swedish society, warranting a cautious optimism regarding the prospects of further activities. The government's policy declaration on 17 September 1996 states that 'Sweden shall be a leading force and an example to other countries in its efforts to create ecologically sustainable development. Prosperity shall be built on more efficient use of natural resources – energy, water and raw materials.' The implementation of this ambitious statement is largely decentralized to the local level of state and society. Thus all of Sweden's 288 municipalities have started a practical implementation of the Rio agreements and have developed Local Agenda 21 (LA21) plans.

The government has proposed to develop the first stages of a broad-based, long-term investment programme (5.4 billion Swedish Kronor [SEK], i.e. 800 million US$) for ecological sustainability and employment. The framework of this programme, as well as the government's overall policy for sustainable development, was announced by the Riksdag in 1997. Sustainable development is the overriding goal both in the Energy Bill that was introduced in the Riksdag in spring 1997, and in the Environment Policy Bill presented in summer 1998. Also, the government's proposal on a new Environmental Code (spring 1998) – where fifteen different environmental Acts are replaced by one all-embracing environmental code – has the aim of making the legislation more lucid, efficient and applicable.

Within the international community, Sweden has played an active role in implementing the global conventions on climate, protection of the ozone layer, biodiversity, transport of hazardous waste, nuclear safety, etc. Within the European Union (EU), Sweden is actively pursuing issues relating to the environment and development co-operation. Regional co-operation in the Baltic region has been broadened and deepened through Swedish initiatives. Considering the fact that the strong decentralist orientation of Agenda 21 has found a fertile soil in Swedish society with its tradition of strong local self-government, policies for ecologically sustainable development in Sweden display a strong multi-level approach.

Aside from this introduction, the chapter is in five parts. Taking its point of departure in a survey of the development of Swedish environmentalism during the twentieth century until Rio 1992, the second part gives an analysis of the environmental goals and instruments as formulated in the flood of official documents that preceded and followed the Rio Conference. The different policy instruments proposed to fulfil the official strategies are also examined with special regard to the role of local government in implementing environmental policies. In the third part, some examples of urban environmental planning and policies are presented and critically discussed. The Swedish post-Rio Agenda is then outlined, highlighting two weak links with regard to its implementation. In conclusion we summarize and highlight some crucial issues of tension and conflict that characterize the processes and outcomes of urban environmental policies and planning in Sweden.

Swedish environmentalism before Rio

The history of Swedish environmentalism during this century describes a route from preservation to sustainability, and covers a number of changes in environmental perspectives and policies which influence the character of the environmental movement as well as its interaction with other spheres in society (Elander *et al.*, 1995; Jamison *et al.*, 1990, Söderqvist, 1986). Until 1960 environmental policy had a very restricted role in Swedish politics, mainly concerning the preservation of specific features of nature, and trying to solve discrete, local and occasional environmental problems, e.g. the pollution of air and water. The organizational basis of environmentalism was dominated by one organization – the Swedish Society for the Conservation of Nature (Svenska Naturskyddsföreningen; SNF), which was created in 1909.

Around the 1960s there was a qualitative shift in the environmental debate. Now it was no longer enough to protect certain places from harmful activities. Already in the 1950s the risks connected to biocides were observed, the effect of mercury on birds was discovered by amateur ornithologists, and both the state and the SNF arranged conferences on these issues. The translation of Rachel Carson's book *Silent Spring* into Swedish in 1962 triggered the first major environmental debate in Sweden (Jamison *et al.*, 1990). It was now recognized that human disturbances spread through ecosystems, causing unforeseen effects, which posed a threat even to humans at the apex of food chains. A holistic approach to the environment was therefore needed.

The environmental debate led to a fourfold increase in membership of SNF (to 20,000 in 1963), a gradual integration of environmental protection into the programmes of the political parties, and the creation of state administrative units responsible for natural resources and natural protection (Lundqvist, 1971). Sweden was the first country in the world to establish a national environment protection agency (1967), and Sweden's Environmental Protection Act (created in 1969) was the most encompassing environmental legislation in the world at that time (Jamison *et al.*, 1990; Weale, 1992: 98). Ecological science and comprehensive planning emerged as dominant instruments for environmental policies. Thus the state was considered to possess the managerial efficiency and the authority necessary to confront environmental problems successfully.

In the late 1960s a global perspective on the environment was gaining ground, manifested by the UN Conference on the Human Environment in Stockholm in 1972 and the creation of the Swedish branch of The Friends of the Earth (FoE). Environmental concern became more far-reaching in the sense that both cases and causes of environmental destruction were attacked. This led to a questioning of industrial society and a search for alternatives to the dominant societal paradigm. The breeding ground for this new kind of critique was the more radical and activist political culture of the late 1960s, coupled with a boom for ecological science. The political answer was more comprehensive planning aided by increasing use of ecological science.

From the mid-1970s until the referendum in 1980 the environmental debate focused on Sweden's plan to build the most ambitious nuclear programme in the world. After a remarkable silence in the 1950s and 1960s, opposition grew fast, leading to the nationwide People's Campaign Against Nuclear Power (Folkkampanjen mot Kärnkraft) in 1978. The campaign organized a broad spectrum of groups and organizations, including environmentalists, peace activists and women's liberationists. Inspired by a merger of socialist and environmentalist ideals, the People's Campaign also spread ideas of a future resource-preserving society with meaningful work for everyone. Ideas for an alternative society thus came to the fore. Organizations like The Friends of the Earth and The Future in Our Hands (Swedish branch in 1976) contributed new values, a focus on lifestyles and a global outlook.

From the mid-1980s, there was a large increase in public concern for the environment, leading in 1988 to parliamentary representation of the Environmental Party (which had been created in 1981). The global character of environmental problems also became more pronounced (e.g. the depleting ozone layer and global warming), and attention shifted to diffuse and prognosticated problems. It was realized that environmental problems cut across sectoral borders in society and demanded the co-operation of state, civil society and the market for their solution. The number of actors involved in environmental problems therefore increased dramatically. 'Sustainability' became the central concept in the environmental debate.

Swedish environmentalism lost its character as a people's movement in the 1980s, as environmental organizations grew more professional and result-orientated, illustrating the tension between pragmatism and idealism. It was the 'deputy-activism' (activism on behalf of a political interest) of Greenpeace (Swedish branch in 1983) and the expert-based pressuring by a revitalized SNF that proved to be organizational models well adapted to the environmental consciousness of the 1980s and the early 1990s. Organizations that were more directly based on member participation, like The Friends of the Earth, The Future in Our Hands and the Environmental Union, experienced a decline in membership. The efficacy of the political branch of the movement, the Green Party, was temporarily disarmed in the early 1990s as older parties recognized the political potential of environmental issues and adjusted their policies accordingly. Efficiently making use of the symbolic dimension in politics (cf. Edelman, 1964) all political parties allegedly became 'environmentalist'. Environmental issues were thus largely absorbed by the traditional left–right dimension in Swedish politics. Nevertheless, in the 1994 elections the Green Party received 7 percent of the votes and was again represented in the Riksdag.

The late 1990s display a dual development. Aside from the success of environmental issues within government and industry, citizens seem today not to give the environmental issue the same priority as at the end of 1980s. The environment was ranked number one by the voters in the election of 1988, but recent surveys (April 1998) show that it has degraded to number six, surpassed by medical care, elderly care, education, employment, and the sustained progress of the Swedish economy (Sifo, 1998). In the 1998 election the environment was not a major issue.

The environmental movement in general has faced a decline in membership since its peak at the end of the 1980s. During the last decade the SNF has lost 20 percent of its members (from 206,000 in 1991 to 168,000 in 1997) and Greenpeace has lost about 60 percent of its financial supporters (from 210,000 in 1988 to 100,000 in 1997).

Goals and instruments in environmental policy

In its preparations for the Rio Conference the Swedish government strongly argued that environmental policy had to be reconciled with democracy, the market economy and economic growth; i.e. environmental policy was regarded as part and parcel of comprehensive welfare policy. This is well in accordance with the Brundtland Commission Report, which defines 'sustainable' as including social, economic and ecological dimensions.

Goals

The declarations and conventions adopted by the Rio Conference were all ratified by the Swedish Riksdag, i.e. the Rio Declaration, Agenda 21, the Forest Protection Principles, the Climate Convention and the Biological Diversity Convention. On the basis of these documents the Swedish government formulated a national strategy for a sustainable society, comprising the following main objectives:

1   to protect human health,
2   to maintain biological diversity,
3   to manage natural resources to ensure their long-term use, and
4   to preserve natural and cultural landscapes.

To reach the stipulated objectives of environmental policy the Swedish Government has adopted a number of fundamental principles:

1   respect for the tolerance limits of humanity and the environment;
2   use of the best available technology and good environmental practice;
3   use of the substitution principle;
4   use of the precautionary principle;
5   use of the polluter pays principle;
6   respect for each country's supremacy over its environment;
7   adherence to Agenda 21.

Taken together the four main goals of environmental policy and the seven principles of environmental protection represent a highly ambitious and comprehensive approach to developing a sustainable society. To what extent these objectives and principles are really adhered to in policies and practices remains to be seen.

Policy instruments

According to the National Environmental Protection Agency, 'a variety of policy instruments and strategies aimed at different players are necessary for the transformation to a more resource-conserving society with environmentally sound habits' (SNV, 1993: 6). Among the actors emphasized are citizens, private companies and the municipalities, with a crucial role assigned to the latter. The state has to create a framework of policies within which this multitude of players can act in a way that supports ecological sustainability.

Policy instruments may be classified in three categories, i.e. administrative-regulatory, financial and social (Elander *et al.*, 1997). *Administrative-regulatory instruments* mean that the state uses its authority by deciding on permission standards, prohibitions and restrictions, different kinds of target levels, etc., and by prosecuting trespassers. *Financial instruments* mean that the state, through the use of economic incentives, creates a situation where action guided by self-interest and rational choice is assumed to favour good environmental solutions. *Social instruments* mean that incentives for environment-friendly action are created through voluntary, pedagogic and informative activities.

Although the central state in Sweden has traditionally used mainly regulatory instruments for environmental protection, there has been a tendency for greater importance to be given to financial and social instruments. The use of fees, taxes, subsidies and other economic rather than administrative instruments is becoming increasingly important, and Sweden seems to have more such instruments than any other country (MoE, 1997). The use of financial instruments in the environmental sector can be traced back to the middle of the 1970s, but was given higher priority at the end of the 1980s (Government Bill 1987/88, No. 85). Financial instruments were now legitimized by their steering capacity, not as a way to increase public revenues which had been the case up to that time.

During the last decade the Riksdag introduced or raised the level of environmental taxes on petrol, oil, energy use, nitrogen oxides and carbon dioxide. The taxes and charges are constructed with the explicit aim of making environmental impact become a matter of monetary calculation (see for example Government Bill 1993/94, No. 100). If the environmental costs are included in the price system, producers as well as consumers are assumed to behave in a way that is more friendly to the environment. In the 1990s Sweden was the first country in Europe to implement economic reforms designed to exchange environmental and energy-related taxes and fees for lower taxes on wage earnings. Evaluations have found that environmental fees and taxes have been efficient both from a socio-economic and an environmental point of view (Government Bill 1993/94, No. 111). The Centre Party and the Green Party especially welcomed this tax reform as being decisive for creating a sustainable society. Aside from their steering capacity, financial instruments have become an important source of public revenues.

In 1998, the revenues coming from taxes relating to environment and energy accounted for 3.2 percent of GDP. The main items of tax revenue are fuels, electricity and the carbon dioxide tax (MoE, 1997: 13).

However, there has not only been a shift from administrative towards economic instruments but also towards a greater emphasis on social incentives, i.e. ways to increase voluntary citizen involvement through changing lifestyles and consumption patterns. Life Cycle Assessment, Environmental Auditing and Environmental Labelling are tools invented in order to make producers and consumers act in a way that is more friendly to the environment. The rationale behind these and other, mainly social, instruments is that citizens and organizations should be educated to act and behave as ecologically enlightened consumers. As the government's Delegation for Ecologically Sustainable Development puts it: 'It is a question of encouraging the driving forces of the market and making all parts of society participate' (DESD, 1997: ch. 2).

Environment and urban planning

Today, the need for more 'sustainable' cities is frequently argued for by researchers, practitioners, policymakers and the public. Aside from the obvious fact that societies like Sweden are urbanized to such a high extent, a major reason for this increasing interest in ecological aspects of city development is the changing nature of what is considered as 'environmental problems'. During the twentieth century the focus of the environmental debate has changed from preservation of specific natural areas over environmental planning to sustainability and ecocycles. It has been a process *from* scattered local problems due to various emissions from industries and municipalities and the preservation of some specific areas from human intrusion *to* diffuse, large-scale, threats like the 'greenhouse effect' or the depletion of the ozone layer. Strikingly, the latter kind of problem is closely linked to production and consumption patterns, and cities are today the major generators of economic activities and social welfare (education, health, etc.) as well as the main consumers of natural resources and the main producers of pollution and waste. Therefore, with regard to many environmental issues of today, both their causes and their remedies are deeply rooted in activities that take place in the city.

In the vast and increasing literature on environmental issues, before Rio comparatively little attention was paid to cities (McLaren, 1992: 56; Stren, 1992). Astonishingly, not even the Brundtland Report (1987), commonly regarded as a turning point for sustainable planning, has much to say about the cities of the industrial world. The Report gives the impression that no special changes of course are necessary for urban development in industrial countries, and that the problem facing the Third World's cities is that they have been without the necessary resources to attain the level of development reached by the cities of the industrial countries today (Naess, 1989: 46).

Nor did the three Swedish Government Commission Reports published in 1990 analysing the city (SOU, 1990a, 1990b and 1990c) regard environmental issues as restricting further urban growth. On the contrary, the dense infrastructure of the city was regarded as a precondition for sustainability. Only in the cities, it was argued, was it possible to build an infrastructure subordinated to principles of ecological balance. Following Olof Eriksson (1991), former head of the Swedish Council for Building Research and environmental adviser to the current government, the environmental chapter in the first of the three reports is fundamentally one-sided, completely dismissing the fact that the city is the main producer of pollution and waste.

However, the publication of the European Community (EC) Green Book on the urban environment in 1990 marks a turning point with regard to bringing sustainability into the urban policy arena in Europe. Today sustainable city development has become commonplace in the set of policy goals given high priority by the political elites in various countries. Strikingly, the focus of the EC Green Book is on the city centres of the West European big cities. But in Sweden we have only three cities of that scale (Stockholm, Gothenburg and Malmö), and their centres are comparatively small, as is still more the case in the Swedish towns and less urbanized municipalities. The urban agglomerations in Sweden are spread out in the countryside, and most of them are still literally 'green' in the sense that their dwellers can always reach parks and green belts at a relatively short distance.

Considering the above-mentioned two circumstances, Swedish towns may have a comparative advantage when it comes to solving the practical problems connected with sustainability. The smaller scale of Swedish towns combined with the more sparsely built environment may make it easier to develop systems of ecocyclical waste management (Bjur and Engström, 1993). Orrskog (1993: 233) even goes as far as concluding that 'Scandinavian cities built on the neighbourhood structure, with green fingers into the city, and with quite good public transport, present relatively good conditions also for sustainability'. On the other hand, the dispersion of residents creates a big problem with regard to transportation. Thus the average Swede travels around 50 kilometres per day, mostly by car.

Although ecological perspectives were not unknown to Swedish policymakers before Rio, they did not enter the field of urban policies until the beginning of the 1970s, and this on two different, although related, paths. The first one was the path of national, comprehensive planning starting with the 1972 Act on National Land Use Planning. The second one emerged from the all-European urban renewal initiatives around 1980. A little later feminist ideas also contributed to the development of ecological consciousness in planning, although they hardly left any mark on planning practice (*Forskergruppen for det nye hverdagslivet*, 1987; Kaul 1991). Indeed, the 1970s represents the high tide of rationalist, public planning in Sweden. Inspired by central place theories with specified demands on the size of various social services, the municipal amalgamation reform had been fully implemented in 1973, the quantitative goal of the Million Dwellings programme of housing had been reached in 1974, and the 1972 'Sweden Plan' symbolized the determination of Social Democracy in Sweden to fulfil long-term planning ambitions. However, as far as ecological care was concerned, no major steps towards its realization had been officially taken. The basic ideology was still mainly one of economic growth and social welfare (Elander, 1978).

By the time stagnation began to haunt the Swedish economy during the 1980s the ideology of rationalist planning was on its way out. Although structure planning was given a prominent position in the Planning and Building Act and the Natural Resources Act, its power could be disputed, as it was only allowed a guiding role, with no decisive power for future land use (cf. Eckerberg, 1995; Khakee, 1995). The property boom coupled with decentralization and deregulation of public services signified a gradual marketization of policy (Elander and Montin, 1990), and the introduction of the concept of negotiative planning. Ironically, just as the ecological dimension was on the threshold of entering municipal planning, planning itself was changing its character in a less demanding direction. Thus, although they became the most important units for public environmental planning in practice (Eckerberg, 1995), the Swedish municipalities have to carry out their task with quite nebulous instruments. Following Ödmann (1987: 263), the municipalities now have to implement 'vague norms in a tough reality'. The 1990s represent an even tougher reality for the municipalities as they have been sternly pressed by central state cut-back policies (Elander, 1996). However, regardless of central state policies, initiatives increasingly taken by municipalities and other local actors today mean that 'sustainability is little by little entering municipal planning and policy' (Malbert, 1992). The Agenda 21 document was the trigger for municipal initiatives on a broad scale.

The post–Rio agenda

In March 1995 a National Committee on Agenda 21 was established. The committee was chaired by the Minister of Environment and included actors representing political parties and various sectors in society. The committee was given three main tasks: (1) to stimulate the Swedish Agenda 21 work, e.g. by the spreading of successful examples; (2) to monitor the national Agenda 21 work and prepare a national report for the UN General Assembly in 1997 (UNGASS); (3) to monitor the Swedish work towards the environmental goals which had been decided. In January 1997, the government appointed a Delegation for Ecologically Sustainable Development, including the five ministers of the environment, education, employment, agriculture and taxes. The delegation was given the task of stimulating the work with Agenda 21 and contributing to the preparation of the Swedish UNGASS report. At the national administrative level the National Environment Protection Agency has a crucial role with regard to the implementation of environmental policies in

Sweden. Thus among its current responsibilities we find the distribution of grants to LA21 projects, and the evaluation of the LA21 work. In addition, because of the explicit lack of national co-ordination, the environmental movement SNF has taken on a co-ordinating role, becoming one of the most important actors concerning the LA21 work (SNF, 1996).

The scope of municipal responsibilities has widened in Swedish society, not least in the environmental field. Thus during recent years there has been an increasing transfer of environmental responsibilities to municipalities and other sub-national actors. In the Government Bill concerning environmental policy for the 1990s (Government Bill 1987/88, No. 85), the government mentions the role played by municipalities in environmental policy only in passing. Six years later, in the Government Bill on the Rio Declaration 1993/94, the municipality is acknowledged as having a key role for the *national* strategy geared at promoting sustainable development. The main reason given for this is that policy and planning to a large extent are implemented at the local level.

This key role of the municipality is strongly emphasized in the Government Commission Report concerning LA21 (SOU, 1994). The municipality is supervisor, planner, producer, entrepreneur, manager, purchaser, educator and information giver, and it is stated that different groups – for example youth, women, voluntary organizations and private companies – should be involved from a bottom-up perspective. The Report does not give any concrete recommendations about what a LA21 should contain, but instead it is stated that every LA21 is presumed to develop according to a democratic process which mirrors the local conditions in each municipality. There seems, however, to be a potential tension between two central policy goals: that of democracy and that of sustainability. On the one hand, the broadly accepted democratic ideology in Sweden says that citizens should be involved in the planning and decision-making processes, but, on the other hand, the goal of sustainability may be difficult to reach without a certain amount of centralized planning and decision-making. Furthermore, the municipalities seem not to have anything to say concerning national goals, but should only feel responsibility for approaching them through local interpretations and constructive strategies (see e.g. Official Letter from the Parliament, Rskr 1994/95, No. 120).

Local Agenda 21

In autumn 1992, the government presented a bill to the Riksdag on the implementation of the Rio proposal, and the Swedish Association of Local Authorities (Svenska kommunförbundet; SALA), the Minister of the Environment and the Minister of Physical Planning sent a joint circular letter to all Swedish municipalities, presenting a summary of Agenda 21 and asking them to initiate local Agenda 21 programmes. During spring 1993, the Ministry and SALA invited municipalities and other actors to a series of regional conferences concerning Local Agenda 21 (LA21). Following the authorities, the major break-through in the LA21 involvement came in 1994, when nearly half of the municipalities reported that work had begun (SOU, 1996, No. 48; SALA, 1995). By July 1995 virtually all Swedish municipalities had taken a decision to initate work on LA21, and about half of them had a special Agenda 21 co-ordinator. Today, all municipalities have initiated LA21 work, and a national survey, carried out in 1996, shows that 3 percent of the Swedish population have been involved in a LA21 project, 20 percent have been reached by written information and 40 percent have heard about Agenda 21 (MoE, 1997: 82).

Local Agenda 21 so far has mainly been used as an instrument for educating citizens and organizations about how to behave in an environmentally responsible way. Citizen involvement has been substantial, and the official picture is that the 'bottom-up' perspective recommended in Rio for subsequent Agenda 21 work has proved successful (MoE, 1997: 12). By international comparison, this is probably true. However, there are also problems, as identified by many Agenda 21 activists in the local communities.

There is a growing insight that the successful implementation of LA21 needs a stronger commitment on the part of the central government. Thus an evaluation carried out by the environmental movement SNF in 1996 found that 50 of the 288 Swedish municipalities perceived that they faced 'national obstacles' in their work with LA21 (SNF, 1996). This critical view is supported by another survey of the municipalities, carried out by the Swedish Association of Local Authorities in 1995, where 24 percent of the Agenda 21 co-ordinators in the municipalities mentioned lack of visible and clear national support and guidance as a strong obstacle in the ongoing Agenda 21 effort (SALA, 1995). Thus, on the one hand national officials declare that the lack of central directives and guidelines is a deliberate approach, where the municipalities are free to act in accordance with their own environmental priorities (SOU, 1996, No. 48: 138; MoE, 1997). On the other hand Agenda 21 activists at the local level ask for a more explicit national policy for sustainable development, including financial and other kinds of support needed to implement this policy.

National policy is also often felt to be inconsistent. It is hard for the municipalities to argue for and implement policies for ecological sustainability when national policies move in a contrary direction. Energy, traffic and infrastructure are areas where national policies are often felt to impede or contradict ecologically sustainable development. Many of the Local Agenda 21 co-ordinators call for a *national* Agenda 21 – a concerted national action programme for sustainable development – to support local activities. Thus, it is not surprising that the national report on Agenda 21 finds that work has been more successful in policy domains where the municipalities have a long tradition of responsibility, e.g. sewerage and waste treatment. On the other hand, in domains where municipal responsibility has traditionally been weaker signs of success are more or less absent (MoE, 1997: 46). This is most noticeable when it comes to energy and transportation. These are domains where the municipality has had limited scope of action. Instead strong networks encompassing national actors within industry as well as in the state have formulated and forcefully implemented policies on energy and transport.

The energy sector

Sweden has one of the world's highest levels of energy use. This can partly be explained by the country's cold climate and energy-intensive industry (Lidskog, 1994). Since the beginning of the 1970s the energy use in Sweden has increased from 457 TeraWatt-hours in 1970 to 633 TeraWatt-hours in 1996 (NUTEK, 1997: 4).[1] Hydropower and nuclear power together produce 25 percent of the total energy supply, and this is part of the reason why Sweden has a relatively low level of carbon dioxide emission. The use of renewable resources has increased during the 1990s, and today accounts for about 20 percent of total energy supply. The country's 7.5 tonnes of carbon dioxide per capita (1996) is a low figure compared to other European nations. However, the Environmental Protection Agency states that the level is too high and has to be reduced by 50 percent to reach a sustainable level (SNV, 1996). The Swedish branch of the Friends of the Earth argues that the shrinkage must be 80 to 90 percent (FoE, 1997).

As described earlier in this chapter Swedish environmental policy has increasingly made use of taxes and other economic instruments, e.g. carbon dioxide tax, electricity tax, differentiated fuel taxes on petrol and oil, sulphur tax and nitrogen oxide charge. However, whereas energy taxes have been gradually raised for households and the service sectors, they have been reduced for industry. In 1988, the Riksdag decided that the discharge of carbon dioxide should not increase. One important argument behind this decision was to make Sweden an example for other nations to follow. In 1991 Sweden was one of the first countries in the world to introduce a specific tax on carbon dioxide. One result of this tax was that district heating to a large degree substituted biofuels for coal, the former being exempted from both carbon dioxide and fuel taxes. The consumption of coal in heating plants fell from 555,000 tonnes in 1988 to 120,000 tonnes in 1994.

In the beginning, the energy-intensive industry was exempted from the carbon dioxide tax with the argument that the tax would damage its international competitiveness. In 1992, the non-socialist government coalition decided to abolish the electricity tax for all industries. It also decided on a substantial reduction of the carbon dioxide tax for the non-energy intensive industry, at the same time as the general carbon dioxide tax was increased. The decrease in energy and carbon dioxide taxes for industry, together with increased production, made the industry's consumption of oil – and thereby the emission of carbon dioxide – increase by 34 percent between 1992 and 1996 (NUTEK, 1997: 18).

At the end of 1996, the government proposed to double the carbon dioxide tax for industry, i.e. to raise the industrial rate to half of the general one (Government Bill 1996/97, No. 29). This proposal seemed to be in line with recommendations given by OECD that Sweden should strive to increase efficiency in its use of energy and to decrease its emission of carbon dioxide (OECD, 1996). The Parliament decided in accordance with the proposal, but it was controversial. As the chairman of the Federation of Swedish Industry put it:

> The emission of carbon dioxide is a global problem. It can only be solved by international agreements. We see no reason for an additional increase of the already high carbon dioxide tax on propellants in Sweden. It will not produce any noticeable effect on the emissions, but deteriorate our international competitiveness.
>
> (Berggren, 1997: 2)

Thus the government's declaration that Sweden should be a leading force and an example for other countries seems not to be agreed upon by all parts of society. Indeed, the government itself vacillates between economic growth and ecological sustainability. However, this is not only visible in the energy sector, but in the transport sector too.

The transport sector

The total emission of carbon dioxide in Sweden decreased by 20 percent during the period 1980–1996, not least because of the development of nuclear power. However, in the transport sector there has been no substitution for oil, which means that the emission of carbon dioxide in this sector *increased* by 27 percent over the same period (NUTEK, 1997: 29). In negotiations that preceded the Rio conference, Sweden was one of the nations that strongly emphasized that the environmental problems of the transport sector had to be included. Officially Sweden boasts of its responsibility to show the world how the transport system should be developed to meet the demands of sustainability (SOU, 1992: 17). However, efforts to walk in that direction face opposition by powerful interests in Swedish society.

The transport sectors accounts for one-fifth of Sweden's total energy use, and 90 percent of this energy is derived from fossil fuels (NUTEK, 1997: 19; MoE, 1997: 64). Aside from the problem of climate-warming gases, the transport sector is one of the biggest sources of emissions of substances causing acidification and eutrophication of soil and water and damage to vegetation. Technical progress has created the potential for cleaner vehicles, especially in the longer term (von Weizsäcker *et al.* 1997: 71–76). However, thus far the positive effects of technical progress have been more or less eliminated by the introduction of heavier and faster vehicles, using larger amounts of fuels.

Some categories of emissions produced by motor vehicles – nitrogen oxides, hydrocarbons, lead – have been substantially reduced by the introduction of better engines, catalytic converters (since 1989 compulsory for new motor vehicles) and improved purification. As in other member states of the European Union, however, carbon dioxide emissions have not declined, and today transportation is estimated to account for one-third of all such emissions (MoE, 1997: 64). Traffic volumes continue to increase – between 1970 and 1990 transportation of persons has increased by 60 percent and transport of goods by 45 percent (NUTEK, 1997: 19) – which causes increased carbon dioxide emissions. In addition, the traffic system impacts the environment by its claim on substantial areas of land, especially in urban areas where green spaces are limited.

Environmental problems connected with heavy road traffic were observed in the OECD review of Sweden's environmental performance, and the country was recommended to introduce cost-efficient measures to restrict the increase in road traffic, and to strengthen traffic planning and the use of public transport in cities (OECD, 1996). In line with these recommendations the national report on Agenda 21 states that 'developments in the transport sector today are not compatible with sustainable development' (MoE, 1997: 9). Thus, changing the transport system into a sustainable one is a major challenge to environmental work in Sweden today.

In the mid-1990s, the Riksdag decided to reduce investments in new roads in favour of more spending on road maintenance and railway investments (Government Bill on infrastructure 1996/97, No. 53). This was criticized by private transport interests, which judge the priority being given to railways as an attempt to 'solve the problems of today with the solutions of yesterday' (Gyll, 1996: 1). The national Agenda 21 report argues that there are tendencies that might change present transport patterns positively. For example, new information technologies may reduce the amount of transport from home to job. The fact that 56 percent of all journeys are shorter than 5 kilometres and yet are mainly made by private cars means that there is a big potential to substitute bicycles and mass transit for car transport. However, so far experience does not indicate a reduction in the overall amount of transportation as a consequence of new information technologies. In addition, the establishment of an increasing number of out-of-town shopping centres, enforced by a coalition of commercial and landlord interests, has contributed strongly to consumers' increasing dependency on the automobile and has caused depletion of city centres.

To summarize, there is no evidence that the amount of transportation will decrease; rather the opposite. In Sweden today there is no overall strategy to introduce biofuels in the transport sector, and the solution to the ever increasing consumption of fuel and the accompanying emission of carbon dioxide remains rather a distant prospect. Since 1987, the average amount of petrol consumption in newly registered cars has not been reduced, despite governmental dialogue with the car industry about a reduction. Using economic instruments to reduce the consumption of fuel (and thereby the emission of carbon dioxide) is something that has been questioned not only by private companies, but also by the the Association of Private Drivers (Motormännens riksförbund), which emphasizes that private cars are intimately connected with the quality of life and that the petrol tax already has reached a level where this quality is threatened (Cornelius, 1997).

Leaving aside the domestic opposition against changes in the transport system, Sweden's membership of the European Union is another important factor that does not necessarily favour the nation's work for sustainable development. In the national report on Agenda 21, the European Union is mainly discussed with regard to the ability to apply environmental requirements to products. The report argues that Swedish environmental requirements concerning products may come into conflict with the single market principle of free movement of goods. In Agenda 21 transport and free trade are touched upon with regard to the importance of liberalizing trade for the benefit of sustainable development. This would give the developing countries better access to export markets, making international trade policy and environmental policy mutually supportive. However, this requires that the environmental problems of the transport sector be solved, especially with regard to the use of fossil fuels. If environmentally friendly ways of transporting do not develop, the increasing transboundary movements will instead increase the destructive pressure on the global environment.

Conclusions

Our conclusions are grouped around three issues. We start by focusing upon the rhetorical dimension of environmental policies, continue by highlighting the tension between economic growth and ecological sustainability in general, and end up by discussing this tension as experienced in Swedish society.

The rhetorical dimension: from sustainability to ecocyclical society

Today's debate on environmental issues is dominated by the global problems such as the greenhouse effect. Comprising ecological, social and economic dimensions, the concept 'sustainability' is commonly stated as an all-encompassing solution to these problems. However, in Sweden sustainability in the 1990s was gradually supplemented by the term 'ecocycle' (*kretslopp*) as a key concept indicating the long-term goal of environmental policies in Sweden (see for example Government Bill 1992/93, No. 180, on guidelines for the ecocyclic development of Swedish society). Creating an ecocyclic society (*kretsloppssamhälle*) means the opposite of the currently dominant principle of linearity. Thus all the material of society's products should be integrated into chains of recovery and recycling, which would diminish the use of raw materials and energy.

Despite the fact that environmental issues have been part and parcel of public debate, there are few signs that there have been major changes in the way society treats natural resources and flows of materials. Instead of minimizing the material used and the production of waste, we recover and recycle waste and, by doing so, encourage more and more material to flow in society (involving for example the transport of waste to the recycling company and new products to the consumer). Today there is a risk that the increasing usage of recycling as the key concept signifying environmental policy ambitions will legitimize further excess exploitation of natural resources (not least fossil energy). At worst this might become just one more example of what Edelman (1977) has coined 'words that succeed – policies that fail'.

Thus the increasing use of the term 'ecocyclical society' instead of 'ecological sustainability' in the official rhetoric is one of many examples of the dynamics of the language of the environment, i.e. the expanding network of environment words that increasingly penetrate texts, words and voices, locally as well as globally (Myerson and Rydin, 1996). Swedish officials have been very sensitive in the use of innovations in the language of the environment, as documented in a number of national reports to the global political community (Rio Declaration, Agenda 21, UNGASS, etc.). However, from this conclusion success in terms of implementation does not automatically follow. Although the rhetorical dimension of environmental policies is important in its own right, one has to be extremely careful when drawing conclusions about its tangible effects.

Sustainability and economic growth

Present-day patterns of production and consumption in Sweden and in other industrialized nations are not sustainable, neither on the input side (the exploitation of non-renewable natural resources) nor on the output side (the environmentally harmful emissions that stem from societal activities). The question to be raised is whether all the activities – government declarations, the creation of public agencies, the creation and dissemination of environmental consiousness – do in fact lead to changes in the present patterns of production and consumption. Of course, there is no simple answer to this question, neither in Sweden nor elsewhere. At all levels of society – international, national, local, individual – we see not one but many answers to the questions of how ecological sustainability should be defined, what priority it should be given compared to other goals, and which means are the most efficient ones to reach this goal. Thus the choir of environmental voices is not an entirely harmonious one; on the contrary it is a rather bewildering mixture. The variety of answers is visible at all levels of society as well as within different sectors.

Environmental policies increasingly emerge from complex coalitions of actors and institutions. Initially, the assumption behind environmental policies was that such issues involved a zero-sum game of social and natural limits to growth. This assumption is visible, for example, in the Government Bill proposing an Environmental Protection Act (Government Bill 1969, No. 28), which emphasizes the importance of not putting too high demands on industry because that would lead to decreasing international competitiveness. Today the ecological challenge is framed in the setting of ecological modernization, where ecological sustainability and economic growth are seen as strengthening each other. Thus in the official rhetoric the zero-sum game of economy versus ecology seems to have become replaced by a positive-sum game.

Most nation states agree that environmental degradation threatens the social and physical resources upon which prosperity and economic development depend, something which is also stated in the European Community's fourth and fifth environmental programmes (1986 and 1992) as well as in the Brundtland Report. The question is what happens when ecological sustainability is interpreted in a way that totally opposes economic growth. As it is now, the underlying assumption is that work for ecological sustainability will not constitute any hindrance to, but can be a trigger for, economic growth. As the Delegation for Ecologically Sustainable Development puts it (DESD, 1997): 'Sweden shall be a driving force and a paragon country for ecologically sustainable development. Ecological sustainability can also contribute to more jobs, economic growth and increasing competitiveness.' Thus environmental concern is more and more regarded as a vehicle for economic growth, although in cases where the two goals seem to be in conflict with one another economic growth is still the heavyweight.

Conflicting coalitions and Agenda 21

With regard to environmental issues on a broad scale we can identify at least three coalitions in Swedish society. The dominating one is constituted by a majority of political parties (including the governing Social Democrats), environmental agencies and to some extent the Swedish Society for the Conservation of Nature (SNF). The approach to sustainability taken by this coalition is more or less identical with the official one as described in this chapter. It comes close to the ecological modernization approach, where all actors through ecological enlightenment will gradually make their activities environmentally adapted. Through knowledge production, pedagogic activities, financial instruments and other steering mechanisms society is assumed to make economic growth and ecological sustainability a happy marriage.

A second coalition, consisting of export-orientated business interests and manufacturing industry, states that Sweden should not try to become a green forerunner. In their latest environmental report *Continued Sustainable Development* (1998), the Federation of Swedish Industry (Industriförbundet) argues that the Swedish state should concentrate its work upon creating international legislation which gives common conditions for all companies in the world, instead of creating national legislation that would cause damage to the country's industrial competitiveness in the international market. Sustainable development is not created, according to the report, by decree but 'through a dynamic process which gives maximum space for creativity and entrepreneurship in the whole society' (Industriförbundet, 1998: 36). Thus, it is assumed that sustainability will be attained through harmonization of international environmental legislation, voluntary agreements (such as the international eco-management system ISO 14000, or the European Union eco-management system EMAS), and ecologically enlightened consumers. Thus, according to the Federation of Swedish Industry, Sweden should first and foremost act in favour of creating international agreements, not trying to become a green island in an archipelago of polluting nations.

A third coalition – which is the weakest one of the three – calls for radical and immediate change of Swedish society. This coalition includes parts of the environmental movement in Sweden, e.g. the Green Party and the Friends of the Earth, a number of organizationally free-standing activists at the community level, and, probably, many young people. This coalition tries to combine the quest for global justice with that of ecological sustainability. Friends of the Earth tries, still with rather limited success, to introduce the concept 'environmental space' in Sweden.

Descending from the national to the municipal level, the overall picture is not unequivocal. One can easily find tendencies going in opposite directions. One urban policy strategy that is popular among many municipalities, towns and cities all over Europe is *the entrepreneurial strategy* aiming at successful competition for capital and workplaces ('city marketing'; 'selling places'). Alongside with this growth-orientated strategy we find the strategy of making municipalities, towns and cities *ecologically sustainable*. Of course, the latter strategy has been triggered by Rio and Agenda 21, although one should not underestimate its foundation in earlier environmentalist ideas and practices, and in locally articulated needs and demands. Looking at the strategies pursued by Swedish municipalities, they do not easily fit into either of these two categories, but one is rather struck by the somewhat contradictory co-existence of the two: they seem to indicate the presence of competing urban coalitions/regimes, i.e. urban growth coalitions and environmental coalitions. Some municipalities seem willing to become number one with regard to entrepreneurial skill and economic prosperity *as well as* sustainability, and they display an impressing variety of policy innovations to approach these goals. Indeed, this dual strategy nicely reflects the growing strength of the ecological modernization perspective.

Swedish municipalities are crucial actors when it comes to sewerage and waste treatment, recycling, green public purchase, green consumption, green accounts, etc., and thus they may initiate and support citizen initiatives leading towards sustainability. However, with regard to energy, traffic, heavy infrastructure, environment protection and agriculture their scope of action is narrower, and the two first mentioned coalitions in this section are difficult to challenge. In these areas business interests, the central state and extra-national actors such as the European Union and transnational companies are actors of a much heavier weight. If these interests move too slowly towards sustainability – or worse, if they step backwards – there is a risk that the relative successes experienced in the Agenda 21 work so far will be followed by a backlash: why should citizens, environmental organizations, municipalities and other involved in LA21 engage in something that lacks support and may even become blocked by elites at the national and international levels?

Note

1   A TeraWatt is one million million watts (1,000,000,000,000).

References

Agenda 21 United Nations Conference on Environment and Development, Rio de Janeiro, 3–14 June 1992. New York: United Nations.

Berggren, B. (1997) Interview with Bo Berggren, chairman of the National Association of industrial companies, pp. 1–2 in *Sverige i rörelse* (Sweden in motion) No. 17 (May 1997).

Bjur, H. and Engström, C.-J. (1993) *Framtidsstaden: diskussion om planering för bärkraftig stadsutveckling* (The city of the future: discussion about planning for sustainable city development). Stockholm: Swedish Council for Building Research.

Cornelius, T. (1997) Interview with Tom Cornelius, managing director of the Association for Private Car Drivers, p. 9 in *My Car. A Magazine from SAAB*, No. 2.

DESD (Delegation for Ecologically Sustainable Development) (1997) *Förslag på investeringsprogram* (Proposal for a programme of investments), 21 March 1997. Stockholm: Delegationen för ekologiskt hållbar utveckling.

EC (European Community) (1986) *Fourth Environmental Action Programme, 1987–92* COM (86) 485 final. Luxemburg: Office for Official Publications of the European Communities.

—— (1992) *Towards Sustainability. A European Community Programme of Policy and Action in Relation to the Environment and Sustainable Development.* Brussels (March).

Eckerberg, K. (1995) 'Environmental planning: dreams and reality', pp. 115–136 in A. Khakee, I. Elander and S. Sunesson (eds) *Remaking the Welfare State. Swedish urban planning and policy-making in the 1990s.* Aldershot: Avebury.

Edelman, M. (1964). *The Symbolic Uses of Politics.* Urbana, Chicago and London: Illinois University Press.

—— (1977) *Political Language. Words that Succeed – Policies that Fail.* New York: Academic Press.

Elander, I. (1978) *Det nödvändiga och det önskvärda. En studie av socialdemokratisk ideologi och regionalpolitik 1940–72* (The necessary and the desirable. A study on Social Democratic ideology and regional policy 1940–72). Lund: Arkiv.

—— (1996) 'Central–Local Government Relations and Regionalism in Sweden', *Österreichische Zeitschrift für Politikwissenschaft* (special issue on Lokale Politik in Europa) 25 (3): 279–294.

Elander, I. and Montin, S. (1990) 'Decentralization and control. Central–local government relations in Sweden', *Policy and Politics* 18 (3): 165–180.

Elander, I., Lidskog, R. and Johansson, M. (1997) 'Environmental policies and urban planning in Sweden. Goals, strategies and instruments', *European Spatial Research and Policy* 4 (2): 5–35.

Elander, I., Gustafsson, M., Sandell, K. and Lidskog, R. (1995) 'Environmentalism, sustainability and urban reality', pp. 85–114 in A. Khakee, I. Elander and S. Sunesson (eds) *Remaking the Welfare State. Swedish urban planning and policy-making in the 1990s*. Aldershot: Avebury.

Eriksson, O. (1991) 'Vetenskap som politiskt alibi. Några reflexioner efter läsning av storstadsutredningen' (Science as political alibi. Some reflexions after reading the investigation on cities), *Tidskrift för arkitekturforskning* 4 (2): 113–117.

FoE (Friends of the Earth) (1997) *Mål och beräkningar för ett hållbart Sverige* (Goal and estimations for a sustainable Sweden). Stockholm: Friends of the Earth Sweden.

Forskergruppen for det nye hverdagslivet (1987) *Veier til det nye hverdagslivet* (Roads to the new everyday life) NORD 1987: 61. Oslo, København, Stockholm & Helsinki: Nordisk Ministerråd.

Government Bill 1969, No. 28 Förslag till miljöskyddslag (proposal for Environmental Protection Act).

Government Bill 1987/88, No. 85 Miljöpolitiken inför 1990–talet (Environmental Policy at 1990s).

Government Bill 1992/93, No. 180, Riktlinjer för en kretsloppsanpassad samhällsutveckling (Guidelines for ecocyclical development of society).

Government Bill 1993/94, No. 100, appendix 1:4 Uthållig utveckling (Sustainable Development).

Government Bill 1993/94, No. 111, Med sikte på hållbar utveckling (Aiming at sustainable development).

Government Bill 1996/97, No. 29, Höjning av koldioxidskatt för industrin och växthusnäringen (Increase of the carbon dioxide tax for the industry).

Government Bill 1996/97, No. 53, Infrastrukturinriktning för framtida transporter (Directions of infrastructure for future transports).

Gyll, S. (1996) Interview with Sören Gyll, director of Volvo, p. 1 in *Sverige i rörelse* (Sweden in motion) No. 4.

Industriförbundet (1998) *Fortsatt hållbar utveckling – industrins miljöarbete* (Continued sustainable development – the environmental work of industry). Stockholm: Industriförbundet.

Jamison, A., Eyerman, R. and Cramer, J. (with Lassoe, J.) (1990) *The Making of the New Environmental Consciousness. A Comparative Study of the Environmental Movements in Sweden, Denmark and the Netherlands*. Edinburgh: Edinburgh University Press.

Kaul, S. (1991) 'Hvor baerekraftig er det byökologiske prosjekt?' (How sustainable is the local ecological project?) *Tidskrift för arkitekturforskning* 4 (2): 25–36.

Khakee, A. (1995) 'Politics, methods and planning culture', pp. 275–296 in A. Khakee, I. Elander and S. Sunesson (eds) *Remaking the Welfare State. Swedish urban planning and policy-making in the 1990s*. Aldershot: Avebury.

Lidskog, R. (1994) 'The politics of radwaste management: civil society, the economy and the state', *Acta Sociologica* 37 (1): 55–73.

Lundqvist, L.J. (1971) *Miljövårdsförvaltning och politisk struktur* (Environmental policy administration and political structure). Lund: Prisma.

McLaren, D. (1992) 'London as ecosystem', pp. 56–68 in A. Thornley (ed.) *The Crisis of London*. London and New York: Routledge.

Malbert, B. (ed.) (1992) *Ekologiska utgångspunkter för översiktlig planering.* (Ecological points of departure for comprehensive planning). Stockholm: Swedish Council for Building Research.

MoE (Ministry of the Environment) (1997) *From Environmental Protection to Sustainable Development. National Report on the Implementation of Agenda 21.* Stockholm: Ministry of the Environment.

Myerson, G. and Rydin, Y. (1996) *The Language of Environment. A New Rhetoric.* London: UCL Press.

Naess, P. (1989) 'Sustainable urban development: The challenges of the Brundtland Commission – A turning point for urban planning', *Scandinavian Housing and Planning Research* 6 (1): 45–50.

NUTEK (Swedish National Board for Industrial and Technical Development) (1997) *Energiläget 1997* (The energy situation 1997). Stockholm: NUTEK Förlag.

Ödmann, E. (1987) 'Vague norms in a tough reality – some remarks on the new Swedish Planning Act', *Scandinavian Housing and Planning Research* 4: 263–268.

OECD (Organization for Economic Cooperation and Development) (1996) *Environmental Performance Review of Sweden.*

Official Letter from the Parliament, *Rskr 1994/95*, No. 120, Stockholm.

Orrskog, L. (1993) *Planering för uthållighet* (Planning for sustainability). Stockholm: Byggforskningsrådet.

SALA (Swedish Association of Local Authorities) (1995) *Agenda 21 i Sveriges kommuner. Redovising av enkätundersökningen 1995.* Stockholm: Svenska kommunförbundet (Swedish Association of Local Authorities).

Sifo (1998) *Svenska folket syn på olika frågor. Viktigaste valfrågorna.* (Swedish people's views on different questions. Most important questions in the general election). Stockholm: Sifo Research and Consulting.

SNF (Swedish Society for the Conservation of Nature) (1996) *Nationella hinder för Lokal Agenda 21* (National obstacles for Local Agenda 21). Rapport 96/9308. Stockholm: Naturskyddsföreningen.

SNV (National Environmental Protection Agency) (1993) *Environmental Protection. The Swedish Experience.* Stockholm: The National Environmental Protection Agency.

—— (1996) *Biff och bil? Om hushållens miljöval* (Beef and car? On the environmental choice of households). Stockholm: Swedish Environmental Protection Agency [report No. 4542].

Söderqvist, T. (1986) *The Ecologists.* Stockholm: Almqvist & Wiksell.

SOU (1990a) Government Commission Report No. 32, Staden (The city).

—— (1990b) Government Commission Report No. 33, Urban Challenges (in English).

—— (1990c) Government Commission Report No. 34, Stadsregioner i Europa (City regions in Europe).

—— (1992) Government Commission Report No. 104, Vår uppgift efter Rio (Our task after Rio).

—— (1994) Government Commission Report No. 128, Local Agenda 21 – en vägledning (Local Agenda 21 – a guidance).

—— (1996) Government Commission Report No. 48, Shaping sustainable homes in an urbanizing world: Swedish national report for Habitat II.

Stren, R. (1992) 'Introduction', pp. 1–7 in R. Stren, R. White and Whitney, J. (eds) *Sustainable Cities: Urbanization and the Environment in International Perspective.* Boulder: Westview Press.

Weale, A. (1992) *The New Politics of Pollution.* Manchester: Manchester University Press.

von Weizsäcker, E., Lovins, A.B. and Lovins, L.H. (1997) *Factor Four: Doubling Wealth, Halving Resource Use*, Sydney: Allen and Unwin.

World Commission on Environment and Development (1987) *Our Common Future. The Brundtland Report*. Oxford: Oxford University Press.

11 Poland

On the way to a market economy

Tadeusz Markowski and Helena Rouba

Introduction

The economic and social well-being of contemporary nations, regions and cities depends more and more on complex interactions framed at the global scale. The influence of these processes, however, depends on the degree of political and economic 'openness' of the countries concerned. In 1945, upon the decision of the victorious powers, Poland had to join the bloc of states which were politically and economically dominated by the USSR. Poland was thus cut off from the processes determining the development of the contemporary world economy: the scientific and technical revolution, and the free flow of information, technology, capital, goods and people. The ineffective system of a central steering of the economy was an additional element which delayed the processes of growth and development. It was only in 1989 that the political turn of events in Poland, followed by other 'socialist countries', opened up for this part of the world the possibility of following the path of democratic development and the market economy. Thus, mainly for political reasons, Poland is now lagging some fifty years behind the developed countries, and trying within a few years to narrow the gap between itself and the developed world, to join the global economy and become an equal partner in resolving global problems.

The coincidence of Polish reforms with the time of preparation for the Earth Summit in Rio de Janeiro and subsequent implementation of the Rio agenda has created a good climate for establishing sustainable development policy in Poland. Agenda 21 is increasingly providing a stimulus for the programmes of sustainable development of many cities and towns.

More than 60 percent of Poland's population live in urban areas which mostly perform industrial functions. The transformation of cities is thus closely related to economic restructuring aimed at reducing the role of industries, especially the ones harmful to the environment. In the field of housing and spatial management there is also a need for reforms and policies allowing the consumption of energy and raw materials to be reduced and the urban environment to be improved. In order to understand the present situation of Polish cities and changes in the direction of a more sustainable form of development it is essential first to consider the political and economic conditions that determined urban development before the introduction of the market economy.

Economic policy and attitude to environmental issues in the period of the centrally planned economy

In the years 1945–1989, within the principles of the socialist system, the direction of Poland's economic development followed the extensive industrialisation model imposed by the USSR. This model was regarded as the driving force of progress, the chief source of new jobs and foreign currency and a means of transforming the class structure of society and of joining in the arms race. The structure of industry was dominated by the mining and processing of coal, sulphur and copper, by the iron and steel industry, and such material-intensive branches of the electromechanical industry as shipbuilding and the production of rolling stock and plant and machinery for the extractive and building industries. Relatively less attention was devoted to the development of a consumer goods industry, which resulted in permanent shortage of basic commodities on the domestic market.

The centralised 'economy of shortage' (Kornai, 1993) did not provide stimuli for raising the quality of products. The state monopoly destroyed internal competition among enterprises, and did not help to reduce the costs of production. Such economies did not develop in total isolation from market economies, which provided some opportunity to test the competitiveness of local industries on international markets. Technological backwardness, which was a permanent structural element of all socialist economies, forced these countries to seek access to new technologies – if only for military purposes – by means of commercial exchange on the global market. However, due to the low technological standards of most goods, participation in international exchange was only possible through the export of raw materials or by reducing prices. One of the ways which allowed the country to attain a certain degree of competitiveness and to obtain foreign currency was excessive, wasteful exploitation of the natural environment. This, in turn, gave totalitarian governments access to strategic goods securing their survival. In almost every centrally steered economy it was common practice not to include the cost of land among the costs of production. Nationalised land was treated as a 'free good'.

The lack of ownership rights and market pricing of land and real property under the previous regime resulted in extravagant use of urban space. Spatial development plans allocated too much space for industrial use and other public purposes, so it happened frequently that for many years these spaces went unused. Housing development was aimed at meeting the quantitative housing needs of the growing urban population. Towns were regarded as housing infrastructure for large industrial plants. Sometimes housing estates were built in zones affected adversely by enterprises harmful to the environment. The standard model of housing development was the gigantic complex of poor quality blocks of flats. These developments formed a ring around most Polish towns and cities – regardless of differences in local conditions. Almost identical blocks of flats were built in all towns and cities, and also even in the villages to serve the needs of state farms. This kind of development resulted in the destruction of social structure and the obliteration of regional differences (Kozłowski, 1994).

Central authorities declared their concern about environmental problems, but this concern remained in the sphere of declarations made chiefly for propaganda purposes, to strengthen official credibility with credit-granting countries and institutions. That is why, despite a wide range of environmental legislation being introduced, the environment, especially in urban and industrial areas, steadily deteriorated (Karaczun and Indeka, 1996). It is not surprising then that environmental problems were one of the reasons for the growth of political consciousness in Polish society. It became clear that without fundamental political reforms nothing could be improved in Poland, neither in the economy and living conditions nor in the state of the environment.

Ecological awareness

Attempts at protecting the natural environmental in Poland date back to the early period of the Polish state, as indicated by royal orders of the eleventh century concerning protection of certain species of animals (notably beavers), as well as numerous later regulations (e.g. the Forest Protection Proclamation of 1778, the 1875 Act on Protection of Birds and relevant school programmes providing knowledge in this field). An Act on Nature Protection was passed on 31 October 1918 (preceding the proclamation of Poland's independence by 11 days). The interwar legislation demonstrates a comprehensive approach to environmental problems which was unprecedented in those times. In the interwar period Poland ratified numerous conventions which created the legal foundation for international co-operation at the global scale. The legislative achievements of the Second Polish Republic in the field of nature protection compared favourably with progressive solutions adopted by other countries (Paczuski, 1994). Many of the pre-war (Second World War) legal acts were still in force in socialist Poland, some of them even for many years.

Progressive degradation of the natural environment during the period of the centrally planned economy, despite fairly good environmental legislation, resulted in growing ecological awareness of the society, particularly in the 1980s, due to publication of real information on environmental pollution and the resulting economic and social losses ('Raport Ligi . . .', 1981). Numerous community-led actions and initiatives gave rise to the formation of new ecological organisations (e.g. the Polish Ecological Club), the reactivation of already existing social organisations (e.g. the Nature Protection League), and incorporation of environmental issues in the activity of political organisations (e.g. Freedom and Peace). Various religious organisations also showed concern for ecological problems. The Chernobyl disaster and subsequent protests against building a nuclear power plant in Poland, and against other investments (e.g. construction of a storage dam on the river Dunajec) contributed greatly to the growth of ecological sensitivity of the society and the mobilisation of social movements.

It has been estimated that the ecological movement in Poland now comprises about two hundred formal and informal ecological organisations formed by various social youth or religious groups (some estimates even suggest there are up to a thousand such groups). They are either solely pro-ecological groups, such as the Nature Protection League, the Polish Ecological Club, the Nature Protection Guard, the Social Committee for the Protection of the Mazurian Lakes, the Green Federation and the Inter-University Ecological Lobby, or they treat ecological problems as part of their activity (e.g. the Polish Sailing Association, scouting organisations). Only some dozen groups, operating mainly in large cities, have permanent qualified staff as well as technical and financial resources sufficient to cover a whole region or the country. None of them is yet so strong as to play an important political role, like, for example, the Green Party in Germany. Local authorities do not yet fully appreciate the integrative role of non-government organisations. However, in 1996 the Ministry for Environmental Protection designed a programme of co-operation with ecological organisations.

Numerous studies (e.g. Kalinowska 1994; Baturo *et al.*, 1997) have shown that sensitivity to environmental issues in Polish society is growing fast. The group mostly concerned about environmental needs comprises young people with education and above average income. Of this group, 67 percent have declared readiness to accept additional taxation for environmental purposes; 52 percent agree to pay up to 10 zlotys {złotych}(about US$3.50) per month; 90 percent are willing to sort their rubbish. The chief motives for increased interest in environmental problems are growing recognition of personal responsibility for health, rise in the prices of energy and waters as well as a fashion for 'environmental correctness'. About 60 percent of those surveyed regard local government as having the greatest role in improving environmental conditions in their neighbourhood.

Ecological awareness is no doubt a result of, among other things, educational progress. Environmental issues have recently been included in school curricula at all levels. Work on the Polish Strategy of Ecological Education, initiated by a resolution of the Polish Parliament in April 1995, is nearly completed. The number of publications and textbooks on ecological issues has increased significantly. Also television, which is the main source of information for a great part of the society, broadcasts serial educational programmes. Yet it is surprising that only 18 percent of the persons surveyed in 1997 had heard about the Rio Summit, only 3 percent knew what Agenda 21 was, and 10 percent knew what the Rio Declaration was. Sixty percent of the leaders of ecological organisations stated that Agenda 21 was not a decisive motive for their activity (Baturo *et al.*, 1997: 63). These results seem to point to insufficient propagation of the Rio documents in Polish society. After 1990 environmental problems lost the political weight which they had in the 1980s – hence their disappearance from the political agenda during presidential and parliamentary elections in recent years.

Economic and environmental policy in the period of system transformation

The effects of the economic reforms

The rapid change in the fundamental principles of the Polish economy compounded by the breakdown of economic ties within the entire socialist bloc brought about economic depression. Industrial production contracted rapidly (by 25 percent in 1990 compared with 1988). A new phenomenon, unknown in socialist Poland, appeared and grew quickly – unemployment. By the end of 1993, 2.9 million people were unemployed, or 16.4 percent of the working population. The real value of average monthly pay dropped in 1990 by a quarter compared to 1989 and decreased steadily until 1994. Prices of goods and services in 1990 increased sevenfold on average compared with those of the preceding year.

The 'shock therapy' of the reforms had a great impact on Polish cities because the collapse of industry resulted in huge unemployment and under-mined the economic base of the cities. Severe economic recession, especially in the initial stage of transformation, affected those cities where mono-branch industry linked to the Eastern bloc prevailed (as in Łódź) and those smaller cities where a single big industrial enterprise provided the main workplace and source of income for the population. The cities with a differentiated economic structure (like Warsaw) were less affected by recession. But every-where new economic principles have created a great challenge for regional and local authorities, urban planners and managers to reconstruct the economic base of the cities and spatial management practice.

In recent years the reforms have started to bring tangible, positive effects. The first symptoms of improvement were a better supplied market, declining annual inflation (to 13 percent in 1997) and decreasing unemployment (to below 11 percent at the end of 1997). The most evident sign of the improving economic condition is the rising rate of economic growth (Figure 11.1). The private sector is rising and consolidating – its share of GNP increased to 60 percent at the end of 1996. Nearly 80 percent of registered business entities were privately owned at the end of 1996. Private companies predom-inate in building, trade and real estate agency. The share of private companies is smaller in sectors such as electricity, gas and water provision and in mining.

In line with world trends the role of manufacturing industry in the economic structure is decreasing. Its share of GNP dropped by 5 percent between 1992 and 1995. Labour productivity in industrial production grew markedly. The continued high rate of investment in enterprises is an indication of the

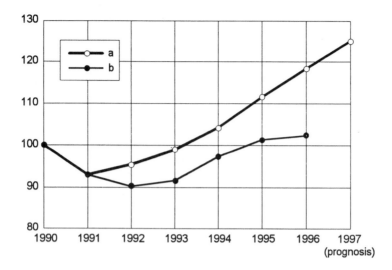

Figure 11.1 Dynamics of GNP (a) and domestic consumption of electric power (b) in Poland (1990 = 100, constant prices)

Sources: *Rocznik Statystyczny 1996* (1997: lxxi); *Mały Rocznik Statystyczny 1996* (1997: 455).

intensified process of modernisation. In 1996 the investment rate was 26 percent higher than in 1995, with the largest outlays on the purchase of plant, machinery and equipment. The inflow of foreign capital, bringing modern technologies of production and methods of management to Poland, is growing. At the end of 1997 foreign investment was over US$17 billion. Significant investment growth has occurred in most sectors of the processing industry, especially in plants producing highly processed goods, such as machines and electrical equipment, medical and precision instruments and motor vehicles. These are favourable tendencies from the viewpoint of environmental improvement.

The process of intensive reform has also caused some adverse phenomena, which may affect further societal assent. One of them is the stratification of society in terms of prosperity. Estimates show that approximately 2.5 million people in Poland live below the subsistence level; that is, their income is about US$36 per month per household member. About 1.5 million families receive social welfare assistance. Particularly difficult is the situation of numerous families experiencing unemployment and suffering from above average social pathology. The housing sector is still in crisis. There are less than 290 flats per 1,000 inhabitants in Poland, which is below the European standard. Housing investment dropped from 6.3 percent of GNP in 1980 to less than 2 percent in 1995. Over the last years we have witnessed a gradual drop in the number of new flats being constructed (from 137,000 in 1991 to 67,000 in 1995). All government programmes designed to improve the housing situation are still beyond the reach of low income families.

Ecological policy

Polish society is now facing the great challenge of catching up with global advancement, which necessitates intensive development based not on the simple exploitation of exhaustible resources but on the use of the latest technologies, modern organisation of work, educational advancement, and careful, economical use of natural resources. Such a model of growth is consistent with the idea of sustainable development, termed 'ecodevelopment' in Poland. According to the Institute for Sustainable Development (Baturo *et al.*, 1997: 101), this term is defined as socio-economic development which, in order to secure equal opportunities for access to the natural environment for contemporary and future generations, integrates political, economic and social activities to ensure balance in the natural environment and continuity of its fundamental processes.

The principle of equal access to the natural environment denotes intergenerational equity, interregional and inter-group equity, balancing local and supralocal interests, and balancing the interests of humans and the natural environment. Ecodevelopment aims to:

- maintain reproduction of renewable resources,
- replace non-renewable resources with substitutes where possible,
- gradually eliminate harmful impacts on the environment so as not to exceed its endurance capacity,
- protect and, if possible, reconstruct biological diversity with respect to: landscape, ecosystem, species and genes,
- create conditions for fair competition between businesses with regard to access to limited natural resources and waste management,
- encourage public participation in decision-making processes, especially those concerning the local environment,
- provide for the ecological safety of human beings, understood as creating conditions favourable to physical, mental and social health (creating and cultivating local ties).

Ecodevelopment, understood as above, represents a considerable extension of the principles of the 'Ecological Policy of the State' (1992) adopted by the Polish Parliament in May 1991 as the official strategy of Poland's human development. This document formulated the main policy directions for all sectors of the economy, specified legal, administrative and economic instruments for nature conservation, and identified short-, medium- and long-term priorities. Equal importance was attached to economic and ecological issues in the country's development. Although accomplishment of these goals is obstructed by economic pressures, the high political status of this document allows its resolutions to be referred to in all work connected with the restructuring of the state and its functions. The implementation of state ecological policy was later the subject of parliamentary debates and in January 1995 the Parliament adopted the 'Programme of State Ecological Policy Implementation Until the Year 2000', which emphasised compulsory observance of the principles of sustainable development by all central government bodies.

In 1995 the Committee for Ecodevelopment was established. This body was headed by the Minister for Protection of the Environment, Natural Resources and Forestry, in order to co-ordinate economic and ecological policy and supervise the implementation of Agenda 21. Thanks to the 1990 Local Government Act, protection of the environment has become an important element of local policy. For many local governments it is the principal guideline of their development programmes. In most cases these programmes follow Agenda 21 principles. A number of cities have joined a pilot project of the National Programme of Actions for Environmental Conservation. Several cities have signed the Charter of European Cities and Towns Towards Sustainability. Some of them (i.e. Gdańsk) have joined the International Council for Local Environmental Initiatives, or have taken other initiatives (i.e. Łódź has joined the European network of 'Healthy Cities' – WHO/Euro).

After 1989 Poland succeeded in creating an extensive system for financing the protection of the environment (including the use of non-budgetary sources). Major components of this system are national, regional and local funds for environment protection and water management, supported by the Environment Protection Bank. These funds collect money from environmental charges and fines, which are then allocated for subsidies and low-interest credits for pro-ecological investments. The funds are also supported by money transferred from the Ecofund, which draws financial means from so-called 'eco-conversion'. This is conversion of part of the Polish foreign debt into subsidising pro-ecological investment projects. In the years 1993–1996, donations from the Ecofund were granted to 240 projects (among other purposes for protection of the ozone layer, the Baltic Sea and species threatened with extinction). Capital outlays for environmental protection, although inadequate to actual needs, are growing continually. Since 1993 they have exceeded 6 percent of total national capital expenditure and 1 percent of GNP (in 1990 the figures were, respectively, 3.6 and 0.7 percent).

Over the last few years Poland's participation in international co-operation in the management of global and environmental problems has increased. Central institutions as well as ecological organisations in Poland participated actively in the 1992 Earth Summit in Rio de Janeiro. A comprehensive 'National Study of Biological Diversity' was elaborated and Poland signed the Conventions on Earth Climate Protection and Biological Diversity. Poland has also signed many other international conventions and protocols on environment protection and has taken part in local and regional environmental initiatives; for example: development of programmes for 'Local Agenda 21', the launch of the 'National Programme of Action for Protection of the

Environment for Central and East European Countries', the institution of foundations for environmental protection, such as the 'Clean Vistula and Seaside Rivers' foundation under the patronage of the Ministry for Environment Protection and the National Fund for Environmental Protection, and the institution of a 'Centre for Environmental Hazards Diminution for Central and East European Countries' (in Katowice).

When assessing the execution of the ecological policy of the state in the transformation period, it is important to take into account not only those solutions which directly relate to environmental protection but also other solutions constituting a basis for restructuring the functioning of the whole state, as they also affect the behaviour of business entities and individual citizens. The introduction of market mechanisms in the economy has increased the sensitivity of business entities to economic incentives, including charges for utilisation of the environment and penalties for disregarding ecological requirements. Combined with higher prices of energy, fuels and other raw materials, this has resulted in more careful and economical use of the environment's resources.

The system reforms which are being carried out also cause certain difficulties in the implementation of ecological policy. The significant growth in the number of businesses makes it difficult to ensure that environmental requirements are respected. Increased importation of foreign goods and the approach of consumption expectations to the model prevalent in Western countries has led to a significant increase in the amount of wrappings, including non-returnable packaging, which has in turn has produced much more communal waste. Growth in the number of cars also means increased air pollution.

Despite Polish society's increased ecological awareness, the high social cost of the reforms sometimes results in ecological considerations being ignored. This is evident in the protests of employees threatened with unemployment resulting from the closure of environmentally harmful factories (e.g. those producing goods containing asbestos), or the protests by entire vocational groups against the restructuring of inefficient industry (e.g. coal mining).

Barriers to a higher rate of progress in the pursuance of environmental policy are subsidised prices of energy; inefficient, environmentally hazardous industrial plants; the exemptions of some such plants from ecological charges and fines; technical and financial constraints on the elimination of environmental pollution; and insufficiently pro-ecological social behaviour. Among the most urgent problems waiting to be resolved is the reduction of energy consumption in the Polish economy (although the growth of GNP in recent years is greater than the rise in the use of electricity, see Figure 11.1). Seventy-five percent of energy produced is based on Polish resources of black and brown coal, the extraction and processing of which are very hard on the environment. Reducing energy consumption is, however, a very costly undertaking. The modernisation of the power industry alone would need an expenditure of $50 billion up to the year 2010, with similar figures for the metals industry, chemical industry and other environmentally adverse branches of industry.

The unit consumption of fuel and energy in industry has dropped markedly only in the case of a few products (e.g. synthetic rubber, pig-iron and steel, and electrolytic copper), whereas for some products it even grew (e.g. brown-coal mining, electrolytic zinc). The 'end-of-pipe' treatment aimed at fulfilling control requirements still prevails in Polish industry. Too many enterprises feel responsible for their products only up to the moment of sale. Little regard is paid to the amount of materials and energy used in production, or to the resulting waste. It is essential to find ways more effectively to enforce environmental protection through tighter controls and appropriate economic instruments (Markowski, 1994).

Changes in the state of the environment

Despite many difficulties, economic reforms and actions aimed at protection of the environment during the period of system transformation are slowly but surely yielding fruit. The forest area is gradually expanding, and now covers 28 percent of the country (while in 1946 it was 20.8 percent). The number and acreage of national parks, nature reserves and protected landscapes is also growing. The total area of land under legal protection constitutes 26.1 percent of the country's area. Harmful emissions to the atmosphere have been substantially reduced in recent years (between 1990 and 1995: dust by 33 percent, sulphur dioxide by 27 percent). Untreated sewage carried away to surface waters has dropped by 54 percent (see Figures 11.2 and 11.3).

The sad heritage of past disregard of environmental requirements has not yet been entirely eliminated. The state of the environment in Poland, particularly in large cities and industrial regions, is still unsatisfactory. Random surveys show that acceptable concentrations of dust in the air are exceeded in all Polish cities – in 15 percent of cases by as much as five times. Concentrations of nitrous oxide exceed admissible norms by two or three times. Despite the construction of many new wastewater treatment plants, only 65 percent of the urban population (and only 3 percent of the rural population) is served by such treatment plants. Even the largest Polish cities, such as Warsaw or Łódź, do not have proper sewage treatment installations.

The management of waste is another difficult problem. The amount of industrial wastes produced every year is not decreasing (Figure 11.3), and the amount of communal waste is steadily growing. Only an insignificant proportion of these wastes is neutralised. Almost all waste is disposed of in dumps poorly suited for this purpose. The appointment of new locations for dumps and waste utilisation plants is very problematic, as it gives rise to violent conflicts between local communities and authorities. Not one large city in Poland has resolved this problem satisfactorily. In extreme cases local governments have to carry away their refuse to locations several hundred kilometres distant (e.g. from Łódź to Upper Silesia). The need to make waste

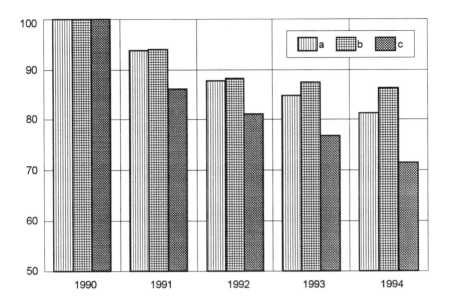

Figure 11.2 Total emissions of main sources of air pollution in Poland (1990 = 100): (a) SO$_2$; (b) NO$_2$; (c) dust

Sources: *Rocznik Statystyczny 1996* (1997: 30); *Mały Rocznik Statystyczny 1996* (1997: 45).

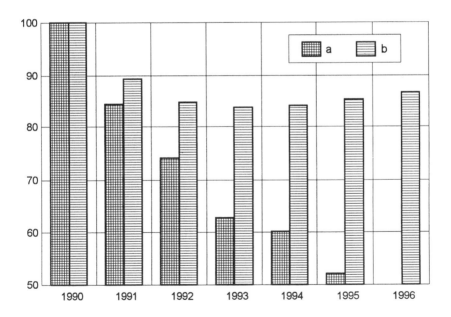

Figure 11.3 Production of wastes and waste water in Poland (1990 = 100): (a) industrial and municipal waste water, not treated, discharged to surface water, produced in the year; (b) industrial wastes noxious to the environment, produced in the year

Sources: *Rocznik Statystyczny 1996* (1997: 35, 27); *Mały Rocznik Statystyczny 1996* (1997: 33, 45).

management more effective is becoming increasingly urgent. The desired changes should include the sorting of waste, recycling and the introduction of modern methods of waste neutralisation. It seems that even a few examples of modern, environmentally friendly waste treatment plants might favourably affect the hostile attitude of residents to locating such land uses in their neighbourhoods.

The problem of waste is compounded by the fact that opening up Poland to contacts with neighbouring countries has brought the *importation* of waste from the West. This waste has been deposited on farms, on the premises of enterprises going bankrupt, and even in forests. In sum, the effective management of waste is one of the priorities to be dealt with on our path to integration with the European Union.

Environmental and structural problems of Polish cities

Urbanisation processes

The 'urban population' in Poland includes residents living in the boundaries of localities with the administrative status of towns or cities. The process of post-war urbanisation in Poland started from a very low level (Figure 11.4). In 1946 only 33.9 percent of the population (8 million people) lived in towns and cities. As a result of urbanisation connected mainly with industrialisation, the size of the urban population rose to 61.7 percent in 1990. In the 1990s the rate of growth of the urban population slowed (in 1996 it comprised 61.9 percent of the total population). The main reasons for this

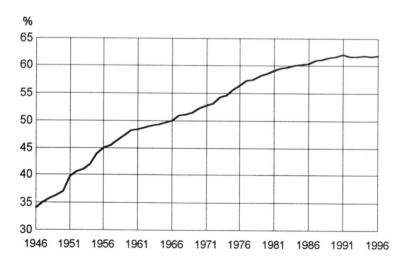

Figure. 11.4 Urban population in Poland as a percentage of total population
Sources: *Rocznik Statystyczny 1996* (1997: 21, 39); *Mały Rocznik Statystyczny 1996* (1997: 55).

were decline in the birth rate (in cities it dropped to 0.03 percent, and in the country to 0.24 percent), decreased migration from the country as a result of shrinking demand for labour in towns, and the crisis in housing. This might suggest the beginning of a de-urbanisation phase in Poland. If this trend persists, one of the most urgent tasks in Polish economic and spatial policy will be acceleration of the restructuring of rural areas. Such restructuring should be targeted at the creation of new jobs outside agriculture and the simultaneous development of technical and social infrastructure which would speed up economic growth and improve the quality of life in rural areas.

A factor conducive to suburbanisation and de-urbanisation in Poland is the growth of motorisation. In the years between 1990 and 1996 the number of cars on the roads rose by 53 percent and now nearly half of all households own a car. A high rate of private car sales (an increase of 33.6 percent between 1995 and 1996) suggests that this trend will persist in the coming years. These processes will contribute to the increase in the cost of the functioning of the increasingly vast urbanised areas and are preconditions for further social differentiation in access to the environment.

The urban settlement network

The present settlement network of Poland, inhabited in 1996 by 38.6 million people, is made up of nearly 58,000 settlements, including 860 towns and cities. In 1996 these towns and cities were inhabited by 26.1 million people. The largest group of towns in Poland is comprised of 450 towns of under 10,000 inhabitants (8.8 percent of the total urban population). Half of Poland's urban population lives in 42 cities of population greater than 100,000. Only 20 of them can be viewed as comprising the metropolitan network, i.e. numbering above 200,000 inhabitants (Figure 11.5). In 1950 there were only six such cities. The biggest is the capital, Warsaw, with 1.635 million inhabitants.

In comparison with many other countries, the urban population in Poland is characterised by only a moderate level of concentration. As Figure 11.6 shows, large cities developed the most quickly. During the so-called 'socialist industrialisation' period large developments, mainly in heavy industry connected with the establishment of the military-industrial complex of the Warsaw Pact states, were concentrated mainly in the areas of hitherto significant urbanisation (i.e. Upper Silesia, the Warsaw district). Although industry developed also in smaller towns, several dozen of which acquired civic rights, the highest population increase after 1945 occurred in towns of above 200,000 population (Parysek and Kotus, 1997).

Apart from industrialisation important factors affecting the urbanisation processes in Poland are administrative and educational functions. The number of *voivodships* ('*voivodship*' is the name of a territorial unit between national state and local commune level) in Poland grew in 1975 from 17 to 49,

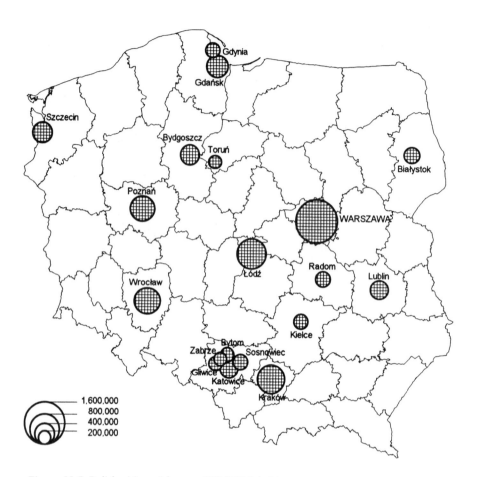

Figure 11.5 Polish cities with over 200,000 inhabitants

which brought about the rapid development of a number of smaller towns that acquired the status of *voivodship* and corresponding administrative functions.[1] System reforms in Poland have also made possible the development of private education at all levels. Owing to this, many small towns have developed not only educational functions but also considerable research opportunities, which have significantly enhanced the importance and attractiveness of these cities.

It can reasonably be expected that structural changes already apparent in the functioning of the Polish economy will expand the functional structure of towns, with a simultaneous reduction of the importance of the industrial function and its related harmful effects on the urban environment.

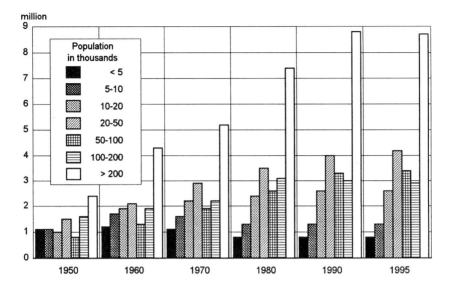

Figure 11.6 Population by size of towns and cities in Poland.
Sources: *Roczniki Statystyczne z lat 1950–1996*, Warszawa: GUS.

The level of development and spatial management of Polish cities

A characteristic feature of many Polish cities is the existence of neglected and degraded areas. They are comprised of housing estates, industrial areas and even town centres with commercial and service functions. The main reasons for this situation seem to be the low standard of buildings already existing at the time of their construction and the faulty management of urban resources and space in the period of the centrally planned economy. Then, no regard was paid either to ownership rights or to the value of land and buildings. The state takeover of the management of most housing resources (which prior to 1945 had been privately owned), and the centralised setting of rents at very low levels, precluded former owners from having any real influence on the use of these resources and their maintenance. As a result even parts of cities formerly regarded as elegant and prestigious were neglected and even fell into complete disrepair. Pre-1945 buildings constitute about one-third of the total housing stock in cities.

Technology used for new buildings and entire housing estates in the centrally planned period was mainly based on the large-panel industrialised construction of block housing. Despite the continued improvement of technological standards, this type of building now represents substandard housing with high costs of maintenance as well as substantial economic depreciation. In some Polish cities this type of high-rise construction is inhabited by 30 to 40 percent of residents. Such areas are now becoming post-socialist clusters of slums with concentrations of lower-income groups. Rehabilitation of these areas will probably take decades.

The supply of new flats in the socialist period lagged far behind the growing demand. In the 1980s the housing shortage became even more acute as the number of new constructions dropped considerably. It is now estimated that about one million Polish families need flats of their own, living up to now with their parents or somewhere else in equally difficult conditions. Another phenomenon not known in the socialist system is the increasing number of homeless people. They are mainly found in the largest agglomerations. The so-called 'housing gap' creates public pressure to work out a successful programme of housing development, consistent with the principles of the market economy but simultaneously accessible to low-income groups in the population. Housing development is one of the principal challenges for Polish cities at the turn of the century (Kozłowski, 1996).

In most large cities, especially in Warsaw, there is a high demand for office premises, modern warehouses, luxury flats and houses. The demand is created by newly established firms, foreigners and the fast developing middle class. New commercial developments and residential estates are appearing, not only in suburban areas but also in the place of degraded pre-war districts of cheap single family housing for workers. In the 'communist' period such areas existed because dislodging the inhabitants was considered too expensive (because of the need to provide replacement accommodation); the most convenient places for gigantic building enterprises were undeveloped ('green-field') suburban agricultural areas (Latour and Orlińska, 1995).

The expansion in the number of individual investors is facilitated by the growing cost of public transport. Low income population groups inhabiting peripheral areas and unable to afford these expenses are leaving the suburban slum areas for centrally located districts and post-socialist blocks of flats. By contrast the former slum periphery is gaining in value due to investments made by rich social groups. This phenomenon might lead to a situation in which income stratification of the society will become apparent in urban space, which means the division into enclaves of wealth and the further deterioration of areas of poverty. Similarly, differential attractiveness of towns for Polish and foreign investors may result in increasing differences between the large, most attractive cities and smaller, underprivileged towns.

Characteristic features of many Polish cities include functional confusion, scarce green spaces (especially in housing estates) and extensive use of areas allocated in spatial development plans for industrial and other public purposes. Industrial areas and unused plots are frequently conveniently located in the vicinity of city centres with good transport connections. Such areas, as well as lands and buildings abandoned by liquidated industrial enterprises after 1989, afford new chances for the restructuring of urban space in Poland. But this opportunity demands regulation of ownership rights following reprivatisation. This problem has not yet been resolved.

Enhancing the status of spatial management might be an important factor in the improvement of urban land management. Under the new Act of 1994, local plans for spatial development have been given the status of local law. They should be prepared with due attention to ownership rights, the value of land and the principle of societal participation. An inherent part of a spatial development plan is a prognosis for its environmental impacts (Rouba and Markowski, 1995). The Act provides that at the central level there should exist a strategic document providing a framework for the country's spatial development. This document should enable integration of spatial planning at the different levels of government. In 1997 a strategy was elaborated called 'Conception of Spatial Development of the Country, Poland 2000 Plus' (*Koncepcja Polityki* . . ., 1997). This strategy refers, both in the name and general idea, to such European conceptions as 'Europe 2000 Plus'. This document has not yet been accepted by the Parliament.

Another important factor, not sufficiently appreciated yet within the new economic reality, is the lack of a cadastral tax relating its level to the value of real estate (Markowski, 1995). New local authorities are rather slow to comprehend the obvious relations between the local development plan and future building rents. Some cities (e.g. Łódź), however, are preparing geodetic and legal documents which will facilitate quick introduction of the cadastral tax after the appropriate resolution has been passed. Cadastral tax, along with enhanced status of spatial development plans and the development of a proper land market, may contribute significantly towards better utilisation of urban space.

Urban environmental management

Changes in the functional structure of cities in recent years, expressed in the significant growth of the role of the service sector and the decreasing importance of industry harmful to the environment in the Polish economy, have already brought about significant improvements in the state of the environment in urban areas. This tendency may be strengthened in the near future by establishing 'special economic zones' in urban areas, where high-tech industry is preferred and the inflow of internal and foreign investments are promoted. This will create new jobs in more environment-friendly sectors and strengthen and differentiate the economic base of the cities.

Another factor which may significantly improve the state of the environment in cities is economical use of resources in the communal economy. The hitherto existing type of multi-family housing (flats) has a heat loss of up to 40 percent. Due to the lack of meters to measure the consumption of water, the use of central heating and even gas, charges for these facilities were assessed on the basis of floor space or the number of occupants. This situation was not conducive to saving resources. In order to reduce resource use in the communal economy, Polish cities have initiated special saving programmes for housing co-operatives (i.e. the Energy Saving Programme), aimed at the improvement of thermal insulation of buildings and supplying them with equipment for measuring the use of thermal energy and gas meters. These programmes are supported by government subsidies, even up to 80 percent of their cost. Measurements have shown that, thanks to the installation of water consumption meters, the use of water has dropped in some households by as much as half, and on average by 20 percent. Similar economic solutions to the consumption of energy or water are increasingly introduced in manufacturing enterprises. As the share of water and energy-intensive industries in the Polish economy is contracting, optimistic forecasts concerning decrease in the use of these resources seem to be justified (see Figure 11.7).

The evident success of energy- and water-saving programmes calls for a new approach to solving the problem of wastes. Following the Agenda 21 resolutions, waste management programmes are being developed at the local level. These are, however, mostly formal documents, prepared with little support from non-governmental organisations or ecological groups. Thus the implementation of the principal postulate of Agenda 21 – that is, social partnership in environmental protection processes – is still distant.

The positive symptoms of transformation are already visible in Polish urban space in the quality of air and water and of the aesthetics of the cities, especially in downtown areas because of newly surfaced streets and well decorated shop windows and advertisements. New office facilities are being put up in the fashionable post-modernist style, signifying the advent of a new era. In contrast are the small positive changes in publicly governed space in the sphere of infrastructure and transport. Integrated revitalisation of urban space in Poland would require undertaking deep legal and financial reforms and

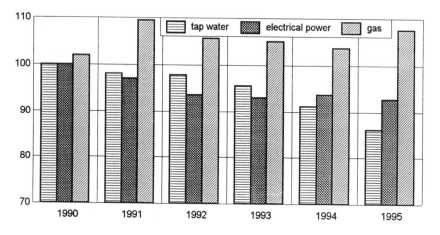

Figure 11.7 Consumption of tap water, gas and electric power in urban households in Poland, per person (1990 = 100)

Sources: *Rocznik Statystyczny 1996* (1997: lxxiii); *Mały Rocznik Statystyczny 1996* (1997: 457).

increasing the level of awareness and competence of political power groups and administration officials.

Cities are places where all the problems of the transforming economy are concentrated. In spite of this obvious fact, Poland is still lacking an official government document which would deal with the problems faced by Polish cities. Such a document should be a statement of the government's urban policy, either in the form of an Act on the Revitalisation of Cities, or within the government's financial programme setting out the directions of modernisation and development of cities. Further economic growth in Poland, a country which aims to catch up with the developed economies, is not possible without a comprehensive vision of economic development which takes account of ecological considerations and the principle of equitable access to the environment. Pro-ecological development of cities as part of the national urban policy should be an integral element of this vision.

Conclusion

Fifty years of the centrally planned economy left behind, among other things:

- a society without a tradition of democratic participation in public life and accustomed to (declaratively at least) a protective function of the state (full employment, low rent-charges, etc.);
- a bad economic structure characterised by raw material-intensive industry, not adjusted to competition on the international market;
- a deteriorated environment in urban and industrialised areas;
- ill-looking cities, chaotically built, mostly with non-economical flat dwellings in blocks, with unsettled properties of land and buildings.

During the transformation period visible changes in Polish cities can be observed. The quality of air and water has been improving, not only due to decreasing industrial production but also as an effect of the dissemination of a preventive approach to environmental protection. The consumption of water and electricity in urban housing decreased significantly due to an increase in prices, the implementation of energy-saving programmes and by supplying buildings with measuring equipment. The progressive development of the commerce and service sector in the cities has resulted in differentiation of the employment structure as well as an improvement in the aesthetics of some quarters and better utilisation of land and buildings in the central areas. The latter problem demands a developed market in real estate, which is, in fact, now already occurring.

The improvement of environmental performance in the cities has a significant impact on decreasing emissions of dust and gases into the atmosphere at the national level. But the development of motorisation may bring new sources of pollution as well as intensive urbanisation of rural areas on the periphery of cities. A seemingly insoluble problem seems to be industrial and municipal solid waste management. Many big cities also lack wastewater treatment plants, and their sludge is drained off into the rivers. In recent years there has been no successful programme of housing development which is both consistent with market principles and simultaneously accessible for less wealthy people.

The ecological awareness of Polish society is rather high, although this concern does not fully coincide with actual performance. There is a large number of ecological organisations, yet their influence in solving environmental problems has been rather limited up to now. The awareness of environmental needs on the part of the government seems to be remarkable, the evidence of which is the legislation on this matter, especially the recognition, even in 1991, of sustainable development as a guiding principle in the nation's economic progress, much in accordance with the definition accepted later in Rio.

The strongest actions according to the Rio Declaration and Agenda 21 seem to be building democracy in Polish society, the creation of a solid legal basis for environmental protection and the active participation of Poland in international co-operation. The requirement to implement the postulates of Agenda 21 is an important factor stimulating local activity. Work on local programmes of sustainable development according to Agenda 21 has recently gained impetus in Poland, particularly in the case of small and medium-sized towns. Numerous international institutions provide assistance in these efforts, for example the UNDP Umbrella Program. Similar programmes are increasingly attracting the interest of large cities, among those being Gdańsk, Warsaw, Łódź and Radom. It is difficult to assess the implementation of the Rio postulates in Polish cities when most of them are in the phase of designing the programmes. It seems that realistic assessment of the participation of local communities in the implementation of sustainable development will be possible only after two years or so from the time the programme commences. The last months of 1997, however, permitted a certain degree of optimism as regards local initiatives.

It is hard to say what would have happened in Poland without the Rio Convention and Agenda 21, because our economic and political reforms coincided with the time of preparation for the Rio Summit. Some of the undertakings discussed above might have been carried out even without Poland's participation in this great event. Past negligence in environmental protection created strong internal demands for improvement in this area. But thanks to the high status of the Earth Summit, ecological problems acquired political significance both on the national and global scales. Thus, state authorities, although very busy with strategic and short-term economic problems, have been forced to pay attention to the environment.

Poland is also benefiting from growing interest in environmental problems at the global scale. Following the assistance and advisory activity of Western countries, Poland is introducing a preventive approach to environmental protection, examples of which are the Cleaner Production programme, promotion of environmental impact assessment, and eco-labelling. The opening up of highly developed countries to the transfer of information, technologies and innovations, has allowed Poland to introduce innovative technological and organisational solutions in manufacturing and environment protection. Eco-conversion of past debt has guaranteed funding for programmes improving the state of the environment. It can be reasonably expected that Poland will soon be able to start paying back its debt to the international community and will be able to provide help to the poorest countries with which it once had friendly political relations and active business contacts.

When considering the impact that the Rio Convention (or internal demand) has had on the directions of change of environmental policy in Polish cities, it is important to underline the increasing role of self-government and the local people in shaping their urban environment. They are supported by the central institutions in greater self-dependence (system of local finance, spatial management) and in the promotion of the sustainable development of the cities. The major discrepancy between the demands of the Rio Declaration and the current situation in the country is between the need to eradicate poverty and to increase the income and wealth stratification of the population. It is strongly visible within the urban population and in the difference in the prosperity of urban and rural populations.

The most important factors for further improvement of the urban environment are the modernising processes of the whole Polish economy and the development of market principles in environmental management. Members of local authorities should have not only environmental consciousness but wider knowledge of the complexity of connections within the urban ecosystems and in the environment as a whole. So, ecological education and professional qualifications are the essential factors in bringing environmental problems to the top of social concerns and solving them successfully.

Due to Poland's geopolitical situation in Europe, the developments in the international situation are also very important for its future prosperity. That is why Poland is making energetic attempts to gain full membership of Western economic and military structures, with full support from most of Polish society. The need to meet the requirements set by these organisations provides an additional stimulus for carrying through the necessary political and economic reforms. Among these requirements, improvement of the environmental aspects of Polish reforms plays a very important role as a stimulus for the implementation of principles of the Rio Declaration. Finally we should say that no part of the Rio declaration is ignored by the Polish authorities or by society, but the pressure of current socio-economic problems and lack of finance creates many obstacles on the quick road to truly sustainable development in our country.

Note

1   Since 1 January 1999 Poland has been divided into 16 *voivodships*, 308 counties (*powiat*) and 2,489 communes (*gmina*). More than 300 small townships will gain new functions and power for their development.

References

Baturo, W., Burger, T. and Kassenberg, A. (1997) *Społeczna Ocena Realizacji Agendy 21 w Polsce* (Social Assessment of the Agenda 21 Execution in Poland), Warszawa: Instytut na rzecz Ekorozwoju.

Karaczun, Z. and Indeka, L. (1996) *Ochrona Środowiska* (Protection of the Environment), Warszawa: ARIES.

Kalinowska, A. (1994) *Ekologia – Wybór Przyszłości* (Ecology – the Choice of the Future), Warszawa: Editions Spotkania.

Koncepcja Polityki Przestrzennego Zagospodarowania Kraju 'Polska 2000 Plus' (A Conception of Spatial Development of the Country 'Poland 2000 Plus') (1997) Warszawa: Rządowe Centrum Studiów Strategicznych.

Kornai, J. (1993) *Anti-Equilibrium*, Warszawa: PWN.

Kozłowski, S. (1994) *Droga do Ekorozwoju* (The Way Towards Ecodevelopment), Warszawa: PWN.

—— (1996) 'Czy transformacja polskiej gospodarki zmierza w kierunku zrównoważonego rozwoju?' (Does transformation of the Polish economy tend towards sustainable development?), in *Mechanizmy i Uwarunkowania Ekorozwoju* (Mechanisms and Determinants of Ecodevelopment), Bialystok: Politechnika Białostocka.

Latour, S. and Orlińska, H. (1995) 'Housing policy and management of resources in urban areas – the past and the future', in T. Markowski and T. Marszal (eds) *Redevelopment and Regeneration of Urban Areas*, Łódź: Łódź University Press.

Mały Rocznik Statystyczny (1996) (Small Statistical Yearbook, 1996) (1997) Warszawa: GUS.

Markowski, T. (1994) 'Environmental policy and regulatory change in Poland', in *Privatisation and Regulatory Change in Europe*, Buckingham and Philadelphia: Open University Press.

—— (1995) 'O spójności uregulowań prawnych w gospodarce gruntami' (For the coherence of legal regulations in land economy), in: *Gospodarka Przestrzenna i Regionalna w Trakcie Przemian* (Spatial and Regional Economy during Transformation), Toruń: Uniwersytet im. Mikołaja Kopernika.

Paczuski, R. (1994) *Prawo Ochrony Środowiska* (Environmental Protection Law), Bydgoszcz: Oficyna Wydawnicza Branta.

Parysek, J. and Kotus, J. (1997) 'Powojenny rozwój miast polskich i ich rola w procesie urbanizacji kraju' (Post-war development of Polish cities and their role in the urbanisation process of the country), *Przegląd Geograficzny* 1–2: 33–54.

Polityka Ekologiczna Państwa (Ecological Policy of the State) (1992) Warszawa: Urząd Rady Ministrów.

'Raport Ligi Ochrony Przyrody o stanie środowiska przyrodniczego w Polsce i zagrożeniu zdrowia ludzkiego' (Report on the state of natural environment in Poland and resulting health hazards) (1981) *Przyroda Polska*, Dodatek no 5/6.

Rocznik Statystyczny (1996) (Statistical Yearbook, 1996) (1997) Warszawa: GUS.

Rouba, H. and Markowski, T. (1995) 'From the past to the future in town planning in Poland', *Scandinavian Housing and Planning Research* 12: 165–172.

12 Recent Australian urban policy and the environment

Green or mean?

Peter Christoff and Nicholas Low

Introduction

In 1997, during the Third Conference of the Parties to the UN Framework Convention on Climate Change at Kyoto, Australia exposed a face to the world which few might have expected to see. In earlier global forums it had professed strong support for international environmental regimes, including on global warming. Here, its representatives claimed that it was unfair and unjust for Australia, whose economy depends so much on the use and export of fossil fuels, to reduce that national dependency. They argued against uniform global reduction targets and demanded significant concessions to permit Australia to increase its greenhouse gas emissions over the coming 15 years by continuing with its fossil-fuel-intensive resource development programme. Australia threatened to withdraw from the Convention if these demands were not met. In the Kyoto Protocol, most developed nations finally agreed to mandated cuts to emissions averaging 5 percent from 1990 levels by 2010. Australia secured agreement to an 8 percent *increase* in its emissions over the next two decades from nations wishing to avoid destabilization of the Convention by its possible secession (see Christoff, 1998b).

This dramatic shift in Australia's greenhouse foreign policy stance and tactics reflects a reaction by the present conservative national Coalition government to the combined, conflicting pressures of globalized environmental regulation and economic deregulation. It also emphasizes the strength of Australia's economic dependence on primary commodity exports (particularly minerals and energy, and predominantly coal) and the influence of this dependence on Australia's domestic and foreign policies. In defending the 'national interest', the policy prioritizes very short-term and selective economic benefits at the expense of other nations, future generations and the environment at large.[1]

Public opinion in Australia predominantly accepts the idea of mandatory greenhouse emission reductions yet, paradoxically, it has been relatively powerless in the face of this policy shift. This suggests a new and significant separation of political action from the mainstream of popular sentiment on environmental matters, occurring at both the national and sub-national levels in Australia's federal system. In particular, Australia's performance at Kyoto points to the substantial institutional barriers which Australia faces in translating ideas of ecological sustainability and environmental justice propounded in the Rio Declaration into an urban setting – the implementation, in short, of a 'brown' environmental agenda.

In this chapter we first review the impact of different forms of globalization on Australian cities and on urban planning for ecological sustainability. We then consider briefly the relationship of Australian environmentalism to urban change and examine changing policy settings and policies at national, State and local levels, particularly in the context of the Rio Declaration and Agenda 21, before discussing the ecological impacts of major Australian cities.

Globalization and Australian cities

Australia has five metropolitan cities, the capitals of its continental states. Widely separated from each other around the coastline, from west to east and then north, these are: Perth in the far west (1.3 million people), Adelaide (1.08 million), Melbourne (3.28 million), Sydney (3.88 million), and Brisbane (1.52 million). Darwin (the small port-capital of the Northern Territory), Hobart (the port-capital of the southern island state of Tasmania), and the national capital, Canberra (inland between Melbourne and Sydney), complete the picture.

Historically, and when compared with many developed countries, Australia evidences a very high urban concentration of population. Some 62 percent of the nation's population live in the five major cities. Melbourne and Sydney together accommodate about 40 percent of the country's population. A further 11 percent of the population inhabit only another 79 regional cities. Indeed the metropolitan cities may be grouped into three tiers. Only Melbourne and Sydney may be regarded as minor 'global cities' in terms of their population, cultural complexity and economic diversity, and the extent and nature of their participation in the global economy. Although Sydney has for decades been creeping ahead of Melbourne in dominance, together they continue to dominate national economic activity (particularly manufacturing) and are increasingly positioning themselves as sites for finance capital operating in the Asian region. Brisbane and Perth are rapidly growing secondary centres for service and resource industry-related activity. The remainder are local settlements predominantly providing focus for domestic economic activity.

Forster (1995) has suggested that the physical development of Australian metropolitan cities during the twentieth century can best be understood in terms of changing transport technologies: first the relatively densely populated 'walking cities' of the nineteenth century, followed by the radial extension of rail and tram networks in the 'public transport cities' of the early twentieth century, then the undifferentiated suburban sprawl of postwar car-born development. While this is substantially true – as it is of cities throughout the developed world – it conceals important differences between Australian, American and European patterns of urbanization. First, Australian cities have never had any tendency to become 'doughnuts' or 'edge cities' in the American sense (Moser and Low, 1986).[2] There has been no persistent hollowing out of the inner city as the middle class leave for the suburbs. On the contrary, and akin to many European cities, the inner suburbs have been repopulated and culturally enriched by successive waves of immigrant settlement, and economically revitalized through a process of gentrification underway since the 1970s. Second, the conventional assumption that Australia's cities are irredeemably 'car dependent' ignores the fact that most peak travel flows are still radial, following paths shaped by earlier linear suburban development following public transport lines. Only a small proportion of travel is now by public transport, with movement concentrated along radial corridors – a pattern which may be explained by the very poor level of service and modal integration provided by the public transport systems of the major cities (Mees, 1999).

Forster comments on two other influential urban tropes derived from North America and Europe: the 'flight to the sunbelt' and the development of 'mega-cities'. In Australia, movement of people and capital from the southern (Australian northerners might say 'Antarctic') cities to those of the subtropical and tropical North has been modest in scale. During the 1980s and 1990s tourism and the retirement to warmer climes of an ageing Australian population brought increased prosperity and population growth to Brisbane and along the 'sunshine coast' south of Brisbane. As for the emergence of a mega-polis, the once-predicted merger of Sydney and Melbourne – connected by a vast south-eastern coastal suburban sprawl – it has not occurred. Such a possibility is made remote by the slowing rate of urban population growth in Australia, the embedded infrastructural strengths of these cities, and the re-centralizing tendencies of post-Fordist cities in general.

Australian cities and their governance have been subject to the complex and sometimes contradictory forces of economic, ecological and cultural globalization. On the one hand, domestic economic investment, and urban development and planning, have been increasingly influenced by a world economy increasingly dominated by liberalized global finance markets. On the other hand, the emergence of global environmental issues in the 1980s and 1990s and the construction of international environmental regimes in response, plus the rise of an international discourse focused on 'ecologically sustainable development' and on regulatory forms of ecological governance, have together guided and channelled Australian domestic environmental policies. These two currents of globalization generally flow in different directions and often produce contradictory vectors of change.

In addition, in Australia as elsewhere, increased capital mobility and the deterioration of the global environment have also generated significant domestic cultural changes, including declining public confidence in economic growth as a vehicle ensuring improvements in quality of life, and increasing public anxiety about the risks associated with technological change and the associated destruction of nature. At the same time, time/space compression and the increasing rapidity of technology transfer and information flows – central characteristics of globalization – have had major implications for the reordering of regulatory regimes (see Gleeson and Low, forthcoming). Together, these transformations have influenced the institutional matrix affecting urban environments at different levels of government and social interaction and fed the political movements and activities which form around particular conflicts.

Over the past four decades, profound changes have occurred in the composition, scale and location of both production and employment in Australia. Broadly speaking, technological change has displaced human labour, intellectual labour has displaced physical labour, the gender composition of employment has been redefined as women have entered the workforce, and employment has been casualized – with permanent employment displaced by contract and part-time work. Unemployment has increased about fourfold since the early 1960s (to around 8.5 percent), but those employed are working longer hours. The industrialization of Asia has transformed the Australian manufacturing sector. These changes have had a profound effect on Australian cities, their internal geography and their environmental impacts.

The Australian manufacturing sector has emigrated, either moving offshore to seek cheap labour in the Asian region or to new high tech establishments built on 'greenfield' sites on the perimeters of the major cities (mainly Sydney and Melbourne). Meanwhile, the rapidly growing service sector has congregated in the inner suburbs. Central business districts have, in effect, developed an extensive business 'penumbra' in the inner suburbs of Australian cities, employing much larger numbers than are employed in ex-urban manufacturing (Mees, 1995). Work locations have spread throughout most suburbs, partly in response to urban economic planning programmes during the 1980s to reduce transport-related pressures and enhance the social stability of suburban regions by creating regional nodes of economic activity. In general, the inner city has been reorganized around consumption, entertainment, and service sector (including education) employment. Growing demand for expensive residential and temporary tourist-oriented accommodation, and the repopulation of the inner city, has driven the refurbishment of the inner suburbs and central districts of Melbourne and Sydney over the past decade. These changes have been strongly supported by state governments.

The increasingly global presentation of cities as cultural commodities has also found focus in Australia. Throughout the twentieth century, cities have become increasingly 'self-consuming'. There is a growing emphasis on the (global) city as a commodity in its own right, competing against other similar commodities in international tourist and investment markets. This transformation of the modern city from location to site, venue and artefact has contributed to the accelerating physical restructuring of Australian cities and a ceaseless refurbishment with bait for custom and capital – casinos, sports facilities and mass entertainment venues, shopping malls, cheap office space and central city accommodation (though see Mees, 1998).[3] Australian state governments, in particular those of New South Wales and Victoria, now spend lavishly on international advertisements packaging their capital city's historic features, ethnic and cultural diversity, infrastructural development and investment advantages, political stability, low crime rate, health and recreational advantages, climate and environmental benefits ('livability'), and major events. The global search for consumers has mixed environmental impacts. Improvements in Sydney's mass transit system and the remediation of a toxic site at Homebush Bay for the 2000 Olympics, or the preservation of Victorian architecture in Melbourne's inner suburbs, must be set against the use of established parkland for Grand Prix racing and sports stadia, the development of new freeways, and the lost opportunities for ecologically sensitive and compact development around Melbourne's docklands and Sydney's Darling Harbour, sacrificed for mega-sports ovals and world exhibition centres.

Finally, Australian cities both contribute to and are affected by the new generation of global environmental problems, such as climate change and ozone depletion, and are influenced by the increasingly international terms of reference of urban environmental governance. The environmental agendas and regulatory frameworks of Australian national and sub-national governments have been influenced by, and articulated through, both formal international regulatory regimes and informal transnational discursive formations which have been generated in response to the two global waves of public environmental concern during the early 1970s and the late 1980s. These two waves – initiated by transnational scientific networks, promoted by international environmental organizations such as Greenpeace, and transmitted through the global media – reshaped domestic environmental consciousness and politics in Australia. The first wave led to the establishment of (sub-national) departments of environment, and environmental protection agencies or related capacities, in a number of (but not all) Australian states, and a greater integration of environmental regulatory capacities focused on urban problems. The second wave fostered an exponential growth in non-government environmental organization membership and in corresponding pressure for further institutional responses and reform by national government (Christoff, 1994; Papadakis, 1993, 1996).

In addition, over the past two decades Australia has been an active – and until recently, proactive – participant in all major international environmental forums, including the 1972 United Nations Conference on Humans and the Environment (UNCHE) in Stockholm and the Earth Summit (UNCED) two decades later at Rio. The Australian government has ratified each of the major environmental framework conventions which evolved during this time and signed environmental treaties governing the conservation of natural and cultural heritage (including world heritage), biodiversity, wetlands, migratory birds, whales and other endangered species, transport, trade in and disposal of toxic wastes, the use and disposal of ozone depleting substances, and the generation of greenhouse gas emissions. These treaties have been acknowledged in national strategies and, more infrequently, have also been the subject of national legislation which in turn gives the Commonwealth government constitutional power to affect the relevant policies of sub-national (state) governments. Significantly, many of these treaties have – or should have – an influence on the governance of urban environments. Finally, the global extension of regulatory frames and relevant policy networks and communities beyond the territorial bounds of any one city, state or nation, has also meant that sub-national environmental standards (e.g. governing air and water pollution, and waste treatment and disposal) have to be set and refined in reference to overseas developments – particularly those in the United States and, to a lesser extent, Europe. Similarly, developments in the OECD's environmental working groups have driven Australian thinking about the use of economic instruments for environmental policy.

Environmentalism and urban Australia

The continent 'discovered' by Europeans in the eighteenth century had already been transformed by over 50,000 years of human occupation in which the indigenous relationship to the land was one of respect and apparent ecological equilibrium. But the Aborigines were driven off the land; many were slaughtered. British settlement brought with it farming techniques which enabled Australia in the short term to produce a vast agricultural surplus, support the nation's industrialization, and make possible the growth of its metropolitan cities. It took a mere two hundred years for colonization radically to transform the continent, strip it of its forests, exploit and deplete its soils, pollute its rivers and seas, and drive its unique flora and fauna toward extinction.

In Australia during the late nineteenth and early twentieth centuries, nature conservation was mainly oriented to meeting the resource, recreational and health needs of the major urban settlements and was largely the work of significant individuals and associations rather than a broadly defined conservation movement. For instance, the National Park (later Royal National Park) near Sydney was proclaimed in 1879, merely seven years after the world's first national park, Yellowstone, in the USA. John Lucas, Member of the NSW Legislative Assembly at the time, commented that the park was created 'to ensure a healthy and consequently vigorous and intelligent community . . . all cities, towns and villages should possess places of public recreation' (in Hall, 1992: 92). Other nineteenth century nature reserves were established largely for similar reasons – including to facilitate appreciation of the novelty of Australia's strange and exotic flora, as with the Fern Tree Gully Reserve (1882) in Victoria. In Victoria, large areas of the upper catchment of the Yarra River were placed in closed reserves to ensure the purity of its waters, the main source for Melbourne at that time. Early legislation, generally initiated by enlightened elites informed of similar developments in Britain and confronted by public outrage at successive epidemics, was enacted to limit disposal of putrescible wastes to urban streams and emissions to air. These laws, and new institutions such as Boards of Works, established to implement major sewerage infrastructural programmes, became part of the new sanitary regime which evolved towards the end of the nineteenth century (Dunstan, 1984, 1985).

In the first half of the twentieth century, the movement for town planning grew with a commitment to improve urban public health, eliminate slum housing and regulate speculative suburban development. This movement began to make a real impact on public policy as the need for reconstruction took hold politically during the war years. In the 1940s, under pressure from the Labor federal government, town planning legislation was introduced in all states. On the whole, the town planning profession developed separately from the environment movement: a gulf separated thinking and action on urban and non-urban policy matters.

Australia was swept up in and transformed by the first wave of the new environmentalism of the 1970s and briefly encompassed both green (nature-oriented) and brown (urban) concerns. During this decade, the emergent Australian environment movement turned its attention to urban issues – energy conservation, transport, destruction of heritage in the built environment, and pollution. Indeed, one of the main foci of the political campaign which brought the Australian Labor Party (at national level) to power in 1972 was the poor environmental condition of the cities after years of market-led development with minimal planning.

The early 1970s were marked by conflicts over the urban environment. The freeing of the global finance industry coupled with the inflow of capital and overseas banks brought with it a massive property investment boom which affected all the major cities.[4] Sydney, in particular, experienced massive redevelopment in its urban core. Development proposals backed by state government planners threatened to destroy some of the older neighbourhoods close to the city centre. Popular movements to protect these neighbourhoods were supported by the building unions who imposed bans on construction – the 'Green Bans' (Manning and Hardman, 1975; Jakubowicz, 1984; Burgmann and Burgmann, 1998). There was an upsurge of interest in the Australian urban and natural heritage, as evidenced by the growth of Australia's pre-eminent urban and nature conservation NGOs. As Davidson (1991: 26) writes, 'from a base of fewer than 200 in the 1960s, the National Trust's membership had by the mid-1970s grown to 48,148 in 1974 and continued to grow steadily to a peak of approximately 75,000 in the early 1980s'. Meanwhile, membership of the Australian Conservation Foundation (ACF) increased fivefold between 1967 and 1971, from 1017 to 5154 (Papadakis, 1993: 152).

Environment movement attention drifted away from the cities during the later 1970s, encouraged partly by the withdrawal of national government involvement in urban policy after 1975, following the election of the conservative Fraser Coalition national government. Instead, forests and wilderness preservation became the foci for public attention. Tasmania became a major site of environmental conflict as its government encouraged logging of forests and damming of rivers in order to attract industry into this remote and sparsely populated island. The drowning of the unique and exquisite Lake Pedder, with its pristine white quartz sand beach, beneath a larger lake created to generate hydro-electricity (in 1972–73) transformed the environment movement. More confrontational and ecocentric NGOs (the Wilderness Society and the revitalized ACF) moved to centre stage, and when the Tasmanian government and its hydro-engineers sought to build a dam on the Franklin River – one of Tasmania's few remaining undammed rivers, set within an area nominated for inclusion on the World Heritage Register – the environment movement was able to mobilize unprecedented metropolitan support nation-wide in a campaign which felled the Tasmanian government, saw the High Court confirm the pre-eminent powers of the national government in this matter, and helped Labor to power in national government in 1983.

The Australian environment movement grew significantly in size during the 1980s[5] but, by contrast with the 1970s, its mobilizing themes and images focused on the destruction or salvation of nature, using sites of symbolic as well as ecological importance – the Franklin River, the old-growth native forests of East Gippsland in Victoria, the Daintree rain forest in Queensland. The divide between urban and non-urban environments remained wide throughout the 1980s. Urban issues, with the marginal exception of toxic waste disposal (protests against proposals for waste disposal plants and a series of accidents at chemical storage depots in both Melbourne and Sydney), were barely considered by the bulk of the environment movement. Partly as a consequence, urban environmental issues were largely left off national and sub-national environmental policy agendas. The environment movement's 'green' focus was allied with constant attempts to 'push' those issues up to the national level, to enable concentration of political effort, encourage greater homogeneity and standardization of administrative and political response reflecting the growing importance of international environmental treaty obligations, and to overcome developmentalist responses at state level.

In the 1990s, global issues such as climate change have again blurred the distinction between 'green' and 'brown', and issues relating to urban use of environmental resources and urban environmental impacts have resurfaced as matters of policy concern. A major national opinion survey in 1993 showed that while unemployment and the state of the economy had overshadowed the environment at this time, 'digging below the surface reveals that the environment is very personally relevant to individuals – almost as much as is unemployment' (ANOP, 1993: 6). Moreover, the survey found 'the main community priorities were increasingly the brown environmental issues': pollution and waste management, water quality, and air pollution, while 'green' issues (concerning forests, plants and animals) were perceived to be less pressing priorities (ANOP, 1993: 8). Interestingly, among global issues only 5 percent of respondents in 1993 mentioned 'the greenhouse effect' – a smaller proportion than in 1991 (ANOP, 1993: 79). The major environment NGOs – the ACF, Greenpeace Australia and the state conservation councils – have also returned to urban concerns, for instance by establishing the Smogbusters campaign and resisting a renewed nationwide bout of freeway construction.

The town planning profession has also shown increased awareness of problems of water supply, waste disposal, and the impact of urban growth on habitat and energy use. Conferences in 1997 emphasized urban ecological sustainability ('Pathways to Sustainability – Local Initiatives for Cities and Towns' which comprised the final phase of a UN review of Local Agenda 21 and follow-up to Habitat II, hosted by the city of Newcastle in New South Wales, and the Victorian State Conference of the Royal Australian Planning Institute. The conference on Environmental Justice of the University of Melbourne, at which this book was workshopped, should also be mentioned). Under an OECD programme to encourage ecologically sustainable urban development, a competition was mounted by the national government to design a new settlement in the Jerrabomberra Valley fringe of Canberra. The winning entry proposed a system of low density urban villages with emphasis on low energy use, the careful management of soils and land form for biodiversity conservation, self-sufficiency and minimal transport demand (Rodger and Schapper, 1994). Another scheme – the Halifax EcoCity Project – proposes a high density ecocity (Ecopolis, 1996) in an inner precinct of Adelaide based on the ideas of Murray Bookchin (1992), Kirkpatrick Sale (1991) and the project's architect Paul Downton (1991). Neither project has been implemented and therefore both remain somewhat utopian.

Changing governmental responses

National government

During the 1970s, the national (federal or Commonwealth) government strongly resisted public pressure to intervene in the affairs of the states on environmental grounds. This resistance was underpinned by a general belief that the Australian Constitution offered few powers for intervention in environmental matters and that resource management (including conservation) was properly the province of the states, to be strongly facilitated by the Commonwealth government where appropriate.

Nevertheless, the strongly reformist Whitlam Labor national government (1972–1975) created the first independent environmental portfolio and passed landmark legislation instituting environmental impact assessment processes and enabling the Commonwealth to act on some of the range of first generation environmental treaties and conventions, including the World Heritage

Convention, Ramsar and CITES, which Australia signed during this period. As a result, and partly in response to international obligations, several key national or regional environmental agencies were established – the Australian Heritage Commission (AHC), the Australian Nature Conservation Agency (ANCA), and the Great Barrier Reef Marine Park Authority (GBRMPA).

When the Hawke Labor government was elected in 1983, its first legislative act was to intervene in the Franklin River dispute on the grounds of Australia's international obligations under the World Heritage Convention. This use of the Constitution's 'external affairs power' sent a signal to the states about the federal government's growing and largely unrealized constitutional capacity in the environmental domain.

Labor remained in government federally and in most States throughout the 1980s. In national government it grew to depend on interest group support as its electoral position became increasingly perilous following the elections of 1984, 1987 and in 1990. The resultant political opportunities were recognized, and the rise of green issues on the national political agenda achieved, by the strong alignment of government and green movement leadership – especially between Prime Minister Hawke, Environment Minister Richardson (also a leading party strategist and Right faction powerbroker), and Phillip Toyne, then executive director of the ACF. The public prominence of environmental issues was further promoted by the emergence of the new generation of global environmental issues, the Brundtland Report and then UNCED, and intense media interest in politically damaging conflict between environmentalists and resource developers (especially in the mining, forestry, and hydroelectric power sectors).

Indeed, in response to media-generated perceptions of political chaos caused by confrontation in the forests and over mines and wilderness, national and state Labor governments sought to regain control of the environmental political agenda by proposing a range of neo-corporatist initiatives to restore order and regulate policy formation. Nationally, between 1989 and 1992, Labor established a plethora of experiments at strategic development, policy integration and consensus-building (see Figure 12.1). Some of these – the Resource Assessment Commission (RAC) and the Ecologically Sustainable Development (ESD) strategy process – were initiated by the Minister for Primary Industries and Energy, seeking to use the 1987 Brundtland Report and its interpretation of the concept of sustainable development, to tame the ecocentric orientation of the environmental debate by drawing the movement's key NGOs into a framework of negotiated solution-making.

The early 1990s may be seen as the high point in environmental capacity building in terms of the creation of new consensus forming institutions, strategic development, and increased capacity for policy co-ordination and implementation. Nevertheless, the environment movement gained few significant outcomes from these processes. But neo-corporatist mechanisms, which had proved relatively effective in delivering union quiescence in industrial relations under the Accord, proved less so in environment policy because of the lateness of the attempt and also the irreducibility of the issues and values at stake.

Of these initiatives, only the ESD and National Greenhouse Response (NGR) processes and strategies, and the InterGovernmental Agreement on the Environment (IGAE) have any significant bearing on the urban agenda. Neither the ESD nor the NGR strategies focused strongly on urban development. Whilst the ESD process included detailed attention to transport and energy production and use (see ESDWG, 1991a, 1991b and 1991c), key urban issues such as population growth, urban expansion, water use and waste disposal figured only marginally. In any case, from 1993 onwards the ESD strategy was given a low priority by the national government and by the States. The NGR strategy, and its subsequent reiterations and transformations, never articulated any clear targets nor was it defined in legislation, despite early commitments nationally and internationally to meet emission reductions to 1988 levels by year 2000 and the additional driving pressure of participation in the evolving international climate change regime.

Table 12.1 Key national environmental institutional initiatives, 1989–1992

Year	Initiative
1989	*Prime Minister Hawke's Statement on the Environment 'Our Country, Our Future'*: the first Prime Ministerial statement to recognize the importance of environmental issues. Major funding allocation to a wide range of programmes and initiating several key institutional innovations, including those mentioned below.
	The 'Decade of Landcare' programme: reflected an alliance between the National Farmers' Federation and the ACF to tackle land degradation.
	Resource Assessment Commission (RAC): established to run independent inquiries and offer recommendations to the Commonwealth government about resource use. Conducted three inquiries – into Forests, Kakadu (mining), and the Coastal Zone.
1992	*Commonwealth Ecologically Sustainable Development (ESD) Strategy and the National Greenhouse Response Strategy (NGRS)*: Nine working groups on Agriculture, Energy Production and Use, Fisheries, Forestry, Manufacturing, Mining, Tourism and Transport, included business, resource sector, environment movement, social welfare, union, and government representatives reported to the federal government in 1991. The resultant strategies were agreed jointly by Commonwealth and state governments late in 1992.
	National Forest Policy.
	Commonwealth Environment Protection Agency (CEPA): established to enable the Commonwealth government to assist in determining national environmental standards and to effect environmental outcomes on Commonwealth land.
	Intergovernmental Agreement on the Environment (IGAE).
	National Waste Minimization and Recycling Strategy.
	Draft National Biodiversity Strategy (finalized in 1996).
	Prime Minister Keating's Statement on the Environment 'Australia's Environment: A Natural Asset'.

After 1992 there was a significant change in the political climate for environmental politics in Australia. Prime Minster Hawke was deposed by his Treasurer, Keating, late in 1991. PM Keating re-established a style of adversarial politics and dismantled most of the consensus-building mechanisms and neo-corporatist forums within the environmental domain established under Hawke. Keating was relatively uninterested in the environment as an issue.[6] His view of cities and their planning was strongly developmentalist, although he was prepared to tolerate a rhetoric of sustainability being written into the feeble attempts of the government to revive the urban agenda of the 1970s.

Despite the publication of a range of national environmental strategies in the period 1992 to 1996 and a significant increase in funding to the environment portfolio, key elements of the experiment in national environmental capacity-building lapsed. The ESD and NGR strategies were effectively emasculated late in 1992, the demise of the RAC announced in 1993, and the newly established Commonwealth EPA absorbed into the national Department of Environment (DASETT) in 1995. These moves were not countered by the IGAE, nor by the creation of a very weak National Environmental Protection Council in 1995. The core issues of nation-building (republicanism and severing residual constitutional ties to Britain), economic modernization, economic *rapprochement* with Asia, and indigenous rights in the domestic context took precedence over environmental concerns.

As opinion polls increasingly demonstrated overwhelming public concern for urban environmental issues, in its final three years the Labor government appealed 'over the heads' of the 'green' environment movement, emphasizing 'brown' issues as a means to capture broad public sympathy and electoral support and recapture the political agenda from the environment movement. The two major initiatives with direct consequence for cities – the IGAE and the Better Cities Program – deserve detailed comment here.

InterGovernmental Agreement on the Environment (IGAE)

The IGAE, signed in 1992 between the Commonwealth, all states and territory governments and the Australian Local Government Association, was the formal expression of the new 'co-operative environmental federalism' sought between state and federal Labor governments in the late 1980s. It was intended to resolve the tensions produced by successive Commonwealth environmental interventions into state affairs. The signatories 'recognize that environmental concerns and impacts respect neither physical nor political boundaries and are increasingly taking on inter-jurisdictional, international and global significance in a way that was not contemplated by those who framed the Australian Constitution' (Preamble). The signatories further agreed that 'the adoption of sound environmental practices and procedures, as a basis for ecologically sustainable development, will benefit both the Australian people and environment, and the international community and environment. This requires the effective integration of economic and environmental considerations in decision-making processes, in order to improve community well-being and to benefit future generations' (paragraph 3.2). Moreover, expressing the precautionary principle, the signatories agreed that 'Where there are threats of serious or irreversible environmental damage, lack of scientific certainty should not be used as a reason for postponing measures to prevent environmental degradation.'

In reality, however, the IGAE served both to reassert state powers over national environmental policy formation by binding the Commonwealth into a framework of consensual macro-policy formation, subject to overwhelming control by a majority vote by the states, and to recognize real problems of compliance and implementation which the Australian federal system generates *vis-à-vis* federal policy. The subsequent evolution of national environmental policy measures (NEPMs) on air, water, and hazardous waste and of the National Pollutant Inventory (NPI), each initiated under the IGAE and promoted within the legislative framework of the NEPC, has been painfully slow.

Better cities programme

The Labor government's major urban programme, called 'Building Better Cities', was initiated in 1991. The programme did not take a systematic approach to ESD but provided central government funding (of relatively small scale – Aus$816 million over five years) to local public and private sector initiatives judged to be 'efficient, socially just and environmentally sustainable'. The principal objectives of the programme were the promotion, planning and support of urban consolidation; efficient provision, charging and financing of urban infrastructure; and improved planning and co-ordination of urban development (Hundloe and McDonald, 1997: 96–7). In co-operation with the federal government, the states created a number of 'area strategies' in the major cities, and in some smaller cities, to co-ordinate the projects.

The area strategies for Victoria included improvements in the public transport infrastructure, construction works at the state's Food Research Institute, public housing projects, development of government-owned land, and flood mitigation and drainage works (Aust. Govt., 1994a). The New South Wales area strategies included redevelopment of the inner city former docklands site at Ultimo/Pyrmont in Sydney, the development of a technology park on former railway land (Eveleigh, in Sydney), public transport infrastructure improvements and a strategy to redevelop dock and railway land in the centre of Newcastle (Aust. Govt., 1994b). These projects are no more than a responsible state government could have been expected to undertake itself, and they must be viewed in the context of about Aus$6 billion cut from central government funding to the states in the late 1980s (Hayward, 1993).

The programme's rhetoric linked social justice and environmental responsibility, thus suggesting an agenda of environmental justice and reflecting the sentiment of the Earth Summit. However, its criteria for environmental justice criteria were far from clearly articulated, and it is also unclear why the selected programmes were thought to achieve the goal. 'Urban consolidation', meaning building flats and terrace houses almost anywhere in the city, was assumed to reduce the city's ecological footprint (Hundloe and McDonald, 1997: 100). But 'ecological footprint' in fact refers to the total level of consumption of environmental resources by a city. The amount of peripheral (agricultural) land saved by urban consolidation is minimal and the level of environmental consumption of a city in other respects is little affected by its density. An evaluation undertaken in 1995 found that 'most of the projects focused on urban form objectives, particularly urban consolidation and higher residential densities, and contribute to ESD to the extent that transport efficiencies and reduced land requirements have energy, air quality and resource [consumption] benefits' (ibid.: 105). But it has never been established that higher residential densities *per se* reduce the consumption of energy or greenhouse gas emissions resulting from travel behaviour. The authors found that, 'One cannot escape the conclusion that from an ESD perspective the Better Cities Program was an *ad hoc* set of projects based on allocating each State and Territory some projects and without any sense of strategic priorities or coherence' (ibid.).

In all, although new regional development and urban renewal schemes (e.g. the Better Cities Program) were proposed under the Hawke–Keating governments, in terms of programme delivery they remained insubstantial for five reasons. First, interventionist federal urban programmes had received strong funding and institutional support under the more radically reformist and controversial Whitlam Labor government (1972–1975). As a result, they were regarded as politically discredited by subsequent, more conservative Labor governments. Second, such schemes were promoted by the left both in the 1970s and the 1980s (under Uren in 1972–1975; under Howe, the Deputy Prime Minister and leader of the left from the late 1980s to 1993). Consequently, because of the dominance of the right-wing factions in the Labor Party, they had little chance of success. Third, support for co-operative federalism under Hawke meant that the national government would not circumnavigate the conservative states by providing significant direct funding for urban development activities to regions and local governments as had the Whitlam government through its Department of Urban and Regional Development. Fourth, because of the environment movement's [inevitable] emphasis on 'green' campaigns, the opportunity was lost for substantial political recognition of and action on brown issues during the window of political opportunity presented by the period 1988–1992. Finally, and most significantly, a range of deregulatory measures, instituted under Labor from the mid-1980s and intensified under subsequent state Liberal governments, have worked strongly against coherent federal and state urban environmental policies or initiatives. Most prominent among these have been the dismantling or incapacitation of various institutions for ecological governance [including local government] and the corporatization and now privatization of energy, water and resource and environmental conservation agencies (including their research and development capacity), and diminution of regulatory control over the new entities.

Since the 1996 national election, under the conservative Howard Coalition government, there has been a slight real growth in funding to environment programmes (with revenue raised from partial privatization of the national telecommunications network *Telstra*). The government has also aggressively worked to undermine Australia's international environmental commitments and performance. Instances include its special pleading against greenhouse gas reduction targets (Christoff, 1998b), weakening of Australia's negotiating position on trade in toxic wastes at recent Basel Convention meetings, permission to excise 20 hectares from a wetland protected under the international Ramsar Convention,[7] and promotion of uranium mining at Jabiluka in the World Heritage listed Kakadu National Park (placing itself in confrontation with the World Heritage Committee which has now demanded a halt to mining at Jabiluka).

The conjuncture, in the 1990s, of neo-liberal federal and state coalition governments has accelerated and radicalized policies for national economic efficiency established and promoted under Labor. For instance, the national competition policy is now enshrined in legislation, and drives an agenda of micro-economic reform and deregulation. Both national and state governments now aggressively espouse developmentalism accompanied by a radical reduction in the size, scope and functions of the welfare state. This, accompanied by state-level privatization of energy and water utilities and pursuit of alleged market efficiencies by the reduction of the public sector, further undermines efforts at implementation and integration of resource conservation and urban environmental measures.

State governments

During the 1970s most states established autonomous departments of conservation, and several, though not all, established environmental regulatory agencies modelled on the United States Environmental Protection Agency. The diversity of institutional forms during the 1980s led to significant differences in the nature and standards of environmental regulation between states. Some leading states also experimented with mechanisms for environmental impact assessment and conflict resolution. Victoria created an independent statutory body to advise on changes in public land use (the Land Conservation Council); New South Wales established the Land and Environment Court – both enacted environmental impact legislation. However, such innovations were constrained in their application by the overwhelming and persistent emphasis by state governments on the facilitation of resource development. There has been a tendency towards institutional convergence in the 1990s, with most states now having established EPAs, adopted mirror legislation to give effect to the IGAE and the related impending creation of nationally uniform environmental measures relating to water, air and selected wastes.

As with central government, there are cross-cutting currents and policy fluctuations at state level. The precautionary principle is now contained in some pieces of legislation in the states of New South Wales, the Australian Capital Territory, South Australia and the Northern Territory, but none in Victoria (Lyster, 1997). Lyster (ibid.: 393) suggests that 'the precautionary principle has arguably acquired the status of an important legal principle' (see also Gullet, 1997; Hannan, 1997).

At present, the two states with the greatest commitment at present to ESD are South Australia and New South Wales. South Australia has produced a co-operative policy with the City of Adelaide for ecologically sustainable development in that city. The New South Wales (NSW) government issues guidelines for ecologically sustainable development to all local councils (NSW Govt., 1998). Under the NSW Local Government Amendment (Ecologically Sustainable Development) Act 1997, 'councils are now expected to adopt a strategic "whole of council" approach toward the recognition of ecologically sustainable development and to respond positively to environmental problems in their area' (NSW Govt., 1997: 1). The intention is to require local councils to include the goal of ESD in their local management plans and in considering any applications – under planning legislation – for development. There is no comparable requirement, however, for the state government to monitor its own activities on a 'whole of state' basis, and the more difficult issues involving conflict between the prevailing model of economic development and ecological goals (such as reducing greenhouse emissions from industry, domestic heating and transport) have not been confronted.

By contrast, Victoria currently has little or no commitment to ESD; indeed, rather the opposite. Under the Kennett Liberal government (1992–present), the most radically neo-liberal of the conservative Coalition governments now in power, Victoria has witnessed a 'degreening' of the state (Christoff, 1998a) – a systematic dismantling of public sector environmental capacity – which includes a 55 percent reduction in the environmental staff of the Department of Natural Resources and Environment, abolition of key consultative and strategic planning bodies (including the LCC), and commercial use of public land once set aside for conservation purposes. The government has adopted a blatantly discriminatory policy by siting toxic waste dumps in the poorer suburbs of Melbourne and has permitted development projects which override state guidelines on biological diversity. The Minister for Planning has repeatedly intervened to set aside key environmental controls in the State Planning Act. The government has cynically co-opted the title 'Agenda 21' to advertise a raft of loosely integrated major infrastructural and construction projects in Melbourne.

Local government

The municipal tier, usually called 'local government' in Australia, has begun to adopt Local Agenda 21 policies slowly. A local government environment network, 'Environs Australia' is promoting and disseminating the concept of ecologically sustainable development. A recent survey of the nation's 770 councils conducted for Environs Australia showed that only 33 councils were actually subjecting their policy processes to Agenda 21 criteria (Whittaker, 1996). However, although this is a small number of councils (4 percent), many were large urban councils and they accounted for about 18 percent of the population of Australia. It seems that the concepts of Agenda 21 are spreading quite rapidly at municipal level. In all, 119 councils (about 15 percent) were showing an active interest in ESD in 1996. Most urban councils now insist on separation of garbage by households, and contract out recycling collection to private firms.

Given the weak constitutional position of the municipal tier in Australia, limits to municipal action may rapidly be reached. Municipal government is not recognized in either the national or state constitutions. Consequently, there are no guarantees that democratic local government will persist. There are many instances in recent history of state governments sacking duly elected local governments which have been politically intransigent or – in the case of the major city councils – have sought to run economic development or conservation programmes which were counter to the policies or political desires of state government (see below). Whittaker notes that, although many of the councils surveyed had adopted environmental indicators and management plans, only 19 percent had established 'sustainability' indicators, and only 14 percent were addressing issues of global sustainability and environmental justice (Whittaker, 1997: 321).

Encouraged by the New South Wales state government, Newcastle City has taken a number of ESD initiatives. It has placed ESD at the core of its economic and employment strategy, aiming 'to become the South-East Asian centre for the sustainable energy industry' (City of Newcastle, 1997: 10). Mention should also be made of successful public transport initiatives in central Brisbane and Perth. Brisbane electrified and extended its rail system, resulting in a doubling in use (albeit from a low base) during the 1980s. Perth electrified its rail system and constructed a new rail line serving the growing residential areas in the northern suburbs. The cities of Sydney, Melbourne and Adelaide have recently joined the International Consortium on Local Environmental Initiatives (ICLEI).

Municipal environmental programmes have concentrated mostly in traditional policy domains such as waste management, town planning, energy and transport. Outside these domains – social service and economic development for example – there is little mention of Agenda 21. Work is also concentrated on council programmes rather than seeking to increase public awareness and consumption behaviour (in contrast, for example, with the approach of Göteborg in Sweden). Whittaker argues that the reason for the slow start is the lack of support for Local Agenda 21 at central and state government levels rather than a lack of capacity or willingness on the part of councils. In Victoria, with the recent sacking of a municipal council opposing the state's plans for a ring road around Melbourne, there is reason to suppose that the state government actively opposes local adoption of Local Agenda 21.

The private or voluntary sector

Government is by no means the only source of activity for ESD in cities. Businesses have scrambled to portray a green image. There is a substantial industry for recycling paper, glass and PET plastics, ink cassettes for computer printers and floppy disks. The Foster Foundation promotes a simple scheme (Greenfleet) for private motorists and the owners of fleets of business vehicles to contribute voluntarily to tree planting programmes and thereby enlarge Australia's carbon sink. In all, the Howard government has placed considerable faith in voluntary agreements with the private sector in which businesses commit themselves to 'action plans' to achieve ecologically sustainable development. Such voluntary programmes are fully in line with the neo-liberal ideological position of the government since they cost the government little, do not involve regulation and place maximum power in the hands of business.

One of these voluntary programmes is called 'Greenhouse Challenge'. Under the programme companies sign co-operative agreements committing them to reduce greenhouse gas emissions.[8] By May 1997 only 42 companies had signed agreements, but this number included some of the largest corporations in Australia such as BHP, Rio Tinto, Shell, the large paper manufacturers, aluminium producers, and a number of electricity companies. However, there are some real and important contradictions in voluntary programmes which may in the end vitiate their success – including the process of privatization of formerly state-owned energy corporations. We consider one example: energy production in the La Trobe Valley in Victoria.

The La Trobe Valley contains enormous reserves of lignite (brown coal) which is burned to produce electricity for metropolitan Melbourne and the rest of the state. Thus the Valley's power stations are among the largest sources of greenhouse gas emissions in Australia. Until 1996 the production and distribution of electricity was wholly in public ownership, incorporated under the State Electricity Commission of Victoria (SECV). The SECV, being accountable to the state parliament, was required to make and implement a plan for the reduction of greenhouse gas emissions. Under the current state coalition government's programme of privatization, however, the industry was split up into a number of corporations and sold to the private sector. The production of electricity was separated from its distribution, with distribution companies operating what are effectively spatial monopolies covering different areas of the state. A co-operative agreement to reduce emissions has been signed by the five major private producers of electricity. This agreement, of course, excludes any additional emissions which may be caused by inefficiencies in distribution, since the producers do not control the distribution networks of power lines. Under the agreement the producers commit themselves to develop common standards and performance measures for emissions, and a common method of verifying performance. Each facility must submit their own greenhouse gas inventories and action plans to reduce emissions at each plant. The aim is to produce electricity by burning coal with greater ecological efficiency, that is to emit less greenhouse gas per quantum of energy produced.

This is a worthy aim which must be part of any attempt to address the greenhouse problem. But there are severe difficulties with such a voluntary, non-regulative strategy. Each distribution company and each private power plant has a big incentive to produce more energy. When the distribution company Eastern Energy was sold to Texas Utilities, Mr Dan Farell, the new company chairman, said he hoped to boost profits by increasing the sale of electricity. Farell said Victoria's electricity consumption was low compared with Texas. Eastern Energy customers use, on average, 5,600 kilowatt-hours per year compared with 14,283 kilowatt-hours in Texas (Walker, 1995). The distribution companies compete to maximize profits. This competition is for investors not for customers, for the customers are mostly tied spatially to a single company. The competition is for the greatest profits rather than the best level of service. Accordingly there is every incentive to reduce the work-force and spend the minimum possible on maintaining the power grid.

The production companies compete with each other to sell energy to the distribution companies. Greenhouse Challenge specifies that all the companies expect their production to increase – as of course they must to make their operation profitable. The technological improvements which are envisaged under the 'action plans' are no more than the companies would carry out under their normal operations to increase the economic efficiency of their production processes. So, for example, the oldest plant in the La Trobe Valley, the Hazelwood power station is owned by one of the signatories of the Greenhouse Challenge voluntary agreement (Hazelwood Power). This power plant ('one of the worst single air pollution sources in Australia' says Strong [1998]), uses old technology due for retirement, but was brought back to peak levels of production by the new owners. It emitted 9.45 million tonnes of CO_2 in 1995–6. By the year 2000 the emissions after the 'action plan' are forecast to be 15.1 million tonnes, an increase of 60 percent. The 'action plan', it is claimed, will save about 0.7 million tonnes, reducing the increase from 67%. Even if the goal of the plan is met it can hardly be claimed as a great achievement. Unfortunately the action plans and agreements are themselves shrouded in secrecy: the detailed draft agreements are 'commercial in-confidence' (secret) and even the identity of the chair of the task force negotiating the agreements cannot be disclosed. Somewhat absurdly, the public profile of the 'co-operative agreement' of Yallourn Energy, one of the private energy producers, claims credit for efficiency measures for the reduction of CO_2 and other greenhouse emissions implemented when the plant was part of the SECV (Greenhouse Challenge Office, 1997).

Consuming cities: concluding remarks

Over the past two centuries, the Australian environment has been refashioned to meet the demands of the nation's urban population. Particularly from the start of the twentieth century, Australia's major cities – their size, the cultural predilections and material demands of their populations, the scale, nature, intensity and complexity of their industrial development – have dominated and 'consumed' the landscape. Their impacts may be considered as occurring over three domains which together define the 'ecological footprint' of a city (Wackernagel and Rees, 1996); namely, the 'localized' impacts of urban production and consumption; the geographically distant impacts arising from urban demands (including direct consumption and the maintenance of urban lifestyles funded by revenue from the export of natural resources); and those impacts which are local in origin but global in extent. These considerations have come into stronger focus with the intensification of regional and global trade: cities are having an increasingly profound effect on spatially distant (and, increasingly, internationally dispersed) environments.

Data on environmental conditions within Australian cities remain incomplete and insufficient for an adequate description of urban environmental conditions because of the poverty of monitoring efforts and also the limited indicators employed – even after twenty years of experience. Nevertheless, it may be said that air quality in all major cities is good by national and international standards, for both Melbourne and Sydney. Trends in atmospheric concentrations of ozone, suspended particulates, sulphur dioxide and nitrogen dioxide have declined over the past two decades but are likely to increase over the next two – the result of policies which prioritize and subsidize road transport over alternatives.

Urban water quality (both potable and ambient) is excellent by international standards (despite Sydney's occasional problems). In addition, demand management in major capital cities has led to a significant decline in per capita water use, although the total demand for water continues to increase. Few accessible rivers remain undammed throughout continental Australia and further withdrawals from major river systems will further imperil aquatic ecosystems already threatened by reduced environmental flows and eutrophication. Continuing increases in total demand represent a significant impending resource allocation problem in several states, involving intensifying competition for limited supplies between agricultural, industrial and domestic users.

Total and per capita energy use – particularly that deriving from fossil fuels – continues to rise. Total and per capita waste generation also continues to increase, despite the targets set in the National Waste Minimization and Recycling Strategy being met. Australia is second only to the USA in the volume of municipal waste generated (kg/per capita), and in both Sydney and Melbourne municipal landfill sites will be saturated by year 2000. Australia's record for recycling is mixed: high by OECD standards for paper, poor for glass and other materials.[9]

As populations in the major capital cities continue to grow, and as urban Australians continue to maintain their collectively lavish and environmentally expensive lifestyle, judged by international standards, the environmental impacts – the ecological footprints – of these cities continue to expand. Two obvious facts stand out from the analysis in this chapter. One is the fluctuating government policy focus, particularly at national and state levels, which corresponds neither with public opinion on the environment nor with global pressures. The other is the widespread failure of government policy to integrate the goal of ecological sustainability with that of economic development, particularly with respect to urban development. The impact of the Rio Earth Summit has, in this respect, been negligible. Environmental management and protection have been viewed as peripheral to, largely separate from and sometimes in conflict with the future of Australia's cities.

Notes

1 For further analysis, see Christoff (1998b).
2 Of course the inner suburbs continued to lose population in the 1980s as Forster (1995: 100) points out. But this was because of the increased affluence of the population which was able to demand and obtain more space in and around the home. This did not leave a 'hole' in the doughnut. On the contrary, what was left was a more prosperous and more physically attractive environment in which house prices consistently rose and to which new investment continued to flow.
3 Mees's analysis of retailing patterns in post-war Melbourne concludes that policy had a positive impact by reducing, indeed in some instances preventing, the development of American-style free-standing malls and the hyper-surbanization of commercial activity (Mees, 1998).
4 Daly (1987) writes: 'The growth of Australia's foreign reserves in 1971 and 1972 and the artificially low value of the Australian dollar, held down by the government under the fixed exchange system, induced a large inflow of speculative capital ($2.0 billion in 1972). At the same time the government relaxed its control of the domestic money supply; this grew at rates exceeding 25 percent in 1973. There were few alternative investments than property for much of this money.'
5 Membership of the Australian Conservation Foundation (ACF) grew from 9,446 in 1981 to 21,400 in 1991 and for the Wilderness Society from 7,332 in 1981 to 16,377 in 1991, (Papadakis, 1993: 150–152).
6 Keating's absence among the cast of national leaders at the Earth Summit in June 1992 signalled a deliberate downgrading of leadership interest in environmental issues, one which persisted over the remaining four years of Labor government.
7 The permit to excise is only the second such instance in the life of the Ramsar Convention. As the intended Victorian chemical storage facility is unlikely to be built, the excision has not occurred.
8 Other organizations such as municipalities may also join the programme.
9 Summaries and data are drawn from SEAC (1996), NSW EPA (1997), and OECD (1998).

References

ANOP (Australian National Opinion Polls) (1993) *Community Attitudes to Environmental Issues*, Canberra: Department of Environment, Sports and Territories of the Commonwealth Government.

Australian Government (1994a) *Better Cities, A Commonwealth Initiative, Victoria Status Report 1994*, Sydney: Commonwealth Department of Housing and Regional Development.

Australian Government (1994b) *Better Cities, A Commonwealth Initiative, New South Wales Status Report 1994*, Sydney: Commonwealth Department of Housing and Regional Development.

Bookchin, M. (1992) *Urbanization Without Cities, The rise and decline of citizenship*, Québec: Black Rose Books.

Burgmann, M. and Burgmann, V. (1998), *Green Bans, Red Union: Environmental activism and the New South Wales Builders Labourers' Federation*, Sydney: University of New South Wales Press.

Christoff, P. (1994) 'Environmental politics', in Brett, J., Gillespie, J. and Goot, M. (eds) *Developments in Australian Politics*, Melbourne: Macmillan, pp. 348–367.

Christoff, P. (1998a) 'Degreening government in the garden state – environmental policy under the Kennett Government', *Environment and Planning Law Journal*, 15 (1): 10–36.

Christoff, P. (1998b) 'From global citizen to renegade state: Australia at Kyoto', *Arena Journal*, No. 10: 113–128.

City of Newcastle (1997) *Initiatives: Steps Towards Sustainability in Newcastle and the Hunter Region*, Newcastle: NSW: Newcastle City Council.

Daly, M.T. (1987) 'Capital cities', in Jeans, D.N. (ed.) *Australia a Geography*, Vol. 2: *Space and Society*, Sydney: Sydney University Press, pp. 75–111.

Davidson, G. (1991) 'A brief history of the Australian heritage movement', in Davidson, G. and McConville, C. (eds) *A Heritage Handbook*, Sydney: Allen and Unwin, pp. 14–27.

Downton, P. (1991) 'Ecopolis now!', *Habitat*, 19/4.

Dunstan, D. (1984) *Governing the Metropolis: Politics, Technology and Social Change in a Victorian City: Melbourne 1850–1891*, Melbourne: Melbourne University Press.

Dunstan, D. (1985) 'Dirt and disease', in Davidson, G., Dunstan, D. and McConville, C. (eds) *The Outcasts of Melbourne*, Sydney/London/Boston: Allen and Unwin, Ch. 7.

Ecopolis (1996) *The Halifax EcoCity Project*, Adelaide: Centre for Urban Ecology.

ESDWG (Ecologically Sustainable Development Working Group) (1991a) *Final Report – Energy Production*, Canberra, AGPS.

ESDWG (Ecologically Sustainable Development Working Group) (1991b) *Final Report – Energy Use*, Canberra, AGPS.

ESDWG (Ecologically Sustainable Development Working Group) (1991c) *Final Report – Transport*, Canberra, AGPS.

Forster, C. (1995) *Australian Cities, Continuity and Change*, Melbourne: Oxford University Press.

Gleeson, B.J. and Low, N.P. (forthcoming) 'Revaluing planning: rolling back neo-liberalism in Australia', *Progress in Planning*, Oxford: Pergamon.

Greenhouse Challenge (1997) *Cooperative Agreements: Loy Yang Power, Hazelwood Power, Yallourn Energy, Edison Mission Energy*, Canberra: Greenhouse Challenge Office.

Gullett, W. (1997) 'Environmental protection and the "precautionary principle": a response to scientific uncertainty in environmental management', *Environmental and Planning Law Journal*, 14: 52–69.

Hall, C.M. (1992) *Wasteland to World Heritage: Preserving Australia's Wilderness*, Melbourne: Melbourne University Press.

Hannan, K. (1997) *ESD in Planning and Environment Act Frameworks*, Conference Papers of the Conference of the Victorian Royal Australian Planning Institute, Melbourne: RAPI, pp. 63–81.

Hayward, D. (1993) 'Dual politics in a three tiered state', *Urban Policy and Research*, 11/3: 166–180.

Hundloe, T. and McDonald, G. (1997) 'Ecologically sustainable development and the Better Cities program', *Australian Journal of Environmental Management*, 4: 88–111.

Jakubowitz, A. (1984) 'The Green Ban movement: urban struggle and class politics', in Halligan, J. and Paris, C. (eds) *Australian Urban Politics: critical perspectives*, Melbourne: Longman Cheshire, 149–166.

Lowe, I. (1997) *Greenhouse, Coping with climate change*, Melbourne: CSIRO.

Lyster, R. (1997) 'The relevance of the precautionary principle: Friends of Hinchinbrook Society versus the Minister for the Environment', *Environmental and Planning Law Journal*, 14: 390–401.

Manning, P. and Hardman, M. (1975) *Green Bans*, Melbourne: Australian Conservation Foundation.

Mees, P. (1995) 'Dispersal or growth? The decentralisation debate revisited', *Urban Futures*, 18: 35–41.

Mees, P. (1998) 'The Malling of Melbourne?', Unpublished paper, Faculty of Architecture, Building and Planning, The University of Melbourne, Parkville, 3052, Australia (available from the author).

Mees, P. (1999) *A Very Public Solution: Transit in the dispersed city*, Melbourne: Melbourne University Press.

Miller, C. (1997) 'Land clearing tie to greenhouse gas', *The Age*, 18 Sept., p. A4.

Moser, S.T. and Low, N.P. (1986) 'The central business district of Melbourne and the dispersal and reconcentration of capital', *Environment and Planning A*, 18: 1447–1461.

NSW EPA (New South Wales Environment Protection Authority) (1997) *New South Wales State of the Environment 1997*, Sydney: NSW EPA.

NSW Govt. (New South Wales Government) (1997) *Local Government Amendment (Ecologically Sustainable Development) Act 1997*, Department of Local Government Circular to Councils 97/75, Sydney: Government Printer.

NSW Govt. (New South Wales Government) (1998) *Environmental Guidelines: State of the Environment Reporting by Local Government Promoting Ecologically Sustainable Development*, Department of Local Government Circular to Councils 98/29, Sydney: Government Printer.

OECD (Organisation for Economic Co-operation and Development) (1998) *OECD Environmental Performance Review – Australia*, Paris: OECD.

Papadakis, E. (1993) *Politics and the Environment, The Australian experience*, Sydney: Allen and Unwin.

Papadakis, E. (1996) *Environmental Politics and Institutional Change*, Cambridge and Melbourne: Cambridge University Press.

Rodger, A. and Schapper, J. (1994) *Sustainable Urban Development in the Jerrabomberra Valley*, Melbourne.

Sale, K. (1991) *Dwellers in the Land: The bioregional vision*, Philadelphia: New Society Publishers.

Strong, G. (1998) 'How is the air up there?', *The Age*, 29 April, p. 15.

SEAC (State of the Environment Advisory Council) (1996) *Australia, State of the Environment, 1996: An Independent Report presented to the Minister for Environment by the State of the Environment Advisory Council*, Canberra: Department of Environment, Sport and Territories.

Wackernagel, M. and Rees, W.E. (1996) *Our Ecological Footprint, Reducing human impact on the Earth*, Gabriola Island, British Columbia, Canada: New Society Publishers.

Walker, D. (1995) 'Victorians should use more electricity, says U.S. buyer', *The Age*, Melbourne, 7 November, p. 3.

Whittaker, S. (1996) *National Local Sustainability Survey*, Occasional paper No. 3, Canberra: Environs Australia.

Whittaker, S. (1997) 'Are Australian councils "willing and able" to implement Local Agenda 21', *Local Environment*, 2/3: 319–328.

13 Jakarta, Indonesia

Kampung culture or consumer culture?

Lea Jellinek

Introduction

Greater Jakarta, the capital of Indonesia with a population of 20 million people, is a city of extremes (Sri Probo Sudarmo, 1997: 231). The city reveals a stark contrast between the individualistic, international, consumer culture and the indigenous, communal, kampung culture. The two cultures stand side by side as if in competition, the consumer culture coming to the fore during the economic boom and the kampung culture becoming more dominant during times of economic stagnation and crisis.

When I first came to the city in the early 1970s, the centre was dominated by kampungs (urban villages) and petty traders. By 1990, the city had been transformed into an Asian Los Angeles. Highways, overpasses, sky-scrapers, hotels and megamalls (self-contained shopping centres) dominated the city. Looking down from the twentieth floor of the Hotel Hyatt many former kampung areas looked like bombed-out sites. The houses of the little people of Jakarta were being replaced by the toll roads and modern multi-storey citadels of glass, cement and marble. Government Ministers told me that the kampung was an anachronism in the modern age, a thing of the past which had to be replaced (Radinal Mochtar pers. comm., 1980; Siswono, pers. comm., 1991).

Many people were being forcefully pushed out of the city centre by modern developments. In the early 1990s, the largest international integrated transport terminal in South East Asia was proposed for Manggarai in Central Jakarta, where an estimated 40,000 people would be affected. The largest International Trade Centre in South East Asia was proposed for Kemayoran, the former airport of Jakarta and 20,000 people would have to move (*Tempo*, 1992: 13–22). A new city modelled on Singapore and Sydney with elegant offices and condominiums, gardens, helipads and the latest in rapid transit systems was proposed for Jakarta Bay – and up to 1.5 million people who occupied and made a living around the bay would be affected (*Majalah Matra*, 1996: 60–70; *Kompas*, 1996: 17). Nobody knew precisely how many people were to be affected because no proper social and environmental impact studies had been done (Bianpoen, 1996, pers. comm.).

Along with the destruction of the kampungs came a consumer culture. The richest built opulent mansions. They shopped at the giant mega-malls and stacked their houses full of consumer goods. In the evenings they went to discos. Entire suburbs were transformed into the equivalent of Beverly Hills. Along with consumption went the destruction of Jakarta's environment. In the 1970s I had been able to see the mountains of Bogor from Hotel Indonesia. By 1990 the mountains were hidden by a constant veil of smog. The city was predicted to run out of groundwater in nine years' time (Bianpoen, pers. comm., 1998).

By 1998, Indonesia (and the Jakartan economy in particular) was in crisis. Jakarta's 'edge cities' account for 70 percent of the population; now 40 percent of this suburban population were unable to feed themselves. Many children had dropped out of school (Marshall, 1998: 22). With the crash of the modern economy, the poor and middle class relied on the kampung social structure and economy for survival. Street vendors dominated the central city streets and many of the vacant building sites were converted into kampungs and market gardens.

Jakarta is the focus of our story because it was at the heart of the economic boom and is now at the heart of the economic crisis. The major impacts of economic development and of the present economic crisis are most strongly felt there. In this chapter I first consider the kampung culture. I then describe the 'consumer culture' that invaded and displaced it in the 1980s and 1990s. I discuss the impact of this culture on the environment. I go on to describe the devastating impact of the economic crisis on the consumer culture, which nevertheless opens up a small window of opportunity upon a more sustainable future society.

Kampung culture

About 80 percent of the urban population have always built their own homes in kampungs and will continue to do so in the future. The kampung areas of the city not only provide most of the housing for the urban poor but also most of the labour needed by the city. They also provide humanity and warmth. Ultimately, they provide the social mesh – the neighbourliness – that holds the city together.

Government authorities tend to notice the less important, highly visible negative physical features of the kampung and fail to notice the kampung's much less tangible but more important social values. Official documents describe kampungs as congested, unsightly, lacking in amenities, disordered, dangerous to health, subject to fire hazard and crime and illegally occupying precious inner city land. By clearing them away and replacing them with multi-storey flats, officials argue that they will be able to create neater, cleaner, legal, organised, high rise, high density settlements complete with modern amenities on precious central city space. But when a kampung is destroyed, not only the improved pathways, drains, mortar and attractive little homes are lost but a whole way of life, work, income, social networks, memories, attachments and accommodation for millions of people. The Kampung culture is one of mutual self-help whose physical fabric has evolved organically – creating a sense of place. High population density is coupled with economic diversity. Low consumption goes with redistribution of resources. Social bonds cannot be separated from economic ties.

'*Gotong royong*' – mutual self-help and exchange – is at the heart of the kampung (village) social and economic relations. From politicians and policy makers to poor kampung dwellers, *gotong royong* is espoused as one of the most positive aspects of Indonesian culture. It means that neighbours know, care for and help one another. Both the urban and rural kampungs have been constructed around the same basic principles which lie at the heart of Indonesian culture: '*makan, tidak makan asal kumpul*' (whether we eat or not does not matter, the important thing is that we gather together), '*bagi, bagi rejeki*' (share our good fortune), '*rukun*' (togetherness), '*terima kasih*' (mutual exchange). Without government intervention this uniquely indigenous welfare system has enabled the redistribution of resources from those who have to those who have not by the people themselves through their own efforts and culture.

The poorest, especially, relied on their neighbours for help in times of hardship. There was no government social security system to pull them through, but they could rely on neighbourly help. If a neighbour lacked food, she could ask her neighbours for assistance. If her children were sick, neighbours visited and tried to help. Kampung mothers talked of growing up together, playing, going to school, raising children, preparing joint festivities (marriages, birth, circumcisions and funerals), and sharing food, resources and worries together. Those who were not related, felt related after having lived for 40 years in such close proximity. People adopted one another's children and formed marital bonds with neighbouring households. Wells, toilets, playing spaces for children, pathways and mosques were built together for communal use.

Until the early 1980s, housing in urban areas was not viewed as a major problem by most kampung dwellers. With Indonesia's gentle climate and the availability of homes in the village, the key problem was living close to income-earning activities in the city. Until the beginning of the 1970s, land and housing were still readily available in Jakarta. In the 1940s and 1950s, the new migrants were even encouraged by the authorities to occupy the many vacant or swamp (wetland) areas in the city. They moved to the city with village neighbours or relatives and clustered together in certain parts of the city. They lived together in simple homes, sharing facilities and learning new trades. When individual breadwinners or couples had accumulated sufficient resources, they bought a plot of land near kinsfolk, friends and work, and built themselves houses. Although many maintained their links with the village, their ties to Jakarta grew stronger until their primary attachment was to the city.

By the early 1980s, kampung densities had increased to over 1,000 people per hectare (Bianpoen, 1983). One person occupied an average space of only 3 to 4 square metres. Spaces were divided and subdivided, creating a honeycomb and a diversity of odd shaped houses, each with its own personality. Some houses were more prosperous and stood proudly above the rest. Others were modest, little shacks barely above the ground. One household could earn ten to thirty times that of a neighbour. This created a network of dependencies with the poor relying on the rich for work and the rich relying on the poor for goods and services (Jellinek, 1991).

Traders moved from house to house delivering fresh vegetables, fruit, ice creams, cakes, kerosene and water. Women washed and sewed clothes for neighbours, prepared cooked food for stalls, cared for their own and neighbouring children. Men rebuilt and upgraded kampung houses with women still living and working inside them. They also carted all types of goods in and out of the community. Within an hour, whatever was needed – a bed, mattress, television, sideboard – was delivered by brokers who operated throughout the community. Many different services were provided. All types of cooked food: porridge, noodle soups, coconut cakes, rice and vegetables could be obtained within minutes. There was a buzz of activity, an excitement that kept everybody busy. It was estimated that 60 to 70 percent of the population earned most or part of their income in these various informal ways in or near the community (Firdausy, 1995: 280–81). Those who went out into the neighbouring city brought income back which again circulated within the kampung community.

Kampungs enabled an economical use of resources. People did not have their own private gardens but used communal space in pathways and community squares. Sleeping, washing and cooking facilities were often shared by many people. Drinking water was bought by the litre and aproximately 40 to 80 litres of water were used per family instead of the 1,000 litres used by each member of a middle class household. People shared electricity connections and spent only Rp.35,000 on fuel per month compared to ten times this amount spent by the middle class. Their houses cost Rp.4,000,000 to 15,000,000 to build instead of ten times that spent by the middle class (Soelistijo, pers. comm., 1998). Table 13.1 provides a rough comparison of consumption levels of kampung and middle class households.

Kampung dwellers mainly walked to work, to school or to the market. Becaks (trishaws), when they were allowed in the city before 1990, were their favoured form of transport along the narrow pathways of the kampung. These man-powered vehicles consumed no fuel and created no air or noise pollution. Kampung dwellers obtained their food from stalls and open-air markets within or near their neighbourhoods – which decentralised trading activities and put them in close proximity to dense residential areas. It also cut costs, encouraged labour creation and avoided the mega-malls' excessive consumption of land and energy for lighting and air conditioning.

Table 13.1 Comparative consumption of kampung and middle class households

	Kampung (Ibu)		Njonja (middle class)	
Basic expenditure	Rupiah value	Percentage of income	Rupiah value	Percentage of income
Food	175,000	54	1,500,000	27
Rent	65,000	20	2,500,000	45
Transport	35,500	11	262,500	4
Water	none	n/a	65,000	1
Kerosene	15,000	5	31,000	0.5
Gas	none	0	n/a	n/a
Electricity	20,000	6	275,000	5
School	17,500	5	275,000	5
Telephone	none	0	350,000	6
Servants	none	0	275,000	5
Rubbish	none	0	20,000	0.3
Total income	R328,000	100	R5,553,500	100

Source: Author's estimates, based on census data (figures are approximate).

Ultimately, the kampungs provided a redistributive mechanism whereby money earned from the city could be channelled to the countryside. For those who were circular migrants, the city was not a place of consumption but rather a place of earning, saving and exchange. They worked hard, lived austerely with relatives or neighbours and sent whatever was left over to the village. During the growing season and dry months, the kampungs provided a refuge for those in the countryside who could not find work and needed to work temporarily in the city. Up to 15 percent of Jakarta's population were circular migrants (Hugo, 1982). The process of movement between city and countryside increased, especially in times of economic hardship.

Members of these communities had a sense of belonging. People knew and greeted each other in the pathways. Any stranger entering the community would be immediately recognised as such and asked where he was going and what he wanted or who he was looking for. Not to ask and not to respond was seen as a sign of arrogance ('*sombong*'). This familiarity with every face and recognition of every stranger was the best protection against crime.

Pride of place was evident from the way the kampung dwellers upgraded and improved their homes and decorated the pathways with greenery and pot plants. Work was regularly done to repair houses – a little improvement here and a little improvement there. The builders in the community were constantly making renovations. As household economies improved, a second storey was added. Glass replaces chicken wire windows. Tiles and cement took the place of earthen floors. Temporary walls give way to brick. Many houses were whitewashed and repainted each year. If left to develop naturally, the kampung became a *permanent* community in the making. With time, it evolved from a rustic village to a shanty town and eventually to an improved inner-city community, especially if it was given legality of tenure.

One should not over-idealise the kampung. As with every community there were conflicts. Neighbours often did not get on with each other. There were jealousies and bitterness. The sanitary conditions – drainage, sewerage and rubbish collection – left much to be desired. Water supply was inadequate and people often had to carry their own water into their homes or wait to buy it from a passing vendor. Conditions were very cramped. But the central city kampung community had many positive intangible values and provided a low cost of living close to work for many poor people.

Consumer culture

During the economic boom of 1975–95 economists spoke and wrote about the massive inflow of foreign investments, average annual growth rates of 7 percent, growth in banking, increase in factory jobs, the fall in annual infla-tion from 600 percent to 6 percent, reduction in poverty from 60 percent of the population to 11.3 percent and the rise in the standard of living from US$50 to US$1000 per capita per annum (BPS, 1998 (July); Schwartz, 1994: 57–58; Booth, 1997; De Tray, 1998). This upward evolutionary devel-opment was assumed to be permanent.

Conspicuous consumption was at the heart of Jakarta's economic boom and it set the trend for the nature and goals of the elite members of society. Although only a minority were able to partake in the most extravagant forms of consumption, the lower and middle class were affected by these patterns and, in their own way, tried to copy them. Houses were cluttered with consumer goods – carpets, couches, sideboards, television sets, video recorders, fans, air conditioners, refrigerators, telephones and rice storage units. Many of these goods were bought on credit. Even though they had more material possessions than ever before, they aspired for more. They constantly compared their lives with those of 'the rich' and felt poor. Relationships between kinsmen and neighbours changed. Whereas formerly, people from the same village or extended family clustered together, now they lived apart. People no longer had time to sit along pathways talking to each other but hurried to work or sat inside and watched television.

The cost of living kept rising, and in 1996 Jakarta was said to have a higher cost of living than New York (Bianpoen, 1997: 8). Life was dominated by envy, keeping up with neighbours and buying the consumer goods that one saw on TV. To accumulate wealth and possessions, needy relatives had to be kept away. Solid doors were shut. Families and neighbours distanced them-selves from one another. Middle class and rich people lived very enclosed lives, moving from their air-conditioned houses to air-conditioned cars and on to air-conditioned offices and mega-malls. They exercised in air-conditioned gyms, ate in air-conditioned restaurants and did not notice the impact they were having on the kampung dwellers or on the city's environment.

The Suharto family set the trend. In 1991, they were estimated to have wealth equivalent to the total amount borrowed from the World Bank – US$40 billion (Aditjondro, 1998: 82). Indonesia's borrowings leaped from US$3.2 billion when Suharto first came to power in 1966 to US$130 billion in 1997. Seventy percent of the country's resources were controlled by 3 to 4 percent of the people (Scott, 1998: 46–47; Greenless, 1998: 30). 'Some economists estimate that Indonesia's private debt of $80 billion is held by at most a few hundred individuals . . . perhaps as few as 50' (Chomsky, 1998: 3). In the 1990s, three conglomerates controlled by the family gained control of 40,000 hectares of land ostensibly to build three mega-cities around Jakarta. Only 10 percent of that land was used (*Kompas*, 1998b: 9). Most of the land lay vacant and was held for speculative purposes.

Pondok Indah (Beautiful Resort), 15 kilometres south of central Jakarta, once a protected water reserve and source of irrigation agriculture, by the 1990s had been converted into a residential area for Jakarta's rich. Farmers, who for generations had cultivated the rich soils, were paid low compensation rates for their land and forced to leave. Indonesian cabinet ministers, generals, police officers and some rich Chinese developers constructed lavish homes in the area. At the entrance to the area, security guards in glass boxes monitored whoever entered and checked that they had a permit and were not taking photographs.

Fortress-like mansions decorated with satellite dishes and surrounded by spiked fences of glass and metal competed with one another for attention. Guards stood at gates, behind high walls in landscaped gardens. One house was surrounded by gardens blended with statues of flying horses and women. Streets were smooth, silent, lined by tall palms and manicured like botanical gardens. Rows of BMWs and Mercedes were parked in the driveways, polished by their chauffeurs. Five cars were visible – two for each member of the family. Families often own at least three other houses and were sometimes away in New York, London, Australia or Paris, staying at their other houses. Servants were left to clean and occupy the 24 rooms of their Pondok Indah house which was decorated with marble, had carpeted floors and chandeliers imported from Turkey, Paris or Italy. Spare houses were often rented to foreigners at exorbitant rates, so there was a close link between the rich of Jakarta and the international capitalist community.

A golf course surrounded Pondok Indah like a protective shield filtering out Jakarta's air pollution – mostly caused by the carbon monoxide fumes from the cars owned by the rich. The golf course was a vast expanse of green, gently undulating hills, ponds, flowers beds and pathways covered by rubber mesh to stop the golfers from slipping. Signs around the golf course instructed non-members to stay out. A Japanese tea house offered expensive cups of tea.

During the economic boom, rich Indonesians preferred to go for weekend shopping trips to Singapore or Bangkok. The Singapore emporia liked Jakartan shoppers, who had unlimited desires and seemingly unlimited budgets. At the beginning of the economic crisis in January 1997, a newspaper reported that a government plane-load of Ministers' wives had gone shopping to Bangkok. While the elite still preferred to shop abroad, the middle classes copied their example in the mega-malls at home buying designer clothes with designer labels. Advertisements on the television and on billboards constantly enticed people to consume more.

The culture of conspicuous consumption bred a culture of want, of individualism, greed and dissatisfaction. One has to question its appropriateness for Indonesia, a country of 202 million people, most of whom are poor. The culture of conspicuous consumption meant that most of Indonesia's resources were being used by an unproductive, selfish minority instead of being shared with the more productive majority. Instead of focusing upon the difficult business of production and saving, the elite were focusing on consumption and it was having a negative impact on Jakarta's environment.

The environment of Jakarta

Greater Jakarta has spread from its origins as the small, coastal fishing village of Sunda Kelapa to Bogor in the hills, 60 kilometres away. It covers the four formerly separate regions of Jakarta, Bogor, Tangerang and Bekasi (now called Jabotabek – Ja Bo Ta Bek). An urban corridor extends along the road from Jakarta 200 kilometres to the inland city of Bandung in West Java (Dharmapathni and Firman, 1997). While the centre of Jakarta has been losing population due to the development of offices and demolition of kampungs, the peripheral areas of Bekasi and Tangerang have been growing at 9 percent per annum (Douglass, 1989: 213).

Between 1975 and 1990, toll roads encouraged the development of many industrial estates, housing estates and new towns around the edges of Jakarta. During the 1950s, 1960s and early 1970s, the bulk of the population had migrated into the city centre. In the 1980s and 1990s, people headed for the periphery hoping to find cheaper land, housing, lower costs of living and more employment opportunities (Dharmapathni and Firman, 1997: 15). By the 1990s, over 70 percent of Jakarta's population was gathering at the periphery, sometimes at a commuting distance of 120 kilometres (Sri Probo Sudarmo, 1997: 231). The city's growth did not follow the numerous Master Plans and Spatial Plans prepared during the 1970s and 1980s which aimed to protect Jakarta's most sensitive environments. Instead, development was *ad hoc* and chaotic following the interests of private enterprise and the market (Douglass, 1989; Sri Probo Sudarmo, 1997).

Jakarta's urban sprawl has meant the loss of precious farming land (see *Jakarta Post*, 1994). Land speculators bought up 60,000 hectares – an area nearly equivalent to the city itself (65,356 ha) – to be used for golf courses, luxury housing estates and gated communities, self-contained with their own shopping malls, restaurants, schools, sports and entertainment centres (Henshall, 1996; Dharmapathni and Firman, 1997: 11). Much agricultural land was lost. Fresh fruit and vegetables became more difficult and expensive to obtain and during the 1980s and 1990s it was more common to see imported oranges, apples (often old and stale) and vegetables from Australia in the supermarkets of Jakarta than good bananas and mangos from Indonesia.

Thousands of weekend bungalows for Jakarta's elite continued to be built around Bogor and into the mountains (Puncak) even though it was against government regulations. This mountain area was supposed to serve as a forest, water reserve, and recharge and recreation area for Jakarta. Farmers were pushed off their land and forced to go to Jakarta to seek work as construction workers, traders or trishaw drivers, or forced to go higher up into the hills to seek new land to cultivate. Fragile, steep mountain slopes and forestry land were converted to small farm plots. Deforestation resulted in rapid water run-off, erosion of top soils and the siltation of Jakarta's thirteen rivers (Douglass, 1989: 220–24).

Water has become a major problem in Jakarta (see *Jakarta Post*, 1993a, 1993b, 1993c). The damage to forests in the mountains in the south of Jakarta means that more water runs off during the wet season and less water is released during the dry. Less water enters the water-table, and pumping of the groundwater by more and more people has caused the water-table to sink. Only 30 percent of the city's population get their water from the municipal supply. The rest rely on groundwater – using wells, hand pumps and electrically pumped bores. All big buildings in the city have their own groundwater supply. Bianpoen, a leading environmentalist and planner in Jakarta, suggests that this water will run out over the next nine years (Bianpoen, 1997: 2). Dams, such as Jatiluhur, which supply water to Jakarta are also running out of capacity. Many people cannot afford piped water connections and are forced to buy their water from vendors, which can absorb up to 8 percent of their income.

The combination of sinking land, dropping water-table and salt water intrusion has made Jakarta's land, particularly in the north, unstable and prone to flooding. Jakarta's land has been sinking at a rate of 34 cm per year in the north and 4 cm per year at the centre (Bianpoen, 1997: 7). By 1997, salt water intrusion from the sea had reached 15 km inland. During the construction boom between the 1970s and 1990s, many wetland areas were built upon, leaving no scope for drainage. The laying of large amounts of concrete covering an estimated 2,621 hectares per year, aggravated the city's drainage and sinking (*Kompas*, 1995a). Some of the multi-storey buildings in the city have shifted off their axes (Douglass, 1989: 218).

The North Jakarta Bay Development (Pantai Utara) was planned and developed in the early 1990s in direct contradiction of the West Java Urban Development Project. The development aimed to reclaim 32 km of land from east to west along the coast and 1.8 km out to sea to house a population of 1.5 million people. The development was financed by international and domestic capital and administered by a Sydney-based firm 'Planning Workshop'. Singapore and Sydney provided the model for this land reclamation scheme.

Reclaiming land from the sea at Rp.1,000,000 per sq. metre was considered cheaper and more efficient than buying land elsewhere in Jakarta (*Kompas*, 1995b: 17; *Majalah Matra*, 1996). Because many city and central government officials had a personal stake in the success of the project, they did not insist on any studies that might advise against the scheme, although by supporting it they were breaking their own planning laws (Laws UU No. 9/1992 about zoning of the sea, UU No. 5/1990 protection of environment and UU No. 10/1992 protection of the people).

Many town planners, city engineers, academics, NGOs and environmentalists opposed the North Jakarta Bay Development. It was feared the 13 rivers which drain Jakarta, and already struggle to get to the sea, would be blocked. All of Jakarta's existing environmental problems: sinking of land and water-table, saline intrusion, water shortages, poor drainage and flooding would be aggravated. Concern was expressed about the impact of this development on Jakarta's remaining 25 hectares of protected mangroves, the fishing villages and the one thousand islands (Pulau Seribu) which form a protective barrier for Jakarta against massive waves hitting the city's coastline. Land for the reclamation of the 2700 hectares was proposed to be taken from the thousand islands. The thousand islands have coral reefs which contain some unique and endangered marine biota. The area was declared a nature reserve and National Park in the 1980s. Much effort and finance was invested to protect it, but the reefs have continued to deteriorate with only 15.4 percent in a good condition in 1995 compared to 80 percent in 1969 (Bianpoen, 1997: 10).

Jakarta's marine waters are said to contain the highest levels of mercury and iodine pollution in the world (*The Age*, 1997; The World Bank, 1995). At the point where some of these rivers meet the sea, I have seen white clouds of foam floating into the bay. One canal comes directly from the Pulo Gadung Industrial Estate and the waters are contaminated (Jacob Kedang, pers. comm., 1998). 'Up to ten kilometres from the shoreline, the sea water does not comply with the standards stipulated by the city's administration' (Bianpoen 1997: 7). In 1997, a fisherman told Louise Williams that he could catch 20 baskets of fish per day when he was young, compared with 10 baskets in 1980 and only two baskets in 1997 (Williams, 1997).

Particulate matter and lead concentrations in Jakarta's air account for 12.6 percent of the city's deaths (Bianpoen, 1997: 9). Traffic contributes over 50 percent of this pollution. Over one-third of Indonesia's total car ownership is concentrated in Jakarta, although less than 20 percent of the city's population have a vehicle. Life on the streets without a car has become increasingly hazardous, especially for kampung dwellers who spend most of their time outside their homes, working on the streets. Streets have become difficult to cross and massive highways divide former kampung communities, making them inaccessible to each other. Moving around the city has become more difficult, not only for the urban poor but for all citizens (Bianpoen, 1998, pers. comm.). The removal of trees and the expanse of concrete and glass, as well as additional heating from air-conditioners and motorcars, have caused external temperatures to rise in some parts of the city.

For how much longer will Jakarta's environment be allowed to deteriorate? What will be the long term impacts of environmental degradation? The economic crisis of 1997 and the halt in large scale developments provide Indonesia with an opportunity to reassess the costs and benefits of economic development and its social and environmental impacts. The government and people no longer have the resources to consume as they did during the 1970s, 1980s and 1990s.

Economic crisis

Suddenly in 1998 Indonesia's economy crashed. Foreign investors and multinational companies fled the country. The annual growth rate has dropped to minus 10 to 12 percent (BPS, 1998; ILO, 1998). The 100-plus banks that flocked to Jakarta after 1987 have collapsed. The international debt cannot be serviced and the country depends on the IMF and World Bank to bail it out. The rupiah exchange rate oscillates from Rp.10,000 to 15,000 to the US dollar when only two years earlier it was steady at Rp.2,500 – a devaluation of around 80 percent. Over half the population (80 million) are said to be below the poverty line, earning less than Rp.5,000 ($US 0.50) per day. An estimated 40 million people are said to be unemployed. Prices have risen so high, that most Indonesians cannot satisfy their basic needs for rice, sugar, salt, kerosene and cooking oil. A large percentage (estimates vary from 10–30 percent) of primary school children have not returned to school because parents need their children to assist with feeding the family and cannot pay the school fees (BPS, 1998; World Bank, 1998; ILO, 1998; *Kompas*, 1998c). Estimates suggest that 60 to 70 percent of the middle to low income communities on the edges of Jakarta have suddenly fallen below the poverty line.

Official figures reveal only a small part of what is happening. Experience and observation show that at the personal and community level a new story is unfolding. Many houses once stocked with consumer goods are now empty. Most possessions have been sold. Mothers complain that they cannot buy milk for their children. They are lucky to obtain a plate of rice each day (Williams, 1998: 12). Mothers bear the burden of finding food for their families. Some turn to household trade, but it is not easy. There are too many traders and too few customers and the cost of wares keeps rising so that traders do not know what price to charge for their products.

Young boys who once went to school now bake corn along the pathways. Some children take turns going to school – one sibling going on one day, the other the next – to save transport costs. Children no longer buy books but gather around one book and study as a group. Children who once bought lunch and snacks at school now carry whatever food they can find from home. The amounts that they have to spend (Rp.300–500, or about 5 cents

US) only buy two sweets or a small cake, leaving them feeling hungry and unable to concentrate for most of the day. Some have shoes on their feet which are held together by elastic bands. One university student could not go to class until a neighbour loaned him a shirt.

People have turned to all sorts of new strategies for survival. Mothers have cut down on family food and are boiling rice and cassava into a porridge so that it expands and can feed more people, like they did during the starvation of the Japanese invasion fifty years ago. People consume only one meal a day. Children have gone out on the streets to work as shoe-shine boys, newspaper sellers, singers, beggars and prostitutes. Those who have land and family have gone back to the village to survive because vegetables, housing, schooling and health care are cheaper there. The number of rubbish recyclers, traders, Otjek (motor-cycle) drivers, becak drivers and prostitutes has expanded.

Only a year ago, street traders had almost disappeared from the central city areas of Jakarta. The middle class patronised mega-malls, restaurants, bars, coffee shops and cafes. The street traders were chased away. Today the central city streets – Wahid Hasyim, Sabang, Tanah Abang, Senayan, Kebayoran, Merdeka Square, Senen – are lined by traders. It was reported by the World Bank, Central Bureau of Statistics and the International Labour Organisation in 1998 that many children were dropping out of school (*Jakarta Post*, 1998b: 3). The middle classes who have lost their jobs in factories, offices, advertising agencies, banks and mega-malls are copying the survival strategies of the poor. Due to this increased competition and drop in consumer demand, 30 percent of the traditional street traders in Jakarta have gone bankrupt in Greater Jakarta (*Kompas*, 1998a).

In one example I know well Saman's kampung house in the middle of the city was destroyed along with hundreds of others to make way for a five-star hotel. The economic crisis has stopped construction and Saman and some of his friends have returned to his former home site, built a small shack and planted bananas, corn and spinach. A cluster of shanties is forming, as people return to the area. It looks as it was in the 1930s and 1940s when it was a market garden (Kebun Kacang). Grass, trees and birds have returned to the centre of Jakarta and 'little people' are making the most of it. They are invading empty blocks of land and using them for housing, growing vegetables and petty trade (Walters, 1998). They are reclaiming the city where they lived before the multinationals and big business evicted them (*Jakarta Post*, 1998a: 3). The government has issued a new decree saying that kampung dwellers who have occupied their land without legal title for many years will be able to get legal title and that vacant land in the city should be utilised (Menteri Negara Agraria [Minister for Lands], Nomor 3 and 6, 22nd and 26th June, Tahun, 1998).

In 1996, kampung dwellers surrounding the old airport of Jakarta at Kemayoran had almost given up hope of being allowed to stay in their area. The project now has no financial support and the key people involved in the project, including the former Secretary of State Minister Moerdiono and Hendro Soemardjan, the head of the Kemayoran Development Agency, have been charged with corruption (*Tajuk*, 1998: 80–87). The community of Kemayoran has gained a reprieve. In April 1998, however, at least 70 houses in Kemayoran were burnt down under suspicious circumstances. Within two months, the community has been almost completely rebuilt, a remarkable feat in this time of economic crisis. These low income people work as drivers, labourers and traders. Where did they get the resources to rebuild? They pooled resources and worked together. Relatives and friends offered assistance. One elderly woman who looks after a number of grandchildren told me 'neighbours each contributed Rp. 20,000 and their labour to help rebuild my house'.

The fishing families who were going to be affected by the development of a new city along the bay are reporting the best fish harvest in years. Land reclamation projects which were to take up 10,000 hectares of land east to west along Jakarta's coastline have been stopped because of lack of funds and because conglomerates acquired the land illegally (*Kompas*, 1998b: 9). Mussels are growing larger than they were a year ago and a new type of fish has come into the bay. Some say it is due to a drop in the level of pollution caused by the many factories which dumped their effluent into the sea. Only two years ago, the bay had been referred to as 'dead' by the developers (Ross, pers. comm., 1996).

Thirty kilometres to the south of Jakarta, ten unemployed youths gather and share food and accommodation. They express the Indonesian philosophy '*makan, tidak makan asal kumpul*' – 'whether we eat or don't eat, the important thing is that we gather together'. The young men come from Sumatra, West and Central Java and Sulawesi. 'We form one large family', the Indonesian concept of '*kekeluargaan*'. They survive because they help one another. When one has money he supports the rest. The ten of them live on Rp.10,000 (one US dollar) per day. When I invited them for dinner at a local stall, they jokingly referred to it as part of a nutritional programme – '*projek perbaikan gizi*'. Under the guidance of an idealistic and dynamic lawyer, the ten young people distribute food parcels of rice, cooking oil and sugar twice a week to the poorest 150 families in their neighbourhood. The money is given by a few members of the Indonesian middle class and an Australian expatriate. The youths and the lawyer also buy and recycle old books and clothing, which helps them make contributions to poor neighbours for the health and schooling of their children (Nainggolan, 1998, pers. comm.). Another group – Suara Ibu Perduli (the Voice of Caring mothers) – is distributing food parcels, free medical care and informal education to families of poor children in six different locations throughout the city and reaching 4,000 people (Dinny Yusuf, pers. comm., 1998).

Help from relatives, neighbours and friends is one of the most important survival mechanisms. Those who have more help those who have less. Some of the middle class and rich are making contributions of rice to people who now cannot eat each day. The NGOs play a critical role in distributing these resources – collecting from the rich to give to the poor. But, not enough of this is going on. Most of the rich are still pretending that Indonesia is not in crisis and are living life as usual, driving big cars, spending up, spoiling their children and going to mega-malls. The gap between those who have resources and those who do not is expanding. The rich are distancing themselves from the problem. The poor are getting more desperate. Crime is growing, and with it fear.

Conclusion: a window of opportunity

Indonesia, and Jakarta in particular, is in a state of shock and panic. The nation is at a major cross-roads and we cannot tell which road it will take. Both co-operation and conflict are evident and have become even more so as a consequence of the economic, political and moral crisis. Kampung communities are surviving because their members are co-operating and supporting one another. At the same time, conflict and violence has become endemic and some say these activities are mainly being orchestrated by the political elite. Jakartan mothers talk of a 'crisis of the kitchen' (Krisis Dapur). While their children starve, their leaders – mainly men – play politics. The clash between power and greed and community, caring and sharing has reached its peak. Which will win, the consumer or kampung culture? How much longer can Jakarta's physical environment deteriorate? Isn't the loss of so much fertile land to construction partly responsible for the famine? Much of the fertile volcanic land of Java has been buried under concrete and the sharing values of the Indonesian people have been shattered by the drive towards individualism and economic prosperity.

Indonesian policy makers were intoxicated by the economic boom of the 1970s and 1980s and wanted to copy everything that was occurring in the West. They felt that the indigenous ways of life of Indonesian society were backward and had to be destroyed. They neglected the important intangible values of kampung society and only focused upon its negative physical shortcomings. They focused too much on the built form and material goods rather than the culture and content of their own society. They were too busy rushing ahead with 'Pembangunan' (development) to notice the loss of jobs, homes, human dignity and meaning for many people. The end result was conspicuous consumption and the destruction of Jakarta's environment.

Indonesia now has the opportunity to reconsider its past 30 years of 'economic development'. It has a chance to remember the social wealth of its value systems of sharing and caring and the wealth of its soil, assets that were being destroyed by unrestrained economic development. Will it take up this opportunity to re-evaluate its past 30 years of development or will political infighting and the panic of immediate survival squander the opportunity?

I have argued in this chapter that kampung communities, with their small-scale, tightly knit community values, once gave low income Indonesian families work, shelter, security, friendship and meaning. Do Indonesians wish to push on as they were before, or re-establish their own indigenous modes of living which may prove more enduring, sustainable and meaningful? I believe that the urban kampungs suggest an indigenous and sustainable way ahead for Indonesia's future. Can a middle way be found to enable the kampung and capitalist cultures to survive and thrive together? The capitalist culture provides the motor and stimulus to the broader economy while the kampung culture provides survival and redistributive mechanisms. Capitalism needs to be curbed so that it benefits the kampung rather than destroys it. Pockets throughout the central city area need to be reserved for kampungs. By comparison, the conspicuous consumption patterns and isolated ways of living of Indonesia's elite are not sustainable – socially or ecologically.

References

Aditjondoro, G. (1998) *Harta Jarahan Harto*, Jakarta: Pustaka Demokrasi.

Bianpoen (1983) *Research and Development for Urban Management. Case Jakarta*, (published report, publisher unknown), Jakarta.

—— (1997) 'Urban Environmental Management (Case: Jakarta)', International Symposium on Saving our City Environment towards Anticipating Urbanisation Impacts in the 21st Century, Universitas Merdeka, Malang Indonesia, Sept. 8–9 (unpublished paper).

Booth, A. (1997) *Poverty in Indonesia*, ILO, South Asian Multidisciplinary Advisory Team, New Delhi.

BPS – Central Bureau of Statistics (1998) Jakarta.

Chomsky, N. (1998) 'Indonesia: master card in Washington's hand', *Le Monde Diplomatique*, Sept., p. 3.

De Tray (1998) *The Australian*, April 11, p. 18.

Dharmapathni, I.A. and Firman, T. (1997) 'Problems and Challenges of Mega Urban Regions in Indonesia: The Case of Jabotabek and the Bandung Metropolitan Area', in McGee, T.G. and Robinson, I.M. (eds) *The Mega-Urban Regions of Southeast Asia*, Vancouver: University of British Columbia Press.

Douglass, M. (1989) 'The Environmental Sustainability of Development, Coordination, Incentives and Political Will in Land Use Planning for Jakarta Metropolis', *Third World Planning Review*, 11/2 (May): 212–235.

Firdausy, C.M. (1995) 'Role of the Informal Service Sector to Alleviate Poverty in Indonesia', *The Indonesian Quarterly* 23/3 (Third Quarter): 278–288.

Greenless, D. (1998) 'No Sanctuary', *The Weekend Australian*, August 15–16: 30.

Henshall, J. (1996) 'Development Planning in Asia's Mega Cities: An Australian Practitioner's Perspective', *Urban Policy and Research*, 14/4: 307–310.

Hugo, G.J. (1982) 'Circular Migration in Indonesia', *Population and Development Review* 8/1 (March): 59–83.

ILO (International Labour Organisation) (1998).

Jakarta Post (1993a) 'Flooding caused by poor drainage', Feb. 10, p. 3.

—— (1993b) 'Java island will face acute water shortage in 2000', August 21.

—— (1993c) 'Waves of flooding hit Jakarta, distrupt traffic', Feb. 9: p. 3.

—— (1994) 'Arable land scarcity haunts East Asia: FAO', June 7.

—— (1998a) 'Slum residents to get ownership of their land' Sept. 10: p. 3.

—— (1998b) 'Pedicabs should have a place in the city: NGOs', Sept. 11: p. 3.

Jellinek, L. (1991) *The Wheel of Fortune*, Sydney: Allen and Unwin.

Kompas (1995a) 'Jadi Bencana, Jika Relamasi Pantura tak Dikaji Secara Ilmiah Tiap Tahun 2,612 ha lahan jadi Hutan Beton' (There will be flood if the Jakarta Bay reclamation project is not done scientifically), Oct. 3, Jakarta.

—— (1995b) 'Reklamasi Pantai akan Tenggelamkam Ibu Kota' (Reclamation will cause flooding in Jakarta, the capital), Oct. 17, p. 11, Jakarta.

—— (1996) 'Ramai-ramai Bangun Kota Pantai Jakarta' (Busy Building Seaside City), June 3, p. 17. Jakarta.

—— (1998) 'Warteg, masih Menjadi Penolong Kaum Bawah?' (Street-side stalls still save the poor), Sept. 22, Jakarta.

—— (1998b) 'Mega Proyek Jonggol, Kapuk Naga dan Pantura Belum Miliki Izin' (Mega-projects and Seaside City Don't have Permits), Sept. 24, p. 9, Jakarta.

—— (1998c) 'Dua dari Tiga Penduduk Indonesia Miskin Sekali' (Two out of three Indonesians are very poor), Sept. 1, Jakarta.

Majalah Matra (1996) 'Kontroversi Reklamasi Pantai Jakarta' (Controversy over Seaside City) No. 119 (June): pp. 60–70, Jakarta.

Marshall, A. (1998) 'Where the young scavenge for a future', *The Age* (Melbourne), August. 22, p. 22.

Ross, P. (1996) Representative of International Planning Workshop Pty Ltd, Sydney (pers. comm.).

Schwartz, A. (1994) *A Nation in Waiting: Indonesia in the 1990s*, Sydney: Allen and Unwin.

Scott, M. (1998) 'Indonesia Reborn', *The New York Review of Books* 45/13, August 13 pp. 43–48.

Sri Probo Sudarmo (1997) 'Recent Developments in the Indonesian Urban Development Strategy', in Burgess, R., Carmona, M. and Kolstee, Th. (eds) *The Challenge of Sustainable Cities, Neoliberalism and Urban Strategies in Developing Countries* Atlantic Highlands, N.J.: Zed Books. pp. 230–244.

Tajuk (Indonesia News, Investigation and Entertainment Magazine) (1998) 'Kasus Kemayoran: pengakuan Moerdiono' (Case of Kemayoran: Moerdiono accused and taken to court), No. 9, June 25, Jakarta.

Tempo Magazine, (1992) 'Geger Proyek Kemayoran' (Kemayaran project emerges), No. 9 (year 12), pp.13–22, Jakarta.

The Age (1997) August 17.

Walters, P. (1998) 'Farmers' lives take root amongst the skyscrapers', *The Australian*, March 25.

Williams, L. (1997) 'Picture postcard look hides Jakarta Bay's toxic Cocktail', *The Age* (Melbourne) August 16–17.

—— (1998) 'Hunger in Indonesia spawning a lost generation', *The Age* (Melbourne), Oct. 20, p. 12.

World Bank Report (1998) 'The Poor in Indonesia's Crisis', Unpublished Report, p. 1.

14 After Rio

Urban environmental governance?

Nicholas Low, Brendan Gleeson, Ingemar Elander and Rolf Lidskog

Introduction

The Rio Declaration of 1992 and its agenda for action in the twenty-first century – Agenda 21 – were bold attempts at steering the nations of the world in the direction of ecologically sustainable development, a direction which includes social and environmental justice on a global scale. Six years later, can we say for the domain of urban policy that the discourse of sustainable development has made progress? Or should we rather say, as one Indian scholar said in 1997: 'five years after Rio we do not have Rio plus five but Rio minus five' (Shiva, forthcoming)?

It has been the intention of this collection to make some assessment of the impact of the discourse of sustainability on the urban environment in different nations. Agenda 21 was a marker in a discourse that by 1992 had already brought together social justice and concern for nature. But it was meant to be more than that; it was meant to represent a commitment to public action, and in this sense Agenda 21 was an act of global governance. In the project which assembled the chapters in this book we posed two sets of questions:

1 Has the Rio Declaration and Agenda 21 had a positive influence – either in the initiation of action or in reinforcing action already under way? Has it had little influence at all? Has it provoked a negative political response?
2 To what extent is urban environmental regulation today a matter for intervention at global level? And what form might the institutions developed for this purpose take?

In the first two chapters in this volume we attempted to set out the wider context of the discourse: first within a framework of political economy, and second within a framework of global environmental governance. In this final chapter we revisit these frameworks in the light of the intervening studies of national urban policy systems. In the next section we consider the first set of questions posed above: concerning the influence of the Rio Declaration and Agenda 21. In the following section we address the second set of questions concerning the future of city governance as a matter for intervention at global level. We then turn to the national studies and consider the varying answers.

The discourse of sustainable development

Capitalism is a material reality today infusing the whole world. By means of technology and science, and in conjunction with the efforts of organized labour, society's creation of wealth has increased enormously. However, during recent decades we have become more and more aware of the dark side of this development. The creation of wealth has been performed through an unlimited exploitation of natural resources with concomitant growth of ecological problems and environmental risks. Whether these problems and risks can be overcome within the framework of a capitalist system is an open question, although the prospects for such a development are bleak.

There are two main discursive 'tropes' competing to shape outcomes which vary depending on the different cultural traditions and political institutions of different places. We call these tropes *raw capitalism* and *environmental governance*. A 'trope' is a literary figure something like a metaphor or template through with which the world is viewed and responses to it constructed. The German word is *leitbild*. Inasmuch as these tropes (and others) vary in impact in different places, there is *variation* rather than uniformity in the actual models of development which are to be found in different regions, countries and cities. Of course there is variation too in the tropes which different scholars have identified. Welford (1997), for example, referring to the work of Johan Galtung and Rudolf Bahro, identifies discursive dimensions between green bioregionalism, socialism, global capitalism and corporatist capitalism (the erstwhile Asian model). We think that the fundamental lines of choice for humankind must now be starkly drawn between just two tropes, and we wish to accentuate that choice, for as Welford (1997: 7) himself says, 'capitalism has pillaged the natural environment'.

Trope A is a particular form of capitalism, or perhaps we should say a reversion to *raw capitalism* from the socially controlled and tamed version which prevailed for some years after the Second World War. It is 'The American Way of Life' redefined for a world of global as opposed to domestic competition. Competition between cities has long been the norm within the United States (Molotch, 1976; Bluestone and Harrison, 1982; Logan and Molotch, 1987), and has also become commonplace in Europe (Jensen-Butler *et al.*, 1997). It is the norm of Trope A and infuses thinking about urban planning wherever it takes root. Trope A praises the holy family of competition, deregulation and privatization. The 'free market' is hailed as a natural necessity (Rothschild, 1990). 'Free trade' which allows technology and investment to penetrate everywhere is a central article of faith. The state's role is to facilitate the operations of the market, to support business and economic growth and to provide the coercive force for and on behalf of capital to keep order and maintain the rule of law supportive of the market: a narrow scope for social and environmental policy, a strong hand for 'law and order' (see Gamble, 1988). Local states become mere corporates competing with other local states to provide a good environment for business within the universal rule of the market. In order to maintain the basic rules of that market and 'free trade', global governance structures are developing: GATT, the WTO, OECD, G7. Technical innovation is believed to solve environmental problems. Ethics is reduced to the simplicities of utilitarian logic. Trope A has been assisted in its drive into the new industrial economies of the East and South by the structural adjustment loans (SALs) offered by the World Bank and structural adjustment packages (SAPs) insisted on by the IMF (Athanasiou, 1998: 148–160).

The alternative discursive trope (Trope B: *environmental governance*) is to be found in the statement to the Rio Earth Summit by Wangari Maathai, founder of the Kenyan Greenbelt Movement (cited in Athanasiou, 1998: 11). It is contained in the environmental justice manifesto adopted by the First National People of Color Environmental Leadership Summit (United Church of Christ, 1991). It was foreshadowed in the World Order Models Project, and is in the principle of 'humane governance' (Falk, 1975, 1995). It is embodied in Factor Four and Factor Ten (von Weizsäcker *et al.*, 1997), and, in embryonic form, in the communitarian traditions of cultures as geographically far apart as Sweden and Indonesia. It is in the emergent global environmentalism of Japanese city governments, and in the popular movements of people against environmental depredation and degradation around the world. While environmental governance is not necessarily inimical to a decentralized market production system – indeed in Ketola's view it leads to an 'ecological Eldorado' for business (Ketola, 1997: 100), it does mean that the market must be returned to democratic, political regulation. Free trade in Daly's view is simply incompatible with ecologically sustainable development (Daly, 1993: 131). In broad terms the components of Trope B are:

1   A value system which prioritizes care for humans and for the earth and espouses the right of all creatures now and in the future to a sustaining, safe, high quality and biologically diverse environment of no lesser quality than the earth provides today (in the year 1998).
2   The regulation of the world economy continuously to achieve the above values, by internalizing within production the environmental and social costs of natural resource use, by adapting production (of both material and informational products) to reduce wasteful consumption, and by engaging in fair and environmentally responsible trade.
3   The elimination of poverty and of all social and environmental discrimination on any basis whatsoever, e.g. by race, ethnicity, class or gender, and especially against those who do not command wealth.

4   Democratization of governance through participation at every level from global to local, and the reassertion of democratic control over the world economy and markets.

Trope A has been carefully cultivated by a world-wide network of well financed discourse-manufacturing institutions: the Right Wing Think Tanks – for example, the Heritage Foundation in the USA, the Centre for Policy Studies in the UK, the Tasman Institute in Australia (Self, 1993: 64–67; Cockett, 1996). These institutions have had a major impact on government. Trope A is dominant at global level in the forums and bureaucracies from which the world economy is governed. But we should also not forget the influence of other existing discursive institutions: universities and schools, environmental and social 'NGOs', local and grassroots organizations. While capitalism is promoted by global hegemons, so is democracy. Civil organizations can have an impact on the publics of democracies whose voting patterns come to influence governments. Recently there has been a surge in publications advocating Trope B (Wackernagel and Rees, 1996; Martin and Schumann, 1997; Athanasiou, 1998). These point to the global dominance of Trope A and portray it as *the* world-wide model of development. They argue apocalyptically – and probably rightly – that the end of the track signposted by Trope A is ecological catastrophe (Wackernagel and Rees, 1996: 55). But this apocalypticism can be self-defeating. People can become discouraged and weary of the doomsters. Agenda 21, however inadequate in the judgement of the strong advocates of Trope B, has played a positive role in promoting a version of Trope B.

There is a power struggle being acted out at every spatial level from the smallest commune or municipality to the United Nations General Assembly and its institutional organs, and in every functional domain. A significant element of that power struggle is the continuous attempt to clothe 'business as usual' in the garments of 'sustainability': 'greenwash' (Athanasiou, 1998: 227–297). It is by no means easy to distinguish genuine attempts by corporations to move towards less damaging forms of production from the representation of 'business as usual' in shades of green. Agenda 21 is a potentially empowering instrument for Trope B. It is not a set of policies to be carried out in a top-down way by an integrated global system of governance. Indeed, while capitalism prefers to impose itself in the same form everywhere under a single rule of minimal law, environmental governance must be interpreted in different ways in different local cultures, even while the principles of justice it embodies are universal.

The changing context of global environmental governance

It is astonishing how accurate in all essentials is the description of capitalism written by Marx and Engels over 150 years ago in the Communist Manifesto

(Marx and Engels, 1967 edn). 'All that is solid melts into air' is given a stunningly contemporary resonance in the light of fears for the atmosphere! There are new institutions like a global finance market, new corporate forms, new technologies, new industries, but capitalism is ever capitalism. What we still have to learn is how to steer the capitalist engine of production and consumption towards ends other than greed and its instant gratification. Although capitalism is still the same, the institutional forms needed for the task of steering have continuously evolved. Capitalism itself constantly puts governance on trial.

Developments like Agenda 21, Local Agenda 21, and Habitat represent the birth of a new model of global governance, a 'partnership' model (Lipschutz with Mayer, 1996), 'globalisation from below' (Falk, 1995), 'glocalisation' (Hempel, 1996), 'cosmopolitan democracy' (Held, 1995; Archibugi, 1998), 'discursive democracy' (Dryzek, 1994). At present the main features of this kind of politics, as it has been realized, are the following:

1   the formation of global networks of 'partners' or 'stakeholders';
2   the partners include organizational manifestations of major interest groups (NGOs), as well as territorial structures of governance – global (e.g. UN), national, regional (e.g. EU), and local governments;
3   the creation of institutions which provide fora for exchange of ideas and negotiation;
4   the use of rhetorical manifestos or 'charters';
5   the open-ended possibility of multiple linkages among actors in the network;
6   the relative absence of coercion of any kind, persuasion being the principle mode of action and 'shame' resting on the articulation of global ethics the main sanction.

It might be appropriate to recall President Lyndon Johnson's words in his State of the Union address of 1967 which seem too hopeful but no less true today: 'We are in the midst of a great transition – a transition from narrow nationalism to international partnership; from the harsh spirit of the cold war to the hopeful spirit of common humanity on a troubled and threatened planet' (Rostow, 1967: 491). The cold war has been left far behind but the world is still struggling towards international – now defined as global – partnership. The partnership model is instantiated in UN conferences (Habitat II), international networks of local governments (ICLEI [International Council for Local Environmental Initiatives], Cities for Climate Protection), and conferences, charters and networks in the European Union. The aim, in the words of Habitat II, is to promote 'civic engagement, sustainability and equity'.

The studies in this volume, as was inevitable, raise many more questions than they answer. There are familiar questions of democracy emerging at the global level. How is accountability to be introduced such that the minuscule number of people actually involved in the global network become answerable to the enormous numbers of the world population? How are people to be selected for inclusion in the network – and how is selection to be connected in some way with accountability? Are the deliberations of the network to result in decisions, and how then are these decisions going to pass into action? Through which institutional arrangements are these decisions to be implemented? Are there any mechanisms created to follow up and evaluate the outcomes? In short, what is to be the relationship in the network between 'policy' and 'action' (Barrett and Fudge, 1981)?

How do the separate currents of global, national and local politics intermingle? While 'cities' have been considered actors in global networks, how are they to be represented – as territorial structures, in which case one would expect to find some system of city-representative government, or as corporate stakeholders of some kind? If we are contemplating a new form of governance shaped as an indeterminate network, then the key questions relate to the thickness and extent of the network – how well does the network integrate vertical and horizontal interactions promoting the emergence of Trope B? How well do institutional structures promote co-operation around the problems posed by the ecological crisis, both among levels from local to global, among policy domains and between state and civil spheres?

In the next section we consider what the national studies in this volume tell us, both about the relative propagation of Tropes A and B and about the growth of 'global partnership'. At national and sub-national levels there are two main questions: how far has the partnership model progressed, and does the political climate and/or institutional framework help or hinder the growth of the model? Most important for the present book is the question of how far Trope B extends across governmental departments, and how well urban planning and management is integrated with the wider environmental problematic. As was observed in Chapter 2, the causes and remedies for many environmental problems are deeply rooted in the kitchens, yards and streets of cities – the ecological 'footprint' of the city (Wackernagel and Rees, 1996). But the problems will not yield to treatment in the usual way: by allocating responsibility to a single profession or department of government.

National studies

The sub-global level has many tiers. We have focused in this volume mainly on the national and sub-national levels but any discussion of Europe has to take account of the influence of the regional governance of the European Union (EU) and other less formal regional associations. The European Community (EC) *Green Book* on the urban environment, published in 1990, brought ecological sustainability into the regional policy arena. The EU has also moved in recent years to develop a spatial regulation regime that will promote sustainability, social cohesion and regional solidarity (Expert Group on the Urban Environment, 1996). But the politics of the EU can work in directions contrary to the thrust of Agenda 21.

Our analysis is structured roughly by the view that the development model tends to propagate outwards from the dominant core economy in any period, today still the USA, but that there are other partners in the core, notably the major European economies and Japan, with considerable power to influence global norms. While America is still 'core', a united Europe with a single currency could prove a major counter-weight. In 1998 Japan appeared to be temporarily out of the game and internally preoccupied. China and India (some would include Indonesia and Brazil) are very populous and receptive to capital and are 'mega-economies in waiting'. There are also a very large number of members of a highly diverse group of nations which are peripheral in the sense that economic and ecological developments within them, while they may play some role, are unlikely to be decisive in shaping global development norms. We turn first, then, to the USA.

The core economy: USA

The discourse of ecologically sustainable development can be read in a variety of different ways. Luke points out in Chapter 3 that in America environmentalism is understood as (a) something everyone likes – like 'Mom and Apple Pie', (b) providing rhetorical support for a variety of 'interest groups' from the poor of the 'Third World' to selfish NIMBY suburbanites, or (c) a threat to the 'American Way of Life'. The neoliberal discourse suppresses understanding of the globalization which follows from its own actuality, and abhors anything governmental. Thus sentiments of antiglobalism can be directed at both economic globalization and ecological globalization in a reading which says that poverty, unemployment, etc. in America are caused by unfair competition by the new industrial economies, orchestrated by the United Nations (playing on the global reading of NIMBY – (b) above). Likewise anti-statism can be twisted around to become anti-environmentalism when those who define themselves as 'pro-nature' are depicted by neo-liberals as 'anti-human'.

The fears of Americans for their economic security can be, and have been, harnessed in support of precisely what is causing both economic and ecological insecurity. Correspondingly it has become necessary for the Clinton–Gore administration, which is, at least superficially, favourable to the ESD discourse, to try to reconnect economic with ecological security, and its local with its international dimensions, and in the process to use this platform to reassert world leadership. The politics which is shaping up in America is one in which global regulation (ecological and economic) in an integrated world is pitted against global free competition (ecological and economic) in an anarchic world (see Wendt, 1992). This is a local politics which has been played out before in America and the results have reverberated around the world. In this politics, Luke claims, talk of 'global partnership' couched in terms of the benefits to 'humanity' is not necessarily helpful. In America it is benefits to America which have to be stressed.

It should not be surprising to learn from Chapter 4 that Agenda 21 has been little taken up at local level in America (about a quarter of one percent of the total population). Lake notes that take-up, where it has occurred, has been mostly in small communities and places of relative prosperity and growth potential. The absence of Agenda 21 in places of economic decline suggests that 'development' is interpreted almost exclusively in terms of raw capitalism (Trope A). Decline is the fault of 'uncompetitiveness' for business. The cultural antipathy to socialism in America, as well as the withdrawal of the federal government from intervention to balance regional economic development and assist declining regions, may account for the seeming difficulty of reconciling environmentalism with the social aspects of Agenda 21. The vertical networks around environmental issues are very weak. State legislatures in particular show little willingness either to encourage local government or to formulate Agenda 21 policies of their own. There is no connecting tissue between tiers of government in environmental matters.

The global aspect of Agenda 21 has made little headway at local level. The story seems to be 'act *locally* and think *locally*'! This peculiarly American political ethos appears in recent works on environmental governance which focus attention on the community, the locality, civic society and the cultural sphere (e.g. Wapner, 1996; Lipschutz with Mayer, 1996). Lipschutz, for example, writes: 'people are more likely to act collectively when their personal experiences and surroundings are implicated in a process than they are to respond to directives from a distance or abstract predictions of future dislocations' (Lipschutz with Mayer, 1996: 39). This is an American reflection which applies most strongly to American people but not to all people.

Agenda 21 has instead been allied with a variety of more 'respectable' causes: local environmental conservation in the face of growth pressures, improving the 'livability' of cities, and environmental improvement to reduce negative attractions for business. And, as Lake points out, there is evidence of greenwash – the 'greening of capitalism' not as transformation of production and consumption but green legitimation of business as usual. Dangerously (for Trope B) Agenda 21 appears to be connected antithetically with environmental justice – as the protection, by exclusion, of prosperous localities from hazardous LULUs (locally unwanted land uses). While the example of Burlington, Vermont, provides a beacon of hope, it seems probable that the USA will be as much the world laggard in adopting the agenda of ecologically sustainable development as it has been the leader in promoting a return to raw capitalism.

Partner economies: Britain, Germany, Japan

The contrast between America and the partner economies is strongly marked. Not only is the advancing model of capitalism according to Trope A under challenge, it conflicts sharply with deeply entrenched traditions of welfare Fordism and socialism (of different kinds) in Britain, Germany and Japan.

In each of these countries, moreover, the new model of ecologically sustainable development (Trope B) has taken root and is struggling into existence. While America is always associated with 'globalization', awareness of the opportunities and costs of the global economy and its effect on the global environment are much more strongly evident in the partner economies than in the American core.

Britain is often linked with America as a member of the Anglophone neo-liberal fan club (Chapter 5). The impact of 'raw capitalism' (Trope A) has indeed been profound, but in the Thatcher years did not result in the abolition of urban planning. The Thatcher government's attempts to dismantle planning in Britain were to some extent frustrated by the opposition of conservative rural constituents fearful of a market 'free-for-all' in the countryside (Allmendinger, 1998). Moreover, EU environmental legislation has, in some instances, encouraged the development of progressive spatial regulation, even in states most resentful of 'Brussels' – notably Britain during the Thatcher–Major era. Cullingworth and Nadin (1994: 140) observe that 'the EC has had a major impact on British environmental policy'. The most notable example is the set of EU regulations on impact assessment. This was introduced into the British planning system as an EU directive. EU regulations also permit actions by third parties against governments and private sector bodies – a case in point being the suit brought by the Friends of the Earth against Britain's Thames Water Utilities in 1991 for allegedly breaching the EU directive on water quality.

Nor did Britain's embrace of Trope A mean the rejection of the agenda of sustainable development – at least in rhetoric. This agenda was contradicted by a massive road building programme from the 1980s which was believed to be a prerequisite of economic growth. The Blair government has begun the task of linking economic and ecologically sustainable development by, for example, organizationally combining the Departments of Transport and Environment. This move, however, contains dangers for the future should transport interests, allied with raw capitalism, become dominant. Local governments have enthusiastically taken up the challenge of Agenda 21. Discussion now focuses more on the question of implementation and its politics, on management tools and the real impact of Local Agenda 21 (LA 21) policies, rather than merely on whether LA21 rhetoric has been adopted into the discourse.

Even the minimum assessment of take-up of LA 21 by local authorities in Britain (15 percent) is hugely greater than that of the USA. Tools for integrative management of local authority policies and programmes to achieve ecological sustainability have begun to be used. Blowers and Young's overall assessment (Chapter 5) is not very positive: steps taken were 'isolated' and 'tentative' compared to what had been envisaged at Rio. The social aspect of sustainability (dealing with inequality) 'received scant attention'. While there may be a better understanding about local–global interactions, this does not necessarily translate into policy decisions. Partnership under the Major government meant partnership between business and government. It remains to be seen whether it will acquire a more inclusive meaning – bringing in NGOs and local citizens– under the Blair government.

Blowers and Young (Chapter 5) refer to the neo-pluralist idea of politics and the imperative for government to do the bidding of business. Yet in the British context we do not find governments at either local or national levels so strictly 'imprisoned' by the market as Charles Lindblom, from an American vantage point, has suggested is universally inevitable under capitalism (Lindblom, 1982). If in Britain the incremental movement towards environmental governance (Trope B) is subject to serious constraints, both structurally (in capitalism) and institutionally (specific to Britain), there has, nevertheless, been progress – not only discursively but pragmatically in terms of 'nuts and bolts'. How much of it is 'greenwash' and how much 'green power' is a question for much further research.

Europe's leading economy has sought political strength and stability in a united Europe. Unity is a historic theme (*leitbild*) for Germany, and reunification has of course dominated recent politics and become Germany's greatest economic challenge. Germany is now caught up in the process of the 'transformation' of Eastern Europe, or, as Athanasiou (1998: 111) puts it, the 'restoration' of capitalism. At National (Federal) level a three-way struggle has been taking place since the 1970s: between Trope A – with its mantra of deregulation and competition – and Germany's post-war statist welfare Fordism (see Cremer and Fisahn, 1998: 59), and also between Trope A and Trope B. Recently social dissatisfaction with the effects of raw capitalism in the East (Germany) has reinforced dissatisfaction with Trope A on other grounds, resulting in a change of government.

While the German federal government takes an outward-looking stance, and several ministries deal with a range of environmental problems (including those arising in foreign aid), there is no national Agenda 21 which sets out to integrate policy between departments and levels of government. Three departments have developed their own response to Agenda 21, but there is little sign of Trope B norms being applied to the main 'economic' ministries. The German urban planning system is a more formally integrated, multi-tiered structure than that of Britain, more decentralized – with considerable power at the *Land* (state) level – and yet steered by a constitutionally embedded commitment to spatial equity at Federal level. The many small municipalities, in the continental tradition, have a permanent place in the constitutional order (see Chapter 6, this volume, and Newman and Thornley, 1996). But city governments are faced with fiscal crisis (increased responsibilities and reduced financial resources) pressure to compete for investment, and reduced autonomy in the face of both multinational firms and an upward shift in governmental power. While Local Agenda 21 thinking has begun to be adopted by growing numbers of municipalities in Germany (2 or 3 percent of the 17,000 municipalities in 1998), cities themselves are in no position to shape the dominant development model. The partnership model is evidenced in the national network of NGOs and inter-municipal co-operative networking. Environmental NGOs themselves are strongly co-ordinated with a view to influencing government policy, and in 1998 the Green Party re-emerged as a major political force in the new coalition government.

Germany faces considerable pressure as the country struggles with its economic problems and the integration of Eastern Germany into the market economy. The chapter on Germany notes the existence of economic, financial, social and legal restrictions on the pursuit of Agenda 21: the short-termism of business thinking coupled with a loss of influence of local authorities in economic planning, fiscal crisis arising from the drain on finances resulting from high unemployment, loss of interest by the general public in environmental matters, and reduced scope for local action on matters such as transport and energy. The struggle between economic and ecological priorities takes place in both the environment and urban policy domains. Pressure to adopt Trope A comes from business which uses rising unemployment figures to back arguments for improving competitiveness by cutting pro-environment measures. In this argument, ' "Location Germany" should be made more attractive and the national economy prepared for competition with other economies' (Cremer and Fisahn, 1998: 61, citing Klodt and Stehn, 1994). But Easterners have become sceptical of the benign effects of globalized capitalism which has become not just 'raw' but 'wild' (beyond even the basic rule of law) in the new territories it invades. Environmentalists continue to argue for reformulation of the economic growth model, finding some encouragement in the growth of new environmental technology.

Environmentalism has seen several different phases in Japan. The enormous 'ecological footprint' (see Chapter 1, and Wackernagel and Rees, 1996) of Japanese cities is well understood, both in terms of the surrounding region (energy, water supply, garbage disposal) and globally (importation of food, use of the atmosphere). Japanese cities have in the past suffered from major episodes of pollution which have stimulated locally oriented environmental activism. In recent years, however, new groups have formed around global issues. Contrary to the USA, and despite the absence of powerful and well financed environmental organizations, public concern for the global issues advanced by the Rio conference is in evidence. The Japanese public, it seems, is much more inclined than the Americans to 'think globally', and this despite little encouragement from national governments distrustful of the anti-developmental stance of environmentalists. We can say perhaps that, as in Germany, not only Tropes A and B are in contention in Japan, but each is contending with Japan's traditional Fordist model.

The Japanese government has introduced important new environmental legislation (an Environmental Basic Law, and Plan) in response to the initiative of Rio, expressing principles of 'recycling, [ecological] coexistence, participation and international cooperation'. This legislation gives new standing for NGOs, and calls for stakeholder partnership and public participation. Cities have themselves responded with considerable enthusiasm and thoroughness to the global call. In fact Japanese cities were already developing some strong environmental policies prior to Rio and well before the national government developed its environmental response (see Chapter 7, p. 137).

Japan provides a good example of the 'discursive change' in civil society discussed by Wapner and Dryzek, and Rio has played a considerable role in this ('consciousness itself can be a form of governance' – Wapner, 1996: 155). Yet when it comes to conflicts between environmental concerns and development projects, the government still pursues construction projects which have a damaging impact on the environment. The Japanese tradition of incorporating the interests of business has meant that these stakeholders are consulted, but some cities have begun to open up to increased public participation in a way which is not traditional for Japan. City governments also have a grasp of the global impact of their ecological footprint, and partnership with stakeholder groups is part of the strategy to contain it. Unfortunately the Japanese attitude to its internal environment, and its rhetoric on global responsibility, is not matched by its actual performance internationally. We cannot ignore the significance of the involvement of Japanese businesses in the destruction of Asian rain forest and the continuing assault on quite threatened whale species.

In the partner economies, as in America, the capacity of cities to develop policies supportive of ecological sustainability is dependent both on the local progress of the discursive trope and on the framework provided by national leadership, financial support, and market regulation. But city governments can also exert upward pressure on the nation state to engage with the ESD agenda and provide important working examples of ESD practices. Municipalities and city governments face increasing pressure to compete for investment which will generate employment. At the same time at local level public attention can be focused on the environmental imperative through both rhetoric and effective policy. The balance between reality and rhetoric varies from place to place and is impossible to discover without detailed local research. A resolution between raw capitalism and environmental governance is temporarily achievable through greenwashing rhetoric, but at the same time rhetoric has a power of its own, setting up public expectations against which the success of governments and corporations will ultimately be measured.

Mega-economies in waiting: China, India

The world's ecological future will be determined by the development model which takes root in those regions of the world which have been – or will soon be – opened up to global capitalism. These are both potentially vast, populous economies, and repositories of much of the world's ecological wealth. With global capitalism comes city growth; with cities, increasing consumption of natural resources and energy, and increasing emission of wastes. These regions include China, India, Russia, and large parts of South

America, East Asia and parts of Africa. This book contains reports on two such 'mega-economies in waiting': China and India. Within ten to twelve years the urban population of China and India (up to 1.3 billion according to estimates in Chapters 8 and 9) will far exceed that of North and South America, Europe and Russia combined.

The ecological impact of urban development is starkly apparent. China's and India's urban environmental problems are not just local air, water, land, and noise pollution, though these are considerable, but the immense drain which consuming cities impose on non-urban areas: productive agricultural land, forests, water, and the atmosphere.

In common with a number of East Asian countries (notably Indonesia – see below), the model of development which has prevailed in China up to now is much more like Trope A than Trope B. The results of massive and sustained capitalist economic growth within the framework of a command economy and polity are everywhere to be seen: loss of agricultural land, severe local environmental degradation from air, water and land pollution, cities congested with traffic and environmentally unplanned construction booms, severe and growing water shortage, and a 'floating' population of 75 to 90 million some of whom are 'environmental refugees'.

India's development is following much the same path as China's, with similar results. Economic growth after opening up to global capitalism has led to massively growing urban populations – a forecast urban population of 500 million by 2010, increase in the numbers of cities, and rapid growth in the size of the leading cities. Major impacts are experienced on the supply of land for agriculture, water supply, fuelwood, and on air quality and traffic congestion. Already inadequate physical infrastructure in cities (for waste disposal, water supply, energy services) is further stressed.

Plainly there are enormous social and environmental injustices emerging in the course of both China's and India's development. Indigenous traditions of environmental care are under threat. Women in particular are 'losing ground' (Ryan and Flavin, 1995: 128). However, while raw capitalism has a somewhat homogenizing impact, there are also major cultural and institutional differences between China and India which may unfold into different futures. The presence of a strong, authoritarian central government is not necessarily incompatible with Trope A – it may even be a precondition. Whether it is compatible with Trope B remains to be seen. In both countries there are reasons for hope that Trope B will make progress in future, but there are also major obstacles.

In China the rhetoric of Agenda 21 has been adopted both at national level and in some city governments. The rhetoric is also being translated into plans and laws. But will the plans be implemented and the laws enforced? There can be little doubt that the government of China is very concerned about the environment of the nation's cities and their wider impact, and appears conscious of its future global role. China has committed 1.5 percent of its GDP to environment protection in accordance with World Bank suggestions (see Chapter

8 this volume, and Ryan and Flavin, 1995: 130). Agenda 21 has been interpreted in terms of *socially* and *economically* as well as *ecologically* sustainable development. In the short run at least there may be contradictions between these terms but the pressing connection between them seems even more obvious in the case of China than elsewhere. While the ecological footprint of Chinese urbanization extends world-wide, the most immediate effect will be felt by the Chinese people themselves. For example, urban development at high densities is made necessary by the diminishing supply of land for food production. Car ownership, ever a sign of status, cannot provide for movement needs in massive, high density cities. The dependence of China on fossil fuel consumption, especially coal, will in future greatly increase carbon emissions. This will contribute to extreme climatic events produced by global warming to which China will be subject, though exactly how and where it is too early to say. For example, the UK Hadley Centre identifies China as an area of greatly increased mean annual 'runoff' as a result of global warming. In 1998 this 'runoff' produced devastating floods (Department of Environment, Transport and the Regions, 1998: 7).

The continuation of growth of Chinese cities in the pattern of the 1980s and 1990s is likely to result in the massive depletion of China's own natural capital: water, air, forests, land. Rapid economic growth in the longer term, along with the widespread distribution of economic wealth as a precondition for further growth of consumption, will require the design, management and regulation of markets to further the ends of ecological sustainability. Development much further along the Trope A track will quickly confront China with its own limits. At the most basic level, neither China nor the world will be able to meet the food needs of the growing population (Ryan and Flavin, 1995: 119–120; Brown, 1995). Future growth along the lines of Trope B, however, will also have repercussions world-wide. As Yin and Wang (this volume, Chapter 8) observe, China's environmental problems stem not just from improving living standards which come with freeing up the economy and allowing markets to develop, but also 'from the way this consumption is stimulated and managed by a capitalist world market . . . it will be increasingly difficult for the Chinese government to resist pressures coming from the global markets'. So if China needs an ecological model of development for national survival and prosperity, then it may have to become a global model.

The Chinese system of government is strong, centralized and integrated. The government has the capacity to implement the necessary market regulation in a 'top-down' manner. There is doubt, however, about the feedback from the 'bottom up'. The lack of openness, pluralism and democracy blocks off multiple potential channels for pressure to be exerted by the groups and publics which feel the effects of ecological and environmental degradation. There are as yet no permitted environmental NGOs (see Lappin, 1994). There is little evidence that markets are being structured to produce environment-favourable outcomes. There is no independent scrutiny of the government's performance at any level. The Chinese government resists any focus for political opposition, and a green movement could rapidly become such a focus. Dryzek (1987, 1994) insists that such feedback is a necessary (if not sufficient) condition of an ecologically 'rational' politics, and that an 'ecological democracy' must be built upon a pluralist civil society. How this civil society can be built in China while maintaining national integrity and order is far from clear. But the lack of the feedback and transparency afforded by democratic freedoms must count as a very serious qualification of hopes of progress towards Trope B in China.

The alliance between strong, centralized governmental authority and raw capitalism is the only kind of partnership we are likely to see in China in the immediate future. However, a Hobbesian breakdown of authority of the kind unfolding in Russia, opening the door to unregulated capital, offers an even more unattractive prospect. In the light of the enormous urban environmental problems facing China in the next twenty years, there is little scope for optimism. There are only two grounds for hope, both of them slim. One lies in the collective wisdom of the Chinese leadership in steering the nation towards a *politically* sustainable system which must necessarily include human rights and freedoms, while at the same time taking strong measures to regulate capitalist growth towards *ecologically* sustainable ends. The other is the strengthening of global institutional structures for the environment to create a new set of global rules for environment-saving production, investment and trade, a development which would of necessity require China's consent and participation.

As with China, the Indian central government has adopted policies to control pollution and conserve the environment. Even before Rio the government had embarked on an agenda of sustainable development. Indian indigenous traditions of environmental care are strong and have not been lost as a result of modernist ideologies. To the contrary, the Hindu ethos and the doctrine of Ghandi are still invoked on the side of environmental protection. Care for nature is a duty enjoined on all citizens under the Indian Constitution of 1949 and on States in 1976. Legislation, therefore, has powerful constitutional support. India has passed laws to control pollution and increase environmental awareness through education. The courts have upheld pollution regulation and closed small polluting industrial plants. Some regions have been successful in limiting the numbers of people migrating to the big cities – mainly as a result of incentives to stay in the village, but Maitra and Krishan (Chapter 9, p. 186) observe that 'Although the system [of land use planning borrowed from the British] has been in operation for forty years, it has failed miserably to solve India's environmental problems.' It is well known also that India has strong NGOs and grassroots organizations devoted to environmental protection and ecologically sustainable development. These have had some notable successes in turning back environmentally damaging development and shifting power to the local level of the Panchayats (local councils; see Shiva, forthcoming). However, while active in rural areas, these NGOs have not generally focused much attention on cities and their impact on the national or global environments.

The continued growth of Trope A capitalism in India's cities will have profoundly detrimental effects on the local and national environment as well as the global. Water-borne diseases are commonplace. Safe water supply and effective waste treatment are among the most pressing urban problems. But also local air pollution from industry and traffic will increase. Likewise while India is at present only a small emitter of greenhouse gases, this will increase rapidly with Trope A economic development. Agricultural land is being lost steadily to urban development and already the footprint of cities shows an ecological deficit (see Chapter 1, p. 22). However, environment-saving technology is being developed to deal with problems such as home cooling and sanitation. And innovative steps have been taken in regional planning.

The Indian political system is less well equipped than China's for strong top-down implementation. The local level is formally responsible for the quality of the urban environment but does not have the financial resources or scope of authority to take the necessary action. But legislation is transparent and open to challenge. The principle of separation of powers enables the judiciary to take a strong role in environmental matters, and there is a grassroots environmental movement with freedom to organize and grow in strength. While top-down power may be lacking, the prospects of bottom-up feedback are good. Local democracy in particular is representative and quite powerful. However bottom-up feedback on the effects of development are likely to be of little use if they do not result in effective top-down regulatory policy being implemented at state and federal levels and adequate financial resourcing for local government.

Peripheral economies: Sweden, Poland, Australia, Indonesia

The progress of *raw capitalism* (Trope A) and *environmental governance* (Trope B) in our examples of peripheral economies is widely divergent. Sweden is perhaps the leading exponent of Trope B, including the partnership model of environmental governance as one crucial element. Poland is preoccupied with its economic transformation to capitalism and therefore might be expected to have absorbed much of Trope A – but has in fact gone some distance towards Trope B. Australia, while once embracing Trope B, is tending in recent years to regress to Trope A. Indonesia is currently devastated by the overenthusiastic embrace of Trope A by a politico-military elite and incumbent middle class.

The growth of Trope B in Sweden coincided with the decay of the Swedish version of the Fordist compact between capital and labour which allocated different 'spheres' of activity to business and the state. Nevertheless the Swedish government has pledged to be a 'leading force' and has embraced ecological sustainability in the form of the 'eco-cyclic' society which encompasses a number of principles central to Trope B: respect for the tolerance limits of both humanity and the environment, the precautionary principle, the substitution principle, the polluter pays principle and 'adherence to Agenda 21'. The national aim is to make manufacturers assume increased responsibility for their products from the cradle to the grave (see Lidskog and Elander, this volume). However, although the Swedish government has very high ambitions with regard to environmental friendliness, one should be careful not to overstate the achievements so far. For example, as in most other EU states during the 1990s carbon dioxide emissions have not declined. Assessing what Sweden has achieved with regard to the Trope B ideal one also has to keep in mind that the country's ecological footprint is still far above a reasonable norm for ecologically sustainable development.

The principle of subsidiarity is embodied in respect for each nation's sovereignty over its environment, and also in the widespread acceptance at local level of responsibilty for implementing Agenda 21. Although municipal planning has had to face cuts in financial grants from central government and is as much subject to the impact of so-called globalization as anywhere else among the developed economies, all 288 Swedish municipalities have started a practical implementation of the Rio agreements and have developed Local Agenda 21 (LA21) plans (Lidskog and Elander, Chapter 10, p. 207). Sweden has embraced the aims of Agenda 21, but it must be said the country was already heading in that direction before 1992. Agenda 21 has nevertheless empowered both government and NGOs in seeking a Trope B model. The hope must be that Sweden, though a small and relatively peripheral economy, will (again) provide a working model for the world of how economic prosperity, a modern lifestyle, social and environmental justice and ecological sustainability are far from mutually exclusive.

The Swedish central government is supportive both of the global aims of Agenda 21 and its local implementation, although the municipalities have complained that central government is not supportive enough. Central government has a tendency to shift responsibility for environmental matters down the scale without giving due financial resources. Thus in Sweden, the municipalities now have to implement a lot of nicely formulated green political goals in a tough reality. In terms of lateral integration, Sweden has a long tradition of interpenetration between government and civil society, and the tradition continues with environmental NGOs. An important element of the Swedish model – characteristic of the Swedish way – is the strong coalition between key NGOs and government aiming to build consensus around the need for economic growth to take a new form: ecological modernization. Green groups have played various political roles: that of popular social movement, deputy activist, and political party. Typically for Sweden, an inclusive National Committee on Agenda 21 was formed to develop national policy, and similar arrangements have been instituted at local level. Much energy in Sweden is put into educating the public so that environmental policy can acquire support at the base of the political pyramid, and even flow from the bottom up.

Poland is an example of a former Eastern bloc country which has sought to combine its transition to capitalism with an ecological sensibility. It has had the difficult task of more or less leaving behind its former development model (command economy), while at the same time tempering resurgent capitalism with the new ideas of ecologically sustainable development. The central planning of Poland's 'partocracy' (centralized party bureaucratic rule) left the country with degraded and irrationally managed urban environments. Towns were 'regarded as housing infrastructure for large industrial plant' (Markowski and Rouba, this volume, p. 220). Every city was ringed by identical blocks of flats – a dreary architecture. While the Polish government declared its good environmental intentions in international fora it cynically ignored their implementation at home, with the result of increasing environmental pollution.

The environment became a cause for social movements opposed to the old regime and hope associated with the new. The new Polish Parliament responded by adopting 'ecodevelopment' as its official model, which according to Markowski and Rouba (this volume, p. 225 and see Baturo *et al.*, 1997) 'in order to secure equal opportunities for access to the natural environment for contemporary and future generations, integrates political, economic and social activities to ensure balance in the natural environment and continuity of its fundamental processes': a clear statement of Trope B. However, 'shock therapy' transformed the social environment, creating unprecedented unemployment, social stratification and insecurity in the cities. Plainly there is immense pressure to succumb to whatever will bring in investment. Foreign capital has indeed flowed to Poland.

In January 1995 the Parliament adopted the 'Programme of State Ecological Policy Implementation Until the Year 2000', which required all central government agencies to observe the principles of ecodevelopment. A test of Poland's resolve on this score will come in the privatization of its electricity production industry. There is no mention of ecological considerations by Poland's task force on the restructuring and privatization of state enterprises; simply: 'A strategy of privatisation for all power plants, distribution units, and thermo-electrical power stations in the sector should be adopted' (Poland: Task Force, 1997: 66).

Some Polish environmentalists have expressed scepticism, perhaps because in the past they have not been able to believe government statements. Athanasiou reports a conversation he had in 1993 with Tomasz Terlecki from the Polish Ecology Club. Asked if Poland will take Brazil's path to capitalism or Sweden's, Terlecki is reported to have answered, 'Brazil, this is totally clear' (Athanasiou, 1998: 114). Terlecki's is only one voice, however. While not ignoring the real difficulties in the way of ecological modernization, Markowski and Rouba paint a different overall picture. Some cities have joined the International Council for Local Environmental Initiatives (ICLEI) and several have signed the Charter of European Cities and Towns Towards Sustainability. At Rio Poland presented significant ideas for environmental conservation and took the initiative in putting the idea of 'Green Lungs for Europe' and other environmental programmes on the agenda for the Baltic and central European states. Poland, as well as Sweden, relevant parts of Germany and Russia, and the other countries in the Baltic Sea Region are all involved in a broad planning initiative titled *Visions and Strategies around the Baltic Sea 2010* (1994). One of four basic values says that the Baltic Sea Region 'shall become a masterpiece for sustainable, environmentally sound development' (ibid.: 4).]

The realities of continuing high unemployment in some regions, and the need for private capital to undertake industrial restructuring, make the environmental agenda tempting to postpone. Yet Poland's new democracy ensures that public concerns will reach the agenda of governments. These concerns fluctuate of course. Environmental participation measured by the membership of environmental groups declined in both Sweden and Poland (not to mention Australia) in the 1990s. But this may signal a shift back to fears for social security rather than new enthusiasm for Trope A. Although the Swedish parliamentary elections in 1998 meant a swing to the right in terms of seats won by the conservative and liberal parties together (from 148 to 159 out of 349 seats), the Social Democratic minority government now prefers to co-operate with the Left Party, which increased its number of seats in the Parliament from 22 to 43, and the Green Party (16 seats). In Sweden, then, as well as in Germany, Britain and France, this will improve the prospects for connecting Trope B more tightly to the older agenda of social justice and security which seems to have been re-ignited all over Europe.

The Polish ecological movement has already had an effect in raising consciousness of environmental issues, and as the movement grows there will be increased pressure to adapt government policy to ecological concerns. As Markowski and Rouba (Chapter 11, p. 238) point out: 'The strongest actions according to the Rio Declaration and Agenda 21 seem to be building democracy in Polish society, the creation of a solid legal basis for environmental protection and the active participation of Poland in international co-operation. The requirement to implement the postulates of Agenda 21 is an important factor stimulating local activity.'

Australia (Chapter 12) is an example of a country with a tradition of environmental awareness and occasional social mobilization, but whose government has swung from firm commitment to an environmental agenda to mere greenwash. Public concern for the environment has waned over recent years, but neither economic nor ecological globalization has been rejected. The public backlash against globalization in the form of new parties of the extreme Right has not gathered sufficient momentum for entry into the parliamentary arena – despite the initial successes of Mrs Pauline Hanson (whose name is probably better know internationally than that of the Prime Minister). Rather the downgrading of ecological concerns has increasingly been seen as the condition of economic globalization. National and state political leaderships have simply overwhelmed what remains of public environmental concern and have expelled environmental issues to the margins of politics.

The present national government does not appear to believe in partnership in any meaningful sense, except with business. It rejects global co-operation on the crucial matter of climate change. Against the world trend, Australia continues to pour money into road-building. Privatization for short-term economic gain has taken precedence over ecological responsibility. Among the public at large there is little awareness of Agenda 21, whose title has even been misappropriated in one State to describe a programme of urban development projects (Europeans may find this hard to believe!). In its publicity during the general election of 1998, the environmental policy of the ruling party was reduced rhetorically to clearing litter and fencing off small sections of unproductive farm property. Public education, seen as so critical to the success of Agenda 21 in many countries, has been largely absent. Nevertheless, earlier environmental initiatives, largely out of the public gaze, continue to bear fruit, and a few local governments have begun Local Agenda 21 programmes. Why so? Perhaps because Australia is economically close to the erstwhile Asian 'Tigers', with which governments have been repeatedly told by Treasury bureaucrats Australia must compete. The same bureaucrats look to the USA as providing the discursive trope for 'reforming' the economy to sustain investment and reduce unemployment.

The proportion of the world's population living in poverty is increasing. But the measurement of 'development' and 'under-development' as welfare differentials between nation states, cities, or even between North and South no longer makes much sense. 'North and South are often taken to denote "rich" and "poor", and this is so misleading as to be dangerous – Southern elites consume, as they pollute, at well above middle class Northern levels' (Athanasiou, 1998: 15). Jellinek's chapter on Jakarta in Indonesia precisely illustrates Athanasiou's point. Jakarta is divided within itself. Indonesia's per capita contribution to global environmental degradation is small only because of the extremely uneven spread of so-called 'development'. Development means the Trope A consumer lifestyle which was seized upon with immense enthusiasm by the new middle class. With 'development' comes the ultimately Western notion that indigenous (e.g. kampung) lifestyles and values are 'poor' while 'rich' means nothing but prodigious consumption for a small minority (see Chapter 13, p. 270).

While this idea is ceaselessly promoted to the 'developing world' by a vast Western-oriented advertising industry owned by Western-oriented elites (Durning, 1993), the idea that indigenous social values, and the environmental merits of traditional lifestyles, can be combined with high quality public services (public health, education, roads, drainage, transport) and labour intensive employment has not been allowed to consolidate into a new model of ecologically sustainable development, a model which is also both *socially* just and *politically* sustainable. The economic devastation in Jakarta, as Jellinek recounts, is today multifaceted. Large parts of the middle class have fallen back into a poverty, which is worse because detached from the social supports of kampung life. It should be remembered that Sweden only developed its socially supportive model as a result of the deprivations of the collapse of capitalism in the 1930s and the continuous struggle of labour. Perhaps Indonesia, drawing on its indigenous culture and looking across to other good examples in the region, can develop its own Trope B model of ecologically sustainable development. Australia, which once looked like taking up that challenge, has unfortunately abdicated from the task to indulge in sporting diversions and capitulation to whatever is good for business as usual.

If we were to scale our case studies in terms of the progress of the partnership model, not just in official rhetoric but in the reality of integration between levels of government and between state and civil society, Sweden would certainly come at the head of the list. A middle level of progress is evidenced in Germany, Britain, Japan and Poland. The USA and Australia would be at the bottom. It is premature to draw conclusions on this issue from the studies of China, India and Indonesia, though India seems to provide more fertile ground for the partnership model than China. In Indonesia the past is not necessarily a guide to the future since Indonesia's political economy has changed radically during the last twelve months and is still in ferment.

The US experience cannot become a normative model for the world, yet the USA is the home of Trope A and of the multinational capital now expanding outward into China, India and Russia. It is this expansion of capital under the aegis of Trope A which poses the maximum danger of ecological catastrophe. The danger of economic collapse is far less significant for the world than the looming ecological crisis. How the questions posed above come to be answered will shape the future of the model of partnership and determine whether it is stillborn, a sickly offspring of the model of development, or a healthy child of democracy. But one thing is certain. The world cannot wait for answers. Like all environmental problems, uncertainty is inherent (see Dryzek, 1987). What then should be done?

Conclusion: the future of urban environmental governance

The tendency of Trope A is to cast cities and planners in the role of corporate actors whose function is to promote economic development – though sometimes environmental quality may be regarded as a means to this end. These actors must not take an interest in the long-term future, nor must they show any interest in their co-competitors except to discover their strategies for attracting investment and match them. Least of all must they *co-operate* with other cities against corporate capital. Bluestone and Harrison (1982: 185) explained some time ago that the end result of institutional competition within this framework in the USA had been a levelling down of *all* regulation and taxation of capital so that *no* city, state or region gains any additional competitive advantage thereby. Institutional competition simply hands an enormous bonus from the citizens to business. What Bluestone and Harrison observed in the USA is now taking place in whatever space capital occupies throughout the globe.

The future for Trope B, however, is not so dark. Much has been made by Marxists of the 'internal contradictions of capitalism', and all the while (in the 1980s and 1990s) capitalism has raged on, apparently overriding all such fears – or hopes. Yet the contradictions are real. The first contradiction between capital accumulation and legitimacy is once again making itself felt, and citizens the world over are demanding change. The contradiction between accumulation and ecology, which feeds into the first contradiction, is also being felt, even though in the short run that contradiction (manifested as between ecology and economic growth) may force citizens to side with capital (see Altvater, 1993). In the longer term a realignment of social movements will only take place when social justice and environmental justice are discursively reconnected in the public sphere (Low and Gleeson, 1998). The impact of gradual ecological modernization yielding to sensitive political leadership and public demand – the reflexive modernity hoped for by Beck and Giddens – should not be discounted. Yet, in the light of today's knowledge, one must hope that *economic* crisis strikes in such a way as to stimulate ecological modernization before it is too late to do anything to stem the gradual but deadly advance of the *ecological* crisis.

City governments can and should take a major role in resisting Trope A and transforming the world's system of city governance in accordance with Trope B. There can be and must be a threefold effort: to transform the discourse of city-marketing and economic competition to one of city-networking and ecological collaboration, to thicken the institutional network of cities and immensely increase the information and communication flows within it, and to create and maintain pressure for democratic institutions at global level.

If Trope B is to make some headway, facilitated by the partnership model, then there must be discursive change. Not only must economic development be subordinated to ecological and social ends (resource conservation and social, environmental and ecological justice), but 'cities' must be viewed primarily as representing people rather than as quasi-firms in a competitive market. That will not mean that economic issues disappear, but that the ends of economic growth are brought to centre stage and debated. To this end, measures of human welfare need to be redesigned, and quickly adopted. The slowing of 'growth' measured by GDP need not mean the end of growth of human welfare; nor does it mean that markets as co-ordinating mechanisms are removed. Rather, it means that markets are designed and regulated to serve social and environmental ends. Dryzek (1996: 29) points out that an evolving ecological rationality has to be able to survive in a world in which individual subjectivities are created by a dominant capitalist-market order. The analysis of Ostrom (1990) and others suggests that such survival is possible, and indeed has occurred, but that effective institutions for governing the commons evolve in accordance with a far different logic than the strategic-instrumental rationality of competing egoists which allegedly governs the behaviour of individuals and firms within markets. We agree with Dryzek and Ostrom that effective environmental governance points to 'a world beyond strategic rationality' (Dryzek, 1996: 29). How is this world to be constituted? What part can urban governance play in its constitution?

Partnership not only *within* cities among different stakeholders, but also *among* cities is indispensable if competitive reduction of environmental regulation and social service provision is to be avoided. Castells's remarks about Europe apply in every region: 'The inter-connection and cooperation between local governments . . . making it difficult for the global economic forces to play one government against the other, thus forcing the cooperation of the global economy and the local societies in a fruitful new social contract' (Castells, 1993: 21). The beginning made by ICLEI and other global networks of cities needs to be taken much further. Capitalism itself has provided the tools, with the immense growth of communications and information technology (Castells, 1989). If Trope A capitalism depends on non-cooperation between its 'prisoners' (in the familiar game of prisoners' dilemma), capital has provided the means to *increase* the chances of co-operation through instant world-wide communication. It remains for city governments to use the tools effectively to thicken the net so as to trap and shape the flows of capital. The aim must be to construct a world-wide network of regulation. City governments will ask each other questions like: which firms have engaged in economic blackmail; which firms are prone to flight; which firms become embedded; what ecotaxation measures have produced good results; which industries are ecologically unsound; what regulatory options exist? The era of 'downsizing and outsourcing' is over. City governments must build up their informational and institutional capacity to meet the needs of the future.

The new skills cities need are those which can rapidly supply answers to these kind of questions and set up new networks of collaboration and new defensive cartels. Moving cities in the direction of Trope B is not a matter for urban planning viewed narrowly as land-use regulation, it is a matter of urban governance – as much across departments as within 'planning' and conservation agencies. The values and skills for that task demand new educational programmes, with an emphasis on democracy, ecology, communication, political economy and the economics of public choice.

City governments seeking to act in the interests of their citizens should exploit what freedom of action they have at global and regional level. But action at the level of the city will hardly be sufficient. City governments need discursive, regulatory and financial support from their national and regional governments. They must demand it. Nation-states are still the principal actors at global level and their interactions will shape the regulatory regimes which must eventually limit the competitive freedom of both cities and corporations. The world created by a thickened network of collaborative cities will be an uneven mesh of regulation. That event will stimulate demands by capital for simplification and governance of the world network at global level. Already we have seen signs of such demands with the ill-fated OECD Multilateral Agreement on Investment – potentially a powerful Trope A instrument for repressing the collaborative regulatory initiatives of city governments. However, the more that capital constructs the statelike apparatus for global governance ('the constitution for the global economy' as the MAI [Multilateral Agreement on Investment] has been termed), the more that demands for democratization of that apparatus will grow, and the more will the focus of public debate turn to the possibility of using the global apparatus for ethical ends: social security, human rights, world peace and ecological sustainability (a political economy anticipated by Falk, 1975, and further pursued by Dryzek, 1994; Falk, 1995; Held, 1995, 1998; Archibugi, 1998).

In summary then, urban planning needs to be redefined as urban environmental governance within a renewed public sphere with an enlarged conception of global, national and local citizenship. This public sphere will increasingly allow access to policy-making processes by citizens, communities and their associations. This enlarged public sphere will encourage critique and debate. Cities will thicken and deepen their networks of co-operation, seeking to demonstrate how economic growth can be subordinated to ecological sustainability without loss of human welfare. Urban governance will thus include supra-city, and sub-city dimensions. The global regulatory apparatus must be strengthened to facilitate action and local level in response to citizen needs.

In some cities, in Europe especially but also in many other parts of the world, city governments have adopted programmes for ecologically sustainable development. Such cities provide one kind of model for the future. However, the world's ecological future will not be determined in those countries which have already made progress towards Agenda 21. It will be determined by the rate of adoption by those great and populous nations whose economies will surge in growth in the twenty-first century. The world's future now depends on how the engine of growth can be steered towards ecological sustainability in those countries and their cities. They will have to take up the leading role. It is in everyone's interests to support and help them find their own particular paths towards Trope B. Seven years after the Rio Earth Summit it is now time to move beyond Rio to make the choice for ecological sustainability by developing the institutions of environmental governance. Indeed, if the planet is to survive then perhaps 'there is no alternative'.

References

Allmendinger, P. (1998) 'Planning and Deregulation: Lessons from the UK', Paper presented at 8th International Planning History Conference, University of New South Wales, 15–18 July.

Altvater, E. (1993) *The Future of the Market*, London: Verso.

Archibugi, D. (1998) 'Principles of Cosmopolitan Democracy', in D. Archibugi, D.Held and M. Köhler, *Re-imagining Political Community: Studies in Cosmopolitan Democracy*, Cambridge: Polity Press, pp. 198–228.

Athanasiou, T. (1998) *Slow Reckoning, The ecology of a divided planet*, London: Vintage Books.

Barrett, S. and Fudge, C. (eds) (1981) *Policy and Action, Essays on the implementation of public policy*, London: Methuen.

Baturo, W., Burger, T. and Kassenberg, A. (1997) *Social Assessment of Agenda 21 Execution in Poland*, Warsaw: Institute for Ecology.

Bluestone, B. and Harrison, B. (1982) *The Deindustrialization of America*, New York: Basic Books.

Brown, L. (1995) *Who Will Feed China?*, New York: Norton.

Castells, M. (1989) *The Informational City*, Oxford: Blackwell.

—— (1993) 'European cities, the informational society, and the global economy', in L. Deben, W. Heinemeijer and D. van der Vaart (eds) *Understanding Amsterdam: Essays on economic vitality, city life and urban form*, Amsterdam: Het Spinhuis.

Cockett, R. (1996) *Thinking the Unthinkable: Think tanks and the economic counter-revolution 1931–1983*, London: Unwin-Hyman.

Cremer, W. and Fisahn, A. (1998) 'New Environmental Policy Instruments in Germany', in J. Golub (ed.) New Instruments for Environmental Policy in the EU, London: Routledge, pp. 55–85.

Cullingworth, J.B. and Nadin, V. (1994) *Town and Country Planning in Britain* (11th edn), London: Routledge.

Daly, H. (1993) 'From Adjustment to Sustainable Development, the Obstacle of Free Trade', in R. Nader *et al. The Case Against Free Trade: GATT, NAFTA and the globalization of corporate power*, San Francisco: Earth Island Press.

Department of the Environment, Transport and the Regions (1998) *Climate Change and Its Impacts. Some highlights from the ongoing UK research programme: a first look at results from the Hadley Centre's new climate model*, Bracknell, UK: The Meterological Office.

Dryzek, J.S. (1987) *Rational Ecology, Environment and Political Economy*, Oxford: Blackwell.

—— (1994) 'Ecology and Discursive Democracy: Beyond Liberal Capitalism and the Administrative State', in M. O'Connor (ed.) *Is Capitalism Sustainable? Political economy and the politics of ecology*, New York: Guilford, pp. 176–195.

—— (1996) 'Foundations for Environmental Political Economy, The Search for *Homo Ecologicus?*', *New Political Economy*, 1/1, pp. 27–39.

Durning, A.T. (1993) 'World Spending on Ads Skyrockets', in L. Brown *et al.* (eds) *Vital Signs*, New York: Norton.

Expert Group on the Urban Environment (1996) *European Sustainable Cities*, Brussels: European Commission, Directorate General XI.

Falk, R.A. (1975) *A Study of Future Worlds*, New York: Free Press.

—— (1995) *On Humane Governance: Towards a new global politics*, Cambridge: Polity Press.

Gamble, A. (1988) *The Free Economy and the Strong State, The politics of Thatcherism*, London: Macmillan.

Held, D. (1995) *Democracy and the Global Order, From modern state to cosmopolitan governance*, Stanford, Calif.: Stanford University Press.

—— (1998) 'Democracy and Globalization', in D. Archibugi, D. Held and M. Köhler, *Re-imagining Political Community, Studies in Cosmopolitan Democracy*, Cambridge: Polity Press, pp. 11–27.

Hempel, L.C. (1996) *Environmental Governance: The global challenge*, Washington, DC: Island Press.

Jensen-Butler, C., Shachar, A. and van Weesep, J. (eds) (1997) *European Cities in Competition*, Aldershot: Avebury.

Ketola, T. (1997) 'Ecological Eldorado: Eliminating Excess over Ecology', in R. Welford, *Hijacking Environmentalism: Corporate responses to sustainable development*, London: Earthscan, pp. 99–136.

Klodt, H. and Stehn, J. (1994) *Standort Deutschland*, Tübingen: Mohr.

Lappin, T. (1994) 'Can Green Mix with Red', *The Nation*, Feb. 14, pp. 193–195

Lindblom, C.E. (1982) 'The Market as Prison', *Journal of Politics*, 44/1–2, pp. 324–336.

Lipschutz, R. with Mayer, J. (1996) *Global Civil Society and Global Environmental Governance: the politics of nature from place to planet*, New York: State University of New York Press.

Logan, J.R. and Molotch, H. (1987) *Urban Fortunes: the Political Economy of Place*, Berkeley and Los Angeles: University of California Press.

Low, N.P. and Gleeson, B.J. (1998) *Justice, Society and Nature: An exploration of political ecology*, London: Routledge.

Martin, H.-P. and Schumann, H. (1997) *The Global Trap: Globalization and the assault on democracy and prosperity*, London and New York: Zed Books.

Marx, K. and Engels, F. (1967) *Manifest der Kommunistischen Partei* (The Communist Manifesto), Harmondsworth: Penguin.

Molotch, H. (1976) 'The City as a Growth Machine: Toward a Political Economy of Place' *American Journal of Sociology*, 82/2, pp. 309–332.

Newman, P. and Thornley, A. (1996) *Urban Planning in Europe: International competition, national systems and planning projects*, Routledge: London.

Ostrom, E. (1990) *Governing the Commons: The evolution of institutions for collective action*, Cambridge: Cambridge University Press.

Poland, Task Force for Structural Policy in Poland (1997) *Sector Programmes for the Restructuring and Privatisation of State Enterprises, Supplementary Report*, Warsaw.

Rostow, W.W. (1967) 'The Great Transition: Tasks of the First and Second Postwar Generations', *State Department Bulletin*, 51, pp. 491–504.

Rothschild, M. (1990) *Bionomics: The Inevitability of Capitalism*, New York: Henry Holt.

Ryan, M. and Flavin, C. (1995) 'Facing China's Limits', in L. Brown *et al.*, *State of the World 1995*, London: Earthscan, pp. 113–131.

Self, P. (1993) *Government By The Market? The Politics of Public Choice*, Melbourne: Macmillan.

Shiva, V. (forthcoming) 'Ecological Balance in an Era of Globalisation', in N. Low (ed.) *Global Environmental Ethics*, London: Routledge.

Taschner, K. (1998) 'Environmental Management Systems: The European Regulation', in J. Golub (ed.) *New Instruments for Environmental Policy in the EU*, London: Routledge.

United Church of Christ (1991) 'Principles of Environmental Justice', *Proceedings of the First National People of Color Environmental Leadership Summit*. Washington, DC (United Church of Christ, Commission for Racial Justice, 105 Madison Avenue, New York, NY 10016, also 475 Riverside Drive Suite 1950, New York, NY 10115).

Vision and Strategies around the Baltic Sea 2010 (1994) Towards a Framework for Spatial Development in the Baltic Sea Region, Karlskrona, Sweden: The Baltic Institute.

Wackernagel, M. and Rees, W.E. (1996) *Our Ecological Footprint: Reducing human impact on the Earth*, Gabriola Island, British Columbia, Canada: New Society Publishers.

Wapner, P. (1996) *Environmental Activism and World Civic Politics*, New York: State University of New York Press.

von Weizsäcker, E., Lovins, A.B. and Lovins, L.H. (1997) *Factor Four: Doubling Wealth, Halving Resource Use*, Sydney: Allen and Unwin.

Welford, R. (1997) *Hijacking Environmentalism: Corporate responses to sustainable development*, London: Earthscan.

Wendt, A. (1992) 'Anarchy is What States Make of It: The Social Construction of Power Politics', *International Organization*, 46, pp. 391–425.

Index